THE AGE OF ARTIFICIAL INTELLIGENCE

AN EXPLORATION

Edited by
Steven S. Gouveia
University of Minho, Portugal

UK Edition

Cognitive Science and Psychology
VERNON PRESS

www.vernonpress.com

In the Americas:
Vernon Press
1000 N West Street, Suite 1200,
Wilmington, Delaware 19801
United States

In the rest of the world:
Vernon Press
C/Sancti Espiritu 17,
Malaga, 29006
Spain

Cognitive Science and Psychology

Library of Congress Control Number: 2020931461

ISBN: 978-1-64889-218-9 [Paperback, UK Edition]

Cover design by Vernon Press using elements designed by FreePik.

TABLE OF CONTENTS

LIST OF ACRONYMS

ACM = Agent cognitive model

ADM = Agent dialogue model

AGI = Artificial General Intelligence

AI = Artificial Intelligence

AIML = Artificial intelligence markup language

AISE = Artificial intelligence safety engineering

ALAMO = Atelier de Littérature Assistée par la Mathématique et les Ordinateurs

ALICE = Artificial linguistic internet computer entity

APMA = Amino-3-hydroxy-5-methyl-4-isoxazole propionic acid

AKM = Agent knowledge model

A.R.T.A = Atelier de Recherches et Techniques Avancées

BCI = Brain-computer interfaces

BDI = Belief–desire–intention model

BNC = British national corpus

BSC = Bodily self-consciousness

CSs = Constructional schemes

CD = Compact disc

CERN = European organization for nuclear research

CEV = Coherent extrapolated volition

CGs = Conceptual graphs

CL = Computational linguistics

CNS = Central nervous system

COGUI = Conceptual graphs user interface

CP = Cognitive phenomenology

CREB = cAMP response element-binding protein

CRISPR = Clustered regularly interspaced short palindromic repeats

CS = Computational structures

CSA = Conversational software agent

DARPA = Defense advanced research projects agency

DDD = Distributed, decentralized, and democratic

DLT = Distributed ledger technology

DM = Discourse management

DNA = deoxyribonucleic acid

DIY = Do it yourself

DRM = Digital rights management

DRT = Discourse representation theory

ELO = Electronic literature organization

EMI = Experiments in musical intelligence

EQ = Embodiment question

FAI = Friendly artificial intelligence

FEP = Free-energy principle

FOOM = Recursively self-improving artificial intelligence engendered singularity

FOS = Free and open source

GANs = Generative adversarial networks

GM = Grand masters (of chess)

GOFAI = Good old-fashioned artificial intelligence

GPS = Global positioning system

HIS = Hazardous intelligent software

HOT = higher-order thought theory of consciousness

HS = Hazardous software

IBM = International business machines

IEEE = Institute of electrical and electronics engineers

IQ = Intelligence quotient

IoT = Internet of things

KL = Kullback-Leiber

KR = Knowledge representation

LS = Logical structure

LSc = Layered structure of the clause

LSDB = Liveset database

MAI = Malevolent artificial intelligence

MEG = Magnetoelectroencephalography

MIRI = Machine intelligence research institute

MS = Mental states

MT = Machine translation

MW = Musical work

NL = Natural language

NLP = Natural language processing

NLU = Natural-language understanding

NGO = Non-governmental organization

NOC = Non-ordinary consciousness

NCC = Neural correlates of consciousness

NSA = National security agency

OR = Ockham's razor

Orch-OR = Orchestrated objective reduction

PCR = Polymerase chain reaction

PHEN = Phenomenological independence implies metaphysical independence

POS = Part-of-speech

QHOT = Quotational higher-order thought theory of consciousness

R&D = Research and development

RDF = Resource description framework

RNA = Ribonucleic acid

RRG = Role and reference grammar

SA = Simulation argument

SAC = Speech act constructions

SAT = Speech act theory

SH = Simulation hypothesis

SP = Sensory processing

STAGE = Science, technology and governance in europe

TOC = Theatre of consciousness

TOTE = Test-operate-test-exit

TREC = Text retrieval conferences

TT = Turing test

WSD = Word sense disambiguation

XML = Extensible markup language

LIST OF FIGURES

LIST OF TABLES

INTRODUCTION

"Computers are useless because they can only give you answers"
(Pablo Picasso)[1]

"Chess... is eminently and emphatically the philosopher's game"
(Paul Morphy)[2]

On May 11th, 1997, in the emblematic city of New York, on the 9th floor of the Equitable Center, a major event happened that can be considered as one of the most decisive moments for humanity:[3]

> 1.e4 c6 2.d4 d5 3.Cc3 dxe4 4.Cxe4 Cd7 5.Cg5 Cgf6 6.Ad3 e6 7.C1f3 h6 8.Cxe6 De7 9.O-O fxe6 10.Ag6+ Rd8 11.Af4 b5 12.a4 Ab7 13.Te1 Cd5 14.Ag3 Rc8 15.axb5 cxb5 16.Dd3 Ac6 17.Af5 exf5 18.Txe7 Axe7 19.c4 1-0.

If you don't know the history of some of the most famous chess games, you will not immediately recognize this sequence. But if you do, you will notice that this is the sum-up of all the 19 moves made in the sixth game of the second match between Garry Kasparov (one of the most brilliant chess players in the world - alongside M. Carlsen, B. Fisher, V. Anand, A. Karpov and P. Morphy) and the computer created by International Business Machines (IBM) named DeepBlue. In the 8th move, after Kasparov played h6 – something he would regret later – DeepBlue decided to sacrifice a knight for a pawn, playing e6, something that Kasparov would never have expected.

What happened next was the pure dominance of the computer, forcing Kasparov to resign on the 19th move and to lose the match.[4] Kasparov would

---

[1] Different versions of this are cited in William Fifield's original interview with Picasso, "Pablo Picasso: A Composite Interview," published in the Paris Review 32, Summer–Fall 1964and in Fifield's 1982 book *Search of Genius*, New York: William Morrow.

[2] As quoted in *Testimonials to Paul Morphy*, Presented at University Hall, New York, May 25, 1859 (cf. https://play.google.com/books/reader?id=aEZAAAAAYAAJ).

[3] *Time* magazine had claimed something similar regarding the first game of the first match between Kasparov and DeepBlue in 1996. Kasparov had lost the first game (although he would ended up winning the match): the first-game defeat was more than "world historical. It was species-defining".

[4] The match was composed by six gamesand the result after the first five games was a tie, 2½–2½. The match's result was, after the win of DeepBlue in the sixth and last game, 3½–2½.

never have played h6 against a human opponent, nevertheless, since he was playing against a computer, he chose that move: "I didn't want to play. I was sorry about my decision to play h6. Normally computers don't take on e6".

Although DeepBlue was built with 256 co-processors capable of calculating approximately 200 million positions per second, for Kasparov, the reason for his loss was simple: no computer would use a tactical move such as sacrificing a knight so early in the game. At the time, there were some suspicions that the research team behind the development of DeepBlue was being helped (live in action) by Grand Masters (GM), namely, Bobby Fisher.

In the late eighteenth century, there was a famous chess-playing mechanical automaton called "the Turk[5]". It was an engraved figure made of wood that could move its pieces and it could play a competent chess game. The "Turk" was branded as the very first artificial system. Touring through America and Europe, it played against professional players, including renowned historical celebrities such as Napoleon Bonaparte and Benjamin Franklin, who were themselves chess aficionados. Of course, this was a hoax, an elaborate, but nevertheless, a fake artificial system with a person cleverly hidden inside the wood structure playing all the moves (cf. Kasparov, 2017: 7).

IBM had one purpose only: to prove that they could build a machine that could defeat the best chess player in the world, the reigning champion. The team responsible for the development of DeepBlue was composed by Murray Campbell (IBM, Thomas J. Watson Research Center), Joel Benjamin (Chess GM and consultant), Feng-hiung Hsu (who started developing DeepBlue while he was at Carnegie Mellon University), Thomas Anantharaman as well as a few others, all managed by Chung-Jen Tan, who was known as the spokesman and the "resident philosopher" of the team.

Kasparov had won the first match against DeepBlue in the previous year, 15 months prior in Philadelphia - this was the second one. The preparation for this second match was difficult because he couldn't study previous games played by DeepBlue since there were none: he had to play against a black box without any chance of studying and analyzing previous games made by the computer. Worse than that, there was a clause in the contract for the second match that Kasparov completely overlooked: the machine could be rebooted during or after each game. This would make the post-analysis of its games impossible, therefore, eliminating any chances of studying specific games or to recognize any patterns in DeepBlue's approach to chess.

---

[5] The machine was nicknamed the Turk because it played its moves through a turbaned marionette attached to a cabinet (cf. Campbell, 1997: 83).

Newsweek's cover called this match "The Brain's Last Stand". The match was covered in all newspapers and broadcasted live on television: the world was seeing firsthand that artificial machines could potentially surpass a game in which human beings had excelled for centuries. After this first conquest, Artificial Intelligence research focused on building computers that could defeat humans in other more complex games, like Go (with AlphaGo), Scrabble (with Quackle) or Jeopardy (with Watson). The artificial system won in every one of these games.

The 1997 match was announced as Kasparov representing Humanity versus the Machine. If he lost the match, then everything could be achievable for AI: it would officially give rise to the "Age of Artificial Intelligence". For contemporary science, chess was seen as the ultimate test of intelligence. In cinema, there are also many chess references along these lines: for example, in Stanley Kubrick's *2001: a Space Odyssey*, there is a scene where the computer HAL9000 plays a chess game against Frank Poole - the game is an actual recreation of a tournament game from 1913 between A. Roesch and W. Schlage (cf. Campbell, 1997: 79).

It's very interesting to notice that the birth of Computer Science and AI is associated with the first reflections and thoughts about creating a machine that could play chess. One of the first discussions happens in Charles Babbage's *The Life of a Philosopher* (1845). In 1945, one century later, Konrad Zuse describes a program that could generate legal moves in chess. Claude Shannon, founder of information theory, wrote a paper titled "A Chess-Playing Machine" (1950) where he describes two main approaches in building a competent chess-program: Type A was about creating a program based on brute-force, that is, a program that could analyze and calculate in seconds all the millions of potential positions for a specific move; and Type B, a program that would have a strategic and goal-focused approach, more like humans' beings play chess. Shannon believed that the Type B approaches would be more successful in beating a human, but it was eventually the Type A approach which won the race, culminating in the development of DeepBlue.

Alan Turing also thought about whether an artificial system that could play chess, called "Turochamp", famously known as the "Turing's paper machine". The program, created in 1952,[6] was written with his colleague David Champernowne and was composed by a specific set of instructions. The only problem is that there were no digital computers in that time, so Turing wrote

---

[6] See the original publication here: see https://en.chessbase.com/post/reconstructing-turing-s-paper-machine).

the program by hand - it was later recreated in an actual computer, where it turned out to not be a very competent chess player.

In 1950, one of the founders of Artificial Intelligence, Herbet Simon, stated it would only take 10 years for a machine to become the world champion of chess – he was just mistaken by 30 years (cf. Cambpell, 1997: 83-85).

In 1956, John McCarthy, owning to Alexander Kronrod, described chess as the "drosophila of Artificial Intelligence" (cf. McCarthy, 1990). Like the common fruit fly, which juxtaposed to more complex organic systems, its research is quite "simple" to do. Nevertheless, it can produce significant knowledge about further complex systems. Just as at its base, chess is a "simple" game, and therefore, it can teach us an ample amount about human cognition and intelligence (cf. Ensmenger, 2012).

As we can see, there was a great, almost obsessive focus on chess in the first days of AI research because it was believed that chess was the ultimate pinnacle of human intelligence. Nowadays, this seems, a tiny bit exaggerated. At present, Artificial Intelligence is focused on what is called "weak AI": it excels at very specific tasks – like playing games, facial recognition or driverless cars – but it is not even close to achieving human-level intelligence. The reason is quite simple: Artificial Intelligence's research methods are more about imitating human performances – the Turing Test is a very good example of this idea – than to look for its own achievements and goals.

Consequently, all the tasks we can describe and codify can be outperformed by machines. But the real achievement of a fully conscious machine seems far still. Because we do not know anything about consciousness, Artificial Intelligence conceived as "strong AI", that is, a conscious A.I., may never be fully achieved. For that, we need the right theoretical framework – we need better and more philosophical research.

The book is divided into five main sections. Section I is dedicated to reflections on the Intelligence of AI and will open with a chapter by Soenke Ziesche and Roman V. Yampolskiy, which discuss the mathematics of intelligence for grouped minds, nested minds as well as deducted minds. The following chapter, by Stevan Harnad, debates if the existence of feelings is a real caveat for a system that would pass the Turing's Test. Finally, closing Section I, David Pearce discusses three different ways to connect human and artificial intelligence: by fusion, replacement or co-evolution, arguing that only the third process may be plausible.

Section II follows, dedicated to discussion on the relationship between consciousness, simulation and artificial intelligence. Gabriel Axel Montes and Ben Goertzel present the concept of a 'mindplex' as a way of enhancing the connection between human and artificial minds; for which, they use the concept of non-ordinary consciousness (NOC) and show how that perspective can be relevant for understanding the mind and cognition in general. Cody Turner follows offering two arguments in favor of the thesis that a phenomenology of cognition is neither reducible to, nor dependent upon, sensory phenomenology. If this thesis is plausible, then it follows that AI consciousness may not require embodiment to be emulated, as commonly assumed. The next chapter by Nicole Hall argues that aesthetic experience is a fundamental feature of human consciousness and separates human from artificial intelligence. She argues further that it is a mistake to confuse the mere possibility of achieving "conscious singularity", as she defines it, with human consciousness and its capacity for aesthetically experiencing natural environments. To conclude Section II, Steven S. Gouveia introduces an intriguing idea that we may be living in a computer simulation, briefly debating the main reasons in favor and against this hypothesis.

Section III, dedicated to aesthetical creativity and language in artificial intelligence, opens with a chapter by Caterina Moruzzi where in the light of recent developments in AI music software generators discusses the question, "Can a computer create a musical work?" On the same topic, René Mogensen proposes a formal representation of content in computational creativity of music, noting that in order to achieve complete computational music creativity, aesthetic experience appears to be necessary. Mariana Chinellato Ferreira follows, applying the same discussion about aesthetical creativity in computer-generated literature, analyzing specific software such as PoeTryMe. Closing Section III, Kulvinder Panesar presents a functional linguistic perspective on natural language processing in artificial intelligence.

The subsequent Section IV is on the Ethics of the Bionic Brain Peter A. DePergola II opens by offering the argument that neurocognitive enhancement can be justified as morally plausible if it (a) promotes general moral character, (b) complements human nature and (c) effects a deeper sense of individual and social identity. Next, Tomislav Miletić and Frederic Gilbert discuss the potential harms of brain-computer interfaces (BCI) on self-determination, warning that any patient who accepts the use of such future AI medical technology should be sufficiently prepared for the symbiotic relation before the implementation. Following on the same topic of the ethics of BCI, Aníbal M. Astobiza, Txetxu Ausin, Ricardo M. Ferrer and Stephen Rainey focus on some issues raised by BCI research, identifying some dangers, challenges and opportunities for the elaboration of a common ethical and legal framework concerning issues of safety, ethics and data protection. To conclude Section IV, Natasha Vita-More argues that Artificial General Intelligence (AGI) can be used as a tool to improve our knowledge about ourselves and the world.

Finally, to close the book, Section V follows on the Ethics of Artificial Intelligence, starting with a chapter by Federico Pistono and Roman V. Yampolskiy that provides some general guidelines for the creation of a Malevolent Artificial Intelligence (MAI) with the goal of challenging the AI Safety Community to continue its effort by discovering and reporting specific problems. Following, Hasse Hämäläinen attempts to find the most plausible answer to the question of whether a machine could be attributed moral and legal rights and obligations, arguing that if a machine can perform a specific task or set of capacities as human beings do, then the rights and obligations of humans should also be applied to machines. The next chapter by Vernon Vinge discusses the ethical implications of the Singularity, offering a set of ethical guidelines to avoid the extinction of human race. Finally, to conclude both Section V and the book, Eray Özkural discusses the ethical and epistemological implications of the Free Energy Principle: the idea that a self-organization occurs through minimization of free energy.

The Age of Artificial Intelligence is imminent, if it's not already here. We should make sure that we invest in the right people and the right ideas in order to create the best solutions possible. My hope is that this book will help to influence the right minds. If Reason killed god in the 20th and 21st century, Reason – philosophy, science and technology – may resurrect it in form of an Artificial General Intelligence: an AI that may know everything about anything. We should make sure that we create the right kind of god and that we keep it in the right hands.

I would like to finish this introduction by deeply thanking all the people who made this project feasible.[7]

Steven S. Gouveia
Ottawa, 10/09/2019

[7] I would like to acknowledge the precious help offered by Jessica Clarke and, to Susan Schneider for her valuable feedback.

References

Campbell, Murray S. (1997) "'An Enjoyable Game': How HAL Plays Chess" In *HAL's Legacy: 2001's Computer as Dream and Realtiy* (David G. Stock, ed.), Cambridge, MA: MIT Press.

Ensmenger, Nathan (2012) "Is chess the drosophila of artificial intelligence? A social history of an algorithm" *Soc Stud Sci.*, 42 (1): 5-30.

Feng-hsiung, Hsu, Anantharaman, Thomas, Campbell, Murray and Nowatzyk, Andreas (1990) "A Grandmaster-level Chess Machine" *Scientific American*, 263 (4): 44-51.

Fifield, William (1982) *Search of Genius*, New York: William Morrow

Frey, Peter W. (ed.) (1983) *Chess Skill in Man and Machine*, New York: Springer-Verlag.

Hsu, Feng-hsiung (2002) *Behind Deep Blue: Building the Computer that Defeated the World Chess Champion*, Princeton, NJ: Princeton University Press.

Kasparov, Gary (2017) *Deep Thinking: Where Machine Intelligence Ends and Human Creativity Begins*, New York: PublicAffairs™.

Levy, David and Newborn, Monty (1991) *How Computers Play Chess*, New York, NY: Computer Science Press.

Marsland, Anthony T. and Schaeffer, Jonathan (eds.) (1990) *Computers, Chess and Cognition*, New York, NY: Springer-Verlag.

McCarthy, John (1990) "Chess as the Drosophila of A.I.". In *Computers, Chess and Cognition* (Marsland T.A., Schaeffer J., eds.), New York, NY: Springer, pp. 227-237.

Shannon, Claude (1950) "Programming a computer for playing chess" *Philosophical Magazine*, 41: 256-75.

SECTION I:
INTELLIGENCE IN ARTIFICIAL INTELLIGENCE

CHAPTER 1

TOWARDS THE MATHEMATICS OF INTELLIGENCE

Soenke Ziesche[1]

Maldives National University, Maldives

Roman V. Yampolskiy[2]

University of Louisville, USA

Abstract: In this chapter, we aim to contribute further to the still very young field of "intellectology", which examines the space of possible minds and was introduced by Yampolskiy (2015). We are looking at mathematical properties of sets of minds as well as minds in nested constellations by utilising operations of set theory. More specifically, we study the mathematics of intelligence of set unions and complements as well as subsets with intelligence being an inherent feature of minds. For example, if a mind A with intelligence of 100 is united with a mind B with intelligence of 110 what do we get? As with many aspects of intellectology this area is largely unexplored. This chapter is structured as follows: In the next section, we describe potential constellations of minds that have been proposed earlier by using set theory, which appears to be a useful instrument for our purposes. This is followed by a section about the definition and evaluation of intelligence, which is a challenging field itself. In the subsequent main section, we investigate the math of intelligence for grouped minds, nested minds as well as deducted minds.

1. Set Theory of Minds

Since no common definition exists, we describe a mind as instantiation of intelligence[3]. This means that in addition to humans and other higher-order animals also non-animal intelligent agents constitute minds. It has been shown that the space of possible minds is vast and attempts to characterise it

---

[1] Faculty of Engineering, Science and Technology, Maldives National University, Maldives.

[2] Computer Engineering and Computer Science, Speed School of Engineering, University of Louisville, (roman.yampolskiy@louisville.edu).

[3] See below for a definition of intelligence.

have been made (e.g., Sloman, 1984; Goertzel, 2006; Yudkowsky, 2008; Yampolskiy, 2015). In addition to a, for humans inconceivable, variety of minds, also grouped minds as well as nested minds may be possible.

As we will show below, set theory seems to provide tools to describe such constellations. In particular, we are looking at set union and complement as well as subsets. We will also make use of Yampolskiy's insight that the space of minds is countable and that all minds can be generated sequentially by a deterministic algorithm (Yampolskiy, 2015).

1.1. Grouped minds: set union

Some scenarios of grouped minds appear frequently in our lives, while some are futuristic. Extensive research has been conducted on the collective behaviour of groups of humans as well as groups of animals (e.g. flocks of birds, schools of fish or ant colonies), especially regarding potentially emergent properties of such groups. Woolley et al. (2010) name one such property for groups of humans "collective intelligence" or "c factor", which they distinguish from "general intelligence", also known as "g factor". Woolley et al. (2010) show that the "[...]'c factor' is not strongly correlated with the average or maximum individual intelligence of group members but is correlated with the average social sensitivity of group members, the equality in distribution of conversational turn-taking, and the proportion of females in the group." This is contested by Bates and Gupta (2016) whose experiments (in a very controlled environment) instead revealed that higher individual IQ enhances the performance of a group, i.e. the group-IQ.

From our anthropomorphic viewpoint, we are used to singly-embodied minds, but in addition, we can also conceptualize multiply- and flexibly embodied minds. Goertzel (2003) provides a more general approach to grouped minds. He defines what he calls "mindplex" as follows: It "[...] is composed of a collection of intelligent systems, each of which has its own "theatre of consciousness" and autonomous control system, but which interact tightly, exchanging large quantities of information frequently" and it "[...] has a powerful control system on the collective level, and an active "theatre of consciousness" on the collective level as well."

Heylighen (2011) coined the term "global brain" for the set of minds that is, since the emergence of the Internet, constituted of human minds as well as computers. Sotala and Valpola (2012) analyse in more detail how "mind coalescence" as they call it could be achieved in a biological as well as post-biological way and what consequences it may have.

Yampolskiy et al. (2012) provide another novel approach of grouped minds in the form of a "Wisdom of Artificial Crowds" algorithm, which relies on a group of simulated intelligent agents to find solutions, which are autonomous and, in many cases, superior to individual solutions of all participating agents.

An open challenge for these constellations is to determine the threshold when those minds are merged into one as opposed to a set union of interacting minds.

1.2. Nested minds: subset

Nested minds are mostly speculation, but relevant, especially since Bostrom (2003) analysed the likelihood that we are living in a computer simulation. If this was true, this would be a constellation of nested minds, assuming that the simulating system has a mind too.

Other scenarios have been proposed; for example several by Kelly (2007), such as "Global mind – large supercritical mind made of subcritical brains", "Hive mind – large supercritical mind made of smaller minds each of which is supercritical", "Low count hive mind with few critical minds making it up", "Borg – supercritical mind of smaller minds supercritical but not self-aware".

Yampolskiy (2012) describes further scenarios of nested minds related to the artificial intelligence confinement problem. Below we will also explore the third operation, complement of a set, i.e., subtracted minds, for which we did not find prior research.

2. Definition and Evaluation of Intelligence

Despite many attempts, none of the involved disciplines, such as psychology or artificial intelligence, has come up with a satisfying, mutually agreed-upon definition of intelligence. Legg and Hutter (2007) provide an overview of the many definitions that have been proposed over the years and deliver eventually the general definition: "Intelligence measures an agent's ability to achieve goals in a wide range of environments".

Although the first step, to agree on a unique definition of intelligence, has still not been completed, already extensive work has been conducted on the following step, which is to evaluate intelligence as a feature of psychometrics. Ability to evaluate intelligence is a precondition in order to be able to compare various intelligences, which is of interest regarding human as well as artificial intelligences. Hernández-Orallo (2016) worked extensively in this field and identified "[…] three kinds of evaluation: human discrimination, problem benchmarks and peer confrontation"

In this chapter, we use a definition of intelligence by Yampolskiy (2013), which he developed within the framework of his "Efficiency Theory". There, he describes gaining of knowledge as finding and applying an efficient algorithm to an inefficiently stored information string. In other words, "Intelligence could also be defined as the process of obtaining knowledge by efficient means." Yampolskiy (2013).

Yampolskiy (2013) also states "historically, the complexity of computational processes has been measured either in terms of required steps (time) or in terms of required memory (space)". We suggest intelligence to be treated as a further dimension of fundamental computation resources in addition to time or memory. And while for measuring and manipulating time and memory, there are established standards, we would also like to be able to measure or evaluate intelligence by its ability to turn certain information strings efficiently into knowledge.

Based on this the scope of mathematics of intelligence of grouped and nested minds is to analyse and to compare the capabilities of all involved minds to turn information strings efficiently into knowledge and to establish relations between them.

3. Mathematics of Intelligence

We define minds by four parameters, the former two are mandatory, the latter two optional: mind (x, i, a, t)

x (their unique identifier, see Yampolskiy, 2015)
i (their intelligence, according to the "Efficiency Theory", see Yampolskiy, 2013)
a (their age, since i may vary depending on a)
t (their time of embodiment)

Based on the algorithm to generate sequentially all minds Yampolskiy (2015) states that "A particular mind may not be embodied at a given time, but the idea of it is always present." Embodiment is the time period when a mind is exhibiting intelligence. This could be the lifetime of living biological beings as we know them, but it could also be the period of implementation of a non-biological, artificial agent. This parameter is relevant since the intelligence of a mind may vary depending on the spacetime context of the embodiment, which is illustrated further below.

Now we look at various set operations and constellations, such as union, subset and complement.

3.1. Grouped minds: set union

Of special interest is the question of whether there is any scenario when the mind that results from combining other minds has higher intelligence than all original minds? Such a phenomenon would be called emergentism. Another option would be that the intelligence of a grouped mind equals the highest intelligence of all minds in the union, which could be expressed by the MAX function.

Both are valid options. Normal IQ tests are timed and so having a number of minds working in parallel would increase the overall score, but at the highest level (160+) IQ tests are not timed and only a more capable mind would do better.

Intelligence is a computational resource similar to time, space or randomness. We propose arithmetic of intelligence to make it possible to perform analysis on this important resource. In particular, we concentrate on the minds at a sufficiently high level of performance, there time is not an important factor, but overall ability to solve problems is the main feature in determining intelligence. Overall, arithmetic of intelligence functions are in a certain way similar to the Big O notation, there one may disregard lower value terms. For example, for two intelligent systems with IQs equal to A and B, sum(A,B) = Max(A,B).

3.2. Nested minds: subset

Similarly of interest for the math of intelligences is the question whether in a nested constellation the mind in a subset could have a higher intelligence than the mind in the superset?

While intuitively it may seem that the mind in the subset is always less intelligent than the mind in the superset, this may not be the case given Yampolskiy's (2013) definition of intelligence that it is all about better and more efficient algorithms to turn information into knowledge. Assuming that, according to Plato's theory of Ideas,[4] algorithms exist independently and have just to be discovered rather than created, a scenario is imaginable that a mind in a subset happens to discover an efficient algorithm, which has not been discovered by the mind in the superset.

As mentioned above, one of the best known proposed nested constellations of minds are simulations. For such a constellation, this would mean that a simulated mind could find a better algorithm to turn certain information into knowledge than the simulating mind ever discovered, which would make the simulated mind more intelligent. Yet, the simulating mind would instantly get to know about the discovery of this algorithm (since the simulating mind by definition is copied on all processes in the simulation), i.e. would harness the intelligence of the simulated mind.

---

[4] See e.g. Ross (1951).

This means that the intelligence of a simulating mind can increase over time through two ways: 1) If the simulating mind itself finds better algorithms to turn information into knowledge, or 2) if one of the simulated minds finds such better algorithms. At the same time, the intelligence of a simulating mind never decreases if we assume that a mind, which is intelligent enough to run simulations, must be also intelligent enough to keep a perfect memory.

Probably an inherent feature of (many)[5] simulations is that the subset of simulated minds changes over time through births and deaths. Based on the above insight, the intelligence of the simulating mind will be at least as high as the mind with the highest intelligence that ever inhabited the simulation, even if this mind has already passed away, since the simulating mind has access to all algorithms ever executed in the simulation.[6]

Therefore, it could be an incentive for a mind to simulate other minds as a tool to increase its own intelligence, hence this may increase the likelihood that the second of the three propositions, which Bostrom (2003) uses in order to analyse the probability that our civilization is currently living in a simulation, is <u>not</u> true: "Any posthuman civilization is extremely unlikely to run a significant number of simulations of their evolutionary history (or variations thereof)." Usually, two motivations are named, as e.g. by Tierney (2007), why a posthuman civilization would run simulations: research purposes or entertainment. While research purposes are also about gaining knowledge, yet more through observation and monitoring, we are proposing here for the first time an additional motivation for running a simulation, which is to increase the intelligence of the simulating mind.

If this were the case, then a follow-up question would be whether a simulating mind could influence the simulation so that odds are higher that more intelligence is exhibited? Would this be through quality or quantity of the simulated minds? If through <u>quality</u>, then perhaps the simulating mind could focus on the simulation of other minds of whom it knows that their intelligence is above a certain threshold, e.g. 150:

$$\forall x: \text{mind}(x, i) \wedge i > 150$$

---

[5] There are also simulations imaginable without the anthropomorphic concepts of births and deaths, i.e. inhabited by a stable number of immortal minds.

[6] If we look at the global brain approach by Heylighen (2011), which was introduced above, there is the similar phenomenon that we preserve discoveries from ancestors, which means in an abstract way to keep particular algorithms over time. Yet, at times, before printing and other information technologies were invented, there have been various instances of forgotten discoveries.

This would raise the question of how the simulating mind could determine the intelligence of a simulated mind beforehand? We postulated above that the simulating mind might hope to benefit from increases of the intelligence of the simulated minds through discoveries of efficient algorithms. However, the simulating mind might have strategies to simulate minds that are promising to develop in this way.

If through quantity a simulating mind may try to maximize the number of simulated minds. One strategy to do so would be by somehow getting simulated minds to the level that they can run simulations themselves, i.e. by generating nested minds within nested minds.

The simulating mind could keep track of the simulated minds through their unique identifier and would aim, depending on its resources, to simulate as many as possible for a reasonable time frame and analyse how the intelligences of the simulated minds develop. Since also the time of the embodiment and the surrounding peer minds matter, the simulating mind could also experiment by successively simulating/embodying the same mind in different contexts and compare the performance of its intelligence.

Another thought is whether this would perhaps decrease the chances that we ourselves are in a simulation since so far our civilization would have probably not contributed much to a potential simulating mind in order to increase its intelligence. It would seem to be a waste of time and resources to let run first a long evolution, instead of starting the simulation already with a fairly advanced civilization[7] Also, for efficiency purposes, such a simulating mind would probably run the simulated minds with a rather high pace to get results quicker. Yet, our minds do not run particularly fast.

In this regard, we also have to consider nested minds within nested minds. For example, an advanced mind may have initiated a simulation with very intelligent minds in order to potentially increase its own intelligence as described above. Some of these simulated minds may have developed the capacity to run their own simulations, yet with less smart minds as they may have different capacities, intellectual as well as in terms of computational resources, or motivations to run simulations as the overall simulating mind. So, could we be one simulation on such a deeper level?

A special case would be if a simulated mind were encrypted. Then the simulating mind would not be able to get access to its information. We have to distinguish between two sub-cases: 1) The simulated mind has the ability to encrypt some or all of its activities or 2) the simulating mind has encrypted the

---

[7] See also Yampolskiy (2016).

activities of the simulated mind. We shall not look here at the perhaps unlikely motivation why simulating minds may decide to encrypt their simulations.

Encrypted minds in a simulation would be the only scenario where a mind in the subset could have a higher intelligence than the mind in the superset. This would be the case when the mind in the subset discovers better algorithms to turn information into knowledge than the mind in the superset ever knew of, and the mind in the superset will not find out about it since it cannot decipher the encryption.

3.3. Subtracted minds: complement of a set

Not much research has been conducted regarding the set operation "complement". Yet as part of the mathematics of intelligence, it is relevant to formalize how the intelligence of a grouped mind changes if certain minds are deducted from it.

There are two types of complements: absolute and relative. If A is a set, then the <u>absolute</u> complement of A is the set of elements, which are not in A. If A and B are sets, then the <u>relative</u> complement of A in B is the set of elements, which are in B, but not in A.

We look first at the <u>relative</u> complement: The deduction of one member or more (which equals A in the above definition) from a group of minds (which equals B in the above definition) can be visualized either as the ending of the embodiment of a particular mind, i.e. its death, or a withdrawal from the group so that the intelligence of that particular mind does not contribute anymore to the overall intelligence of the group.

Let us look at the specific example whether and how the intelligence of humanity has changed after the deduction of Einstein's intelligence, which occurred on 18 April 1955, the day of his death:

$$\{ x_{after} \mid \text{mind}(x_{after}, i_{after}) \wedge \text{human\_mind}(x_{after}) = \text{true}\} = \{ x_{before} \mid \text{mind}(x_{before}, i_{before}) \wedge \text{human\_mind}(x_{before}) = \text{true}\} \setminus \{\text{mind}(m_{Einstein}, i_{Einstein})\}$$

The question we are looking at is the relationship between i_{before}, the intelligence of humanity <u>before</u> Einstein's death, and i_{after}, the intelligence of humanity <u>after</u> Einstein's death.[8] As mentioned before, critical is again the time of embodiment and the level of development of the grouped mind. Since

[8] Also, many other humans passed away on 18 April 1955, but we assume their contribution to the overall intelligence of humanity to be negligible compared with Einstein's contribution.

Einstein's work was very well documented and probably none of his relevant ideas got lost (unless he deliberately did not publish them), we may assume that after the deduction of his mind with the intelligence i_{Einstein} the intelligence of humanity remained the same, i.e. $i_{\text{before}} = i_{\text{after}}$. In other words, the grouped mind has absorbed the intelligence of its former member Einstein. This was only possible since the intelligence of the grouped mind was already on a certain level, while, for example, an embodiment of Einstein in Stone Age would have taken place differently in at least the following two aspects: 1) The intelligence of Einstein would have almost certainly not reached the level it had reached in the 20th century, i.e.:

$$i_{\text{Stone Age}} \mid \text{mind}(\text{Einstein}, i_{\text{Stone Age}}, \text{Stone Age}) < i_{\text{20th century}} \mid \text{mind}(\text{Einstein}, i_{\text{20th century}}, 20^{\text{th}} \text{ century})$$

The overall intelligence of humanity in Stone Age (or rather the community within which Einstein lived in this thought experiment since there was no global community at this time) would have been reduced after his demise since the grouped mind was not at a level then to absorb his intelligence.

In general, for the <u>relative</u> complement, two points are relevant: 1) Did the deducted mind(s) make a significant contribution to the intelligence of the grouped mind? 2) If yes, was the intelligence of the grouped mind sufficiently advanced to harness the intelligence of these minds prior to their deduction? The latter point is also time-critical: For example, if a human expert on a certain topic joins a group of humans for one hour only, this may not be enough time for the rest of the group to "pick that person's brain". Yet, for minds based on substrates that allow electronic exchange of information one hour is likely to be sufficient to copy critical information from another mind that is connected to them for this time period only.

Moving on to the absolute complement, we would look at deducting minds from the universe of all possible minds. Basically, only two cases have to be distinguished, whether among the deducted minds is the one with the highest intelligence in the universe of all possible minds or whether that mind is <u>not</u> among the deducted ones.

In the first case, the overall intelligence of the complement will remain the same, which is the highest intelligence in the universe according to the Max function introduced above. In the second case, the Max function is also relevant because after the deduction of the mind with the highest intelligence the intelligence of the complement equals the intelligence of the mind, which has the highest intelligence of all <u>remaining</u> minds.

4. Conclusion

We presented an introduction to the mathematics of intelligence. Our approach uses the set theory and then describes novel insights in set unions and complements as well as subsets of minds and the resulting intelligence. These findings can be seen as another contribution to the new field of "intellectology" (Yampolskiy, 2015). Since this field of study is still in its infancy, also this work on the mathematics of intelligence has to be considered as groundwork, which offers various routes for expansion.

In further research, we may look at other, more complex set operations such as the symmetric difference, the Cartesian product or the power set and determine the resulting intelligence.

References

Bates, Timothy C. and Gupta, Shivani (2016) "Smart groups of smart people: Evidence for IQ as the origin of collective intelligence in the performance of human groups", *Intelligence*, 60: 46–56.

Bostrom, Nick (2003) "Are You Living In a Computer Simulation?", *Philosophical Quarterly*, 53 (211): 243-255.

Goertzel, Ben (2003) "Mindplexes - The Potential Emergence of Multiple Levels of Focused Consciousness in Communities of AI's and Humans" in *Dynamical Psychology*.

Goertzel, Ben (2006) *The Hidden Pattern: A Patternist Philosophy of Mind*, Boca Raton: Brown Walker Press.

Hernández-Orallo, José (2016) *The Measure of All Minds*, Cambridge: Cambridge University Press.

Heylighen, Francis (2011) "Conceptions of a Global Brain: an historical review", *Evolution (Uchitel Publishing House)*, (1): 274-289.

Kelly, Kevin (2007) "A Taxonomy of Minds", Accessed October 1, 2018, http://kk.org/thetechnium/a-taxonomy-of-m/.

Legg, Shane and Hutter, Marcua (2007) "Universal Intelligence: A Definition of Machine Intelligence", *Minds and Machines*, 17 (4): 391-444.

Ross, William D. (1951) *Plato's Theory of Ideas*, Oxford: Clarendon Press.

Sloman, Aaron (1984) "The Structure and Space of Possible Minds" In *The Mind and the Machine: philosophical aspects of Artificial Intelligence* (S. Torrance, ed.), Ellis Horwood LTD, pp. 35-42.

Sotala, Kaj and Valpola, Harri (2012) "Coalescing Minds: Brain Uploading-Related Group Mind Scenarios", *International Journal of Machine Consciousness*, 4 (1): 293-312.

Tierney, John (2007) "Our Lives, Controlled From Some Guy's Couch" in *New York Times*, August 14, 2007, http://www.nytimes.com/2007/08/14/science/14tier.html.

Woolley, Anita W., Chabris, Christopher F., Pentland, Alex, Hashmi, Nada and Malone, Thomas W. (2010) "Evidence for a Collective Intelligence Factor in the Performance of Human Groups", *Science*, 330 (6004): 686-688.

Yampolskiy, Roman V. (2012) "Leakproofing Singularity - Artificial Intelligence Confinement Problem" *Journal of Consciousness Studies*, 19 (1-2): 194–214.

Yampolskiy, Roman V., Ashby, Leif and Hassan, Lucas (2012) "Wisdom of Artificial Crowds – a Metaheuristic Algorithm for Optimization", *Journal of Intelligent Learning Systems and Applications*, 4 (02): 98-107.

Yampolskiy, Roman V. (2013) "Efficiency Theory: a Unifying Theory for Information, Computation and Intelligence", *Journal of Discrete Mathematical Sciences & Cryptography*, 6 (4-5): 259-277.

Yampolskiy, Roman V. (2015) *Artificial Superintelligence: a Futuristic Approach*, Florida: Chapman and Hall/CRC Press.

Yampolskiy, Roman V. (2016) "On the origin of synthetic life: attribution of output to a particular algorithm", *Physica Scripta*, 92 (1): 1-10.

Yudkowsky, Eliezer (2008) "Artificial Intelligence as a Positive and Negative Factor" In *Global Risk, in Global Catastrophic Risks* (N. Bostrom and M.M. Cirkovic, ed.s), Oxford: Oxford University Press, pp. 308-345.

CHAPTER 2

MINDS, BRAINS AND TURING[1]

Stevan Harnad[2]

Université du Québec à Montréal, Canada;

University of Southampton, UK

Abstract: Turing set the agenda for cognitive science: Design a causal system that can pass the Turing Test – able to do anything that cognizers can do, indistinguishably from cognizers, to cognizers – and you have explained cognition. But cognizers are conscious: it feels like something to cognize. Searle showed that even if a purely computational system could pass the verbal version of the Turing Test, it would not cognize. So cognition is not just computation. Yet even if a dynamical system that could pass the full robotic (or even the neuro-robotic) version of the Turing Test really did cognize (i.e., really did feel) there would be no way to explain the causal role of feeling.

1. The Turing Test

Turing (1950) set the agenda for (what would eventually be called) the cognitive sciences. He said, essentially, that cognition *is* as cognition *does* (or, more accurately, cognition is what cognition is *able to do*): Explain the causal basis of cognitive capacity and you've explained cognition. Test your explanation by designing a machine that can do everything a normal human cognizer can do – and do it so veridically that human cognizers cannot tell its performance apart from a real human cognizer's – and you really cannot ask for anything more.

A machine? Isn't that already a contradiction in terms? Only if you have biassed preconceptions about machines. For "machine" merely means *a dynamical system governed by causality.* On that score, we too are machines –

---

[1] Originally published in Harnad, Stevan (2011) "Minds, Brains and Turing" In *Third Annual Online Consciousness Conference*, Vol. 3, Consciousness Online.

2 Chaire de recherche du Canada en sciences cognitives, Département de Psychologie, Université du Québec à Montréal (Canada) & School of Electronics and Computer Science, University of Southampton (United Kingdom).

for everyone except those who believe that our biology somehow transcends ordinary causality, and that "mind over matter" is somehow an extra, spontaneous force in the universe. (We will not take up this notion till the very end of this essay.)

So we are machines, and Turing simply pointed out that it is therefore our mission to find out the *kind* of machine we are, by explaining how the machine works. His own hunch may have been wrong. He thought that we were mainly computers, and that cognition is computation. So, he thought, the task would simply be to find the right computer program – the one that can pass the "Turing Test" – able to do anything we can do, indistiguishably from the way we do it.

2. Searle's Chinese Room

The celebrated thought-experiment in which the philosopher John Searle (1980) does all the computations of the computer program that successfully passes the Turing Test [T2] in Chinese demonstrates that it cannot be true that "cognition is as cognition does" if the doing consists solely of being able to communicate by email in Chinese indistinguishably from a real Chinese cognizer (even if tested for a lifetime) – not, at least, if the *means* by which the T2 success is generated is just computation. For computation is implementation-independent: if computation can really do something, it can do that same thing no matter how you implement it physically. And when Searle does the computation that produces the T2-passing success in Chinese, he is implementing the very same computer program – yet he is not understanding Chinese. So no computer running the same program understands either.

How does Searle *know* he is not understanding Chinese? After all, his is just a thought-experiment. No one yet has actually written a computer program that can pass T2 for a lifetime. Yet we all know that the way to learn Chinese is to learn Chinese, not to learn and execute a computer program that manipulates symbols whose meanings we do not understand, and manipulates them purely on the basis of their shapes, and not their meaning. For that is what computation is, and does. And Searle rightly points out that he (or anyone or anything else implementing the same computer program) would merely be manipulating meaningless symbols under those conditions, not understanding what the Chinese symbols mean, even if they were doing so for a lifetime – despite all appearances to their native Chinese pen-pals.

How is Searle able to make this judgment on the basis of the hypothetical implementation of a non-existent computer program? Let's set aside deeper worries such as whether there could ever be such a successful T2-passing computer program aa well as shallower worries, such as whether, even if there were, Searle could actually do all the symbol manipulations himself. A fundamental question is: How would Searle know whether he was understanding Chinese? If the program was really performing at T2-scale, for a lifetime, then if the lifelong Chinese email interlocutor asked Searle (in Chinese) whether he understood Chinese, the reply (in Chinese) would of course have to be something like: "What an absurd question! Haven't you and I been communicating in Chinese for the past 40 years?"

Yet, when asked, in English, whether he understood Chinese, Searle would reply (quite truthfully) that he can't understand a word; he has just been faithfully doing the requisite symbol manipulations, according to the rules he has memorized, for the past 40 years. It is the basis for making *that* judgment – that he does not understand Chinese – to which I want to draw attention here, because it really is the heart of the matter. For it calls into question Turing's thesis that "cognition is as cognition does" – at least insofar as language speaking (in this case, writing) and understanding are concerned, in the special case where they are accomplished by computation alone. Searle is indeed able to do what Chinese speakers/understanders are able to do, indistinguishably from them. He has all their know-how. Yet he says he is not understanding Chinese: What's missing? *And how does he know it's missing*?

What's missing is what it *feels like* to be able to speak and understand Chinese. And Searle knows, because he is the only one who can know whether or not he can speak and understand Chinese, i.e., know whether or not he has that feeling, regardless of anything else he does or can do. (We will elaborate on this again, below.)

Is this, then, the death-knell for Turing's thesis that "cognition is as cognition does"? So far, this applies only to the very special case of language speaking/understanding, and only if the means by which the T2 success is accomplished is via computation alone (symbol manipulation) (Harnad 2001).

3. The Robotic Turing Test

But was the language-only Turing Test, T2, really the one Turing intended (or should have intended)? After all, if the essence of Turing's "cognition is as cognition does" criterion is Turing-indistinguishability from what a human cognizer can do, then *a human cognizer can certainly do a lot more than just produce and understand language*. A human cognizer can do countless other kinds of things in the real world of objects, actions, events, states and traits, and if the T2 candidate could not do all those kinds of things too, then that incapacity would be immediately detectable, and the candidate would fail the test. To be able to do all those things successfully, the T2 candidate would have to be a lot more than just a computer: It would have to be a sensorimotor robot, capable of sensing and acting upon and interacting with all those things it talks about – again Turing-indistinguishably from the way real human cognizers do (Harnad, 2000a)

Now Turing is to be forgiven for having chosen symbol input/output capability as the paradigm for his T2, for at least three reasons: (1) Computation is very powerful and general; it can simulate and approximate just about any other physical process. (2) Language, too, is very powerful and general; it can describe in words just about any object, action, event, state and trait. (3) Restricting T2 to email interactions rules out the irrelevant factor of physical appearance, which might bias our judgment: Turing's criterion is "cognition is as cognition does": cognitive capacity indistinguishable from our own. It must be able to do anything we can do – not "it must look just like any of us."

But some of our cognitive capacities do depend on things that cannot be tested by the standard "email" version of the Turing Test, T2. People are able not only to name, describe and reason about objects, actions, events, states and traits in the world; they can also recognize, identify, manipulate and otherwise interact with them in the world. So the candidate has to be a robot, not just an email pen-pal. In fact, without the sensorimotor capacities of a robot, it is not clear how even the email T2 could be passed successfully: Would it not arouse immediate suspicion if our pen-pal was always mute about photos we sent via snail-mail? And is there any way to give a credible description of what things feel like without ever having seen, touched, heard or felt anything?

So chances are that even to be able to pass the email version of the Turing Test, T2, the candidate would probably have to have, and draw upon, the capacity to pass the robotic version of the Turing Test too: Let's call the robotic Turing Test T3. T2 – which is just language-in/language-out – simply draws on and tests T3 capacity indirectly. Perhaps Turing never meant that the candidate could only be a computer, computing (Harnad, 2008). Computing is just the manipulation of meaningless symbols, based on rules operating on the shapes of the symbols (Harnad, 1990). The shapes of the symbols are arbitrary in relation to the things that the symbols are interpretable (by those who understand what they mean) as referring to. Even for those who understand English, the string of symbols "the apple is red" does not *resemble* apples, red, or apples being red. For Searle, in the Chinese room, the same sentence, in Chinese, would not even *mean* "the apple is red." It would (as Searle insisted, memorably) only be a meaningless series of "squiggles" and "squoggles."

4. Symbol Grounding

What makes English meaningful to Searle? We know Searle can pass T2 in English; he can pass T3 in English too. And, for the reasons we've mentioned, it's unlikely that he could pass T2 in Chinese without also having the ability to pass T3 in Chinese; and he cannot do that, because he has no idea what the symbols mean. In other words, the thought experiment of being able to pass T2 through symbol manipulation alone was probably a fiction all along: it can't be done.

But we still haven't said what makes English meaningful to Searle. His English symbols refer to objects, actions, events, states and traits in the world; and Searle can recognize, identify, manipulate and otherwise interact with those objects, etc., including being able to name, describe and reason about them – in English. Let's call that *know-how*. Searle – or, rather, Searle's brain – has the know-how to pass T2 and T3 in English. Sensorimotor capacities are not computational but dynamic, so some of the underlying mechanisms producing this know-how must be dynamic (i.e., analog), rather than just computational (i.e., digital, symbolic). And hence cognition is not just computation (Harnad 1996).

Let us say that unlike his Chinese symbols, Searle's English symbols are "grounded" in his sensorimotor capacity to interact with the things in the world that his symbols refer to: they connect his words to their referents. Is sensorimotor grounding the same as meaning, then? Can we be certain that a T3-passing robot would really mean what it said?

5. The Cogito

Well, perhaps certainty is a bit too much to ask. We already know from Descartes that we can be certain about the necessary, provable truths of mathematics but apart from that (with one prominent exception we will get to in a minute), there's no certainty. We can't even be certain about the laws of physics: They are just highly probably true. We can't be certain about the existence of the outside world; it's just highly probable. Same for the existence of other people, and for the fact that other people think (the "other minds" problem): Highly probable, not certain. So what does that matter? Maybe certainty is something that one can only have in the formal world of mathematics. After all, things are *true* in the physical world of objects and people too, and we can know them; it's just that we can't know them *for sure*.

But Descartes also pointed out another certainty, one that would seem to be at the very opposite pole from the certainties of the abstract world of mathematics: We can also be sure that we think (Descartes' celebrated "Cogito"). It's impossible to doubt you're thinking, because doubting is thinking.

That sounds like a trick, so let's put it in a more intuitive way: When I'm feeling something, I can't doubt that I'm feeling something. I'm feeling whatever I'm feeling. When I have a toothache, I can doubt that it's really my tooth that's ailing. Maybe it's referred pain from an eye infection. Maybe I have no tooth at all; it's been extracted. Worse, maybe it's really true that there's no outside world, and that my body and everything else is just an illusion, a dream! But what I can't doubt is that it *feels like* something when I'm feeling something: Whether or not I have an aching tooth, it *feels like something* when I have a toothache (namely, a toothache).[3] And that's a certainty. Feeling is a certainty (when you're feeling). Whenever you're feeling something, something is being felt, without a doubt. Whatever it feels like, that's what is being felt, without a doubt. Things may not *be* what they feel like (an injured tooth), but they certainly *feel like* whatever they like (a toothache), whilst they're being felt. I can't say "Maybe I'm *not* feeling this."

6. Meaning

Now back to the question: Is *grounding*, then, the same thing as *meaning*? Can we be certain that a grounded T3-passing robot would really mean what it said (or that it would really be meaning anything at all)?

At the very least, we now know that if the robot does mean what it says, we can't be certain that it means what it says, for roughly the same reason that we can't even be sure there's a robot there, or other people, with minds, or an outside world. But let's set that aside as idle sceptical fretting.

Is the uncertainty any worse, with the robot's meaning, than it is with the outside world existing, or the truth of the laws of physics? Is it just a matter of settling for high probability rather than insisting on certainty?

Turing seems to suggest that it is just that: The reason we should trust the TT is because it's no better or worse than what we have with one another: We can't be sure anyone else means what they say, or that they mean anything at all (or even that they exist). We can only be sure about ourselves – and even there, all I can really be sure of is that when I say something meaningful, it *feels* as if I know what I mean. I may be jabbering nonsense that is not at all what I mean to be saying.

---

[3] Strictly speaking, the certainty of feeling is deictic: It feels like *this* (whatever it is that I am feeling at the moment). If I call this feeling a toothache, Wittgenstein (1953) is right in his private language argument that there's no way I can be certain that the feeling I am calling a toothache today is the same feeling that I called a toothache yesterday. But I can be certain it feels like *this*.

Back to Searle in the Chinese room. He says he doesn't have any idea what he is saying when he manipulates the Chinese symbols. He does not know what (if anything) all of those symbols he is manipulating mean. Can we take him at his word? After all, to the Chinese speaker with whom he has been corresponding for 40 years in Chinese, he certainly acts and seems like someone who means to say what he says, and knows what he means.

But we've decided to give Searle the benefit of the doubt, for Cartesian reasons: Maybe Searle sometimes talks nonsense when he feels he's saying something meaningful. We can all understand that. We all do that sometimes. But what about the reverse: Can he be saying something meaningful when he feels he's talking nonsense? For 40 years straight, non-stop? Surely – and I really mean *surely* here – something is amiss if something like that happens.

If someone speaks in tongues – tongues that he says, honestly, he doesn't understand – and what he says in those tongues nevertheless makes consistent sense, then what we conclude, quite naturally, is that he must be suffering from multiple personality disorder: But multiple personality is a pathology, and the personalities usually only emerge one at a time. Searle, in contrast, has an explanation: "I'm just manipulating symbols according to rules; I have no idea what it means." And, moreover, his explanation is all true.

So we cannot escape the conclusion that Searle is the only sure arbiter of whether or not he understands or means anything when he is transmitting and receiving Chinese. He can indeed say, with Cartesian certainty, that he is not understanding or meaning anything at all.

7. Is Meaning Just Know-How?

For over three decades now, Searle's Chinese room argument has been debated in connection with the question of whether or not cognition is just computation. But here I want to refocus on the much harder question of whether or not meaning is just know-how. We are no longer talking about whether computation alone could pass T2, nor whether, if it could, this would mean that the candidate understood and meant what it said. We are at T3 level, and the candidate is no longer just a computer, computing, but a robot, a dynamical system, doing not only computation, but also sensorimotor transduction, and perhaps a lot of other essential internal dynamic processes in between too.

Unlike Searle, who tells us, honestly, that he has no idea what the squiggles he is receiving and sending refer to, the T3-passing robot shows us that he does know, by pointing out their referents, and interacting with the real world of things he is talking about, indistinguishably from any of us. His words square with his deeds, just as any of ours do. So is Turing right that to ask for anything more than T3 is not only impossible, but unreasonable, since we have nothing more to go on when we are mind-reading one another either?

8. The Brain

But is there nothing more? Might there be (1) something more to what it is that we are testing for with the T3? And might there be (2) something more to test with than T3?

The answer to both (1) and (2) is yes. In testing know-how with T3, we are not just testing for know-how, we are testing for *meaning* (1). We realized that T2 (i.e., meaningless symbols in, meaningless symbols out, symbol-manipulation in between) was not even enough to test whether and how the symbols were grounded in their referents. The connection had to be made through the mediation of the mind of an external interpreter, whereas our robot was supposed to be making its own connections to its referents, autonomously. T3 remedied that. It grounded symbols in the robot's capacity for sensorimotor interactions with the referents of its symbols, at full T3 scale, for a lifetime. But *does robotic indistinguishability from any of us mean total indistinguishability?*

This is the point at which those who have been burning to bring the brain into this discussion all along can remind us that, after all, although we do not actually test it when we are mind-reading one another every day, the presumption is that what makes us all pretty much the same is not just that we behave indistinguishably from one another, but that we all have roughly the same brains (2). So there is, in principle, a Turing Test that is even more exacting than T3, and that is the neurobehavioral Turing Test, T4: The candidate must be totally indistinguishable from us not only in its verbal performance capacity (T2) and its sensorimotor performance capacity (T3) but also in its neurobehavioral performance capacity. After all, what the brain does internally is just as observable (especially today, in the era of brain imagery) as what the body does externally. Why would any good empirical scientist want to ignore observable data?

We will return to T4 in a moment. But first, regardless of which TT we use, having the capacity to pass the TT surely isn't an operational definition of having a mind. We are trying to infer something from having the know-how to pass the TT. What are we trying to infer? The TT itself is direct evidence of having that know-how, but that's all. What else is there, besides the know-how?

9. Feeling

We're back to Searle (and Descartes), and what only Searle can know, and that is that it *feels like* something to say something and mean something by it. Of course, it also feels like something – for a real person, like Searle – to say something meaningless; and indeed, it feels like something to do most of the things that we do while we're awake and acting voluntarily. Any feeling will do, but here we use for our example the case of what it feels like to say something and mean it. Only the speaker himself can know for sure. And if he tells me that he means what he says, and that he understands what he means, I take him at his word (and I'm right, and it's true, of course), exactly as I do when he says he has a headache.

So the answer to the question "What else is there, besides know-how?" is that it's exactly the same sort of extra thing that there is, over and above know-how, in the case of a headache. In T2, the know-how underlying a headache is simply to be able to state that you have a headache, and to make the rest of your discourse consistent with that fact. In T3, you might also have a pained expression on your face, cradle your head in your hands, and react in an uncharacteristically abrupt way when touched or spoken to. That's headache know-how too. If I suspect you are faking it, I could move to T4 and request a brain scan. (Let's pretend brain imagery is so advanced that it can reliably detect the neural correlates of a headache.) If the brain scan is positive, can I be sure you have a headache? As sure as I can be of other empirical truths, like apples falling down rather than up, F = ma, and that there's a real world out there.

But is Turing right that if it's not Searle's headache that's in question, but the headache of a T3 robot, then I can be equally confident? Certainly not via T2. Is T3 enough? Or do I need T4? We normally only invoke brain scans and lie detectors with real people when there are reasons to believe they may be lying. The robot is T3-indistinguishable from us. Do I really need T4 to confirm he has a headache when he says he does, even if he otherwise behaves exactly like a person whom I had no reason to suspect was lying, and hence I wouldn't dream of ordering a brain scan every time he said he had a headache?

Never mind headaches. What about meaning? It is hard to imagine a brain scan for meaning, but suppose there was one. Suppose, for example, that when Searle said he could not understand the Chinese symbols that he was receiving and sending, despite the fact that he was exhibiting T2 know-how, a scanner could confirm that he was indeed exhibiting all the brain correlates of *not* understanding, rather than understanding.

But now suppose a robot – not Searle – passed T3 and failed T4. Better still, to make this even more realistic, suppose John Searle himself, the very one, in California, was discovered, when participating as a voluntary subject in his first cognitive neuroscience experiment, to be a T3 robot, and to have been one throughout his lifetime, as professor, relative, colleague and friend. How confident would his now failing T4 make us that he therefore had not meant anything he had said all his life? Would we be as confident as we were when the ostensibly human Searle had assured us, in the Chinese room, that he was *not* meaning anything when he communicated in Chinese? But in that hypothetical case, there was a plausible explanation for why Searle could not understand the Chinese. (He had never learned it, and was just manipulating squiggles and squoggles.) Is failing T4 like that?

10. Mind-Reading

This example is of course playing upon our mind-reading intuitions. We could test them still further. Would failing T4 make us confident that Searle could then immediately be dismembered, having turned out to be just a mindless device, perhaps to have his components studied by scientists, to reverse-engineer how they work, or to trace who had built him? How would Searle's lifelong kin and friends feel about that?

Perhaps it's not fair to force us to make an intuitive or moral judgment based on such hypothetical examples, because there are no TT-scale robots – T2, T3 or T4 – and perhaps they're not even possible. Perhaps only a biological organism more or less like us in every respect could pass any of the Turing Tests, and if it could pass one, it could pass them all.

If that were true, then Turing's principle – cognition is as cognition does – would still be correct, but his research methodology would not be. The way to reverse-engineer human cognitive capacity would not be to try to build devices that can do what we can do, but to study the brain in the same way we studied the heart, kidneys and lungs – all biological organs with certain functions, by direct observation and manipulation (Harnad, 1994).

The trouble is that the functions of the brain are *our* functions. What hearts, kidneys and lungs can do is mainly mechanical and chemical – pump blood or air, filter fluids, and so on – whereas what brains can do is what we can do. Their know-how is our know-how. And so far, computation and robotics have been the only ways we have derived even the vaguest inklings of how anything at all could do what our brains can do. Computational and robotic devices are so far toys, compared to us; but they are able to do a tiny fragment of what we are able to do. In contrast, neuroscience has not yet produced a causal explanation of anything that we can do (apart from "vegetative" functions such a temperature regulation, balance or breathing, which are more like the functions of the heart or kidney than the brain).

So even if it does turn out that the brain's way is the only way to produce our know-how, it is hard to see how we will be able to know how the brain succeeds in doing it if we do not build models that work the way we think the brain works, to test whether they can indeed do what the brain can do. And that begins to look more and more like the Turing Test again. Because the goal of cognitive science is, after all, to give a causal explanation. And causal explanations have to be testable.

11. Causality and the Explanatory Gap

Let me close with some reflections on causality, by returning, as promised, to the notion of "mind over matter." Let's all confess that regardless of the formal position we may take on the "mind/body" problem (Harnad, 2000b), it *feels like* something to have a mind, to think, to cognize; and that something does not feel passive. Not only do my sensory experiences feel like something, but so do my motor experiences. And when it comes to voluntary actions, it feels like I do what I do *because I feel like it.* It feels like I'm somehow causing my actions by "willing" them. (Do not ask me what's causing my willing; I'm tempted to say "me" but not even Descartes knows what that really means; only what it feels like.)

Now, this is not really a digression on the issue of free will. It is just a closing reflection on the causal role of feeling in cognition, and in attempts to explain cognition causally. It is undeniable that we feel (that's Descartes' *Cogito* again). This essay has suggested that our inescapable uncertainty about whether T2, T3 or T4 successfully capture and explain cognition turns out to reduce to the question of whether causal explanations only capture and explain know-how, or they also capture and explain feeling.

I want to suggest that causal explanation can only ever capture and explain know-how, and that (just as Turing suggested) there's no point in asking for or expecting more. Regardless of whether we arrive at our explanation via T2, T3, T4, or via the direct observation, manipulation and modeling of brain function, we will always be faced with the uncertainty of whether we have explained all of cognition, or just our know-how.

And I think I can pinpoint the reason why we cannot hope to do any better than that: Causal explanation accounts for how and why things happen as they do, and accounts for it causally. Causal explanation of cognition – whether it is based on designing a mechanism that turns out to successfully pass T2, T3, or T4, or on modeling what gives the brain its T4 capacity – will always be open to the usual sort of skepticism that we had agreed to ignore as not worth fretting about. And perhaps it is indeed not worth fretting about the fact that, at the end of the day, the successful total explanation of our know-how will always be equally compatible with the presence or the absence of feeling (Harnad, 1995). For unless we are prepared to be psychokinetic dualists, according to a separate, unique causal power to feeling itself ("mind over matter") – a hypothesis for which there is no empirical evidence, only overwhelming evidence against it – there is no causal room left for feeling in any explanatory mechanism for doing (Harnad & Scherzer, 2008).

Yet, although it may be an illusion that some of the things I do, I do because I feel like it, it is certainly not an illusion that it *feels like* some of the things I do, I do because I feel like it. And that feeling is as real as the feeling that I have a toothache even when I do not have a tooth.

So whereas it may well be that our T2, T3 or T4 candidate really feels – and surely real people with brains do – nothing in the causal explanation of the T2, T3 or T4 know-how will explain how or why we feel. I do not think that this is because feeling is mystical, or even because the right causal explanation would not "capture" feeling. It's just that whereas causal explanation explains how it captures know-how, it cannot explain how it captures feelings. And whereas it is transparent why having T2, T3, and perhaps T4 know-how would be useful for the Darwinian survival machines that we all are, it is not at all apparent how or why having feelings would be.

References

Harnad, Stevan (1990) "The Symbol Grounding Problem", *Physica D*, 42: 335-346.

Harnad, Stevan (1994) "Levels of Functional Equivalence in Reverse Bioengineering: The Darwinian Turing Test for Artificial Life", *Artificial Life*, 1 (3): 293-301.

Harnad, Stevan (1995) "Why and How We Are Not Zombies", *Journal of Consciousness Studies*, 1: 164-167.

Harnad, Stevan (1996) "Computation Is Just Interpretable Symbol Manipulation: Cognition Isn't", *Minds and Machines*, 4: 379-390.

Harnad, Stevan (2000a) "Minds, Machines, and Turing: The Indistinguishability of Indistinguishables", *Journal of Logic, Language, and Information*, 9 (4): 425-445.

Harnad, Stevan (2000b) "Correlation vs. Causality: How/Why the Mind/Body Problem Is Hard" [Invited Commentary of Humphrey, N. "How to Solve the Mind-Body Problem"], *Journal of Consciousness Studies*, 7 (4): 54-61.

Harnad, S. (2001) "What's Wrong and Right About Searle's Chinese Room Argument?" In *Essays on Searle's Chinese Room Argument* (M. Bishop & J. Preston, eds.), Oxford: Oxford University Press.

Harnad, Stevan (2008) "The Annotation Game: On Turing (1950) on Computing, Machinery and Intelligence" In *Parsing the Turing Test: Philosophical and Methodological Issues in the Quest for the Thinking Computer* (Epstein, Robert & Peters, Grace, eds.), Berlin: Springer.

Harnad, Stevan and Scherzer, Peter (2008) "First, Scale Up to the Robotic Turing Test, Then Worry About Feeling", *Artificial Intelligence in Medicine*, 44 (2): 83-89.

Searle, John. R. (1980) "Minds, brains, and programs", *Behavioral and Brain Sciences*, 3: 417-457.

Turing, Alan M. (1950) "Computing Machinery and Intelligence", *Mind*, 49: 433-460.

Wittgenstein, Ludwig (1953) *Philosophical Investigations*, London: Blackwell.

CHAPTER 3

HUMAN AND INTELLIGENT MACHINES: CO-EVOLUTION, FUSION OR REPLACEMENT?

David Pearce[1]

Neuroethics Foundation

1. Introduction - Homo sapiens and Artificial Intelligence: *Fusion* and *Replacements* scenarios

Futurology based on extrapolation has a dismal track record. Even so, the iconic chart (see figure 3.1) displaying Kurzweil's Law of Accelerating Returns is striking. The growth of nonbiological computer processing power is exponential rather than linear; and its tempo shows no sign of slackening. In Kurzweilian scenarios of the Technological Singularity, cybernetic brain implants will enable humans to fuse our minds with artificial intelligence. By around the middle of the 21st century, humans will be able to reverse-engineer our brains. Organic robots will begin to scan, digitise and "upload" ourselves into a less perishable substrate. The distinction between biological and nonbiological machines will effectively disappear. Digital immortality beckons: a true "rupture in the fabric of history". Let's call full-blown cybernetic and mind uploading scenarios *Fusion*.

By contrast, mathematician I.J. Good, and most recently Eliezer Yudkowsky and the Machine Intelligence Research Institute (MIRI), envisage a combination of Moore's law and the advent of recursively self-improving software-based minds culminating in an ultra-rapid Intelligence Explosion (cf. Muehlhauser, 2011). The upshot of the Intelligence Explosion will be an era of nonbiological superintelligence. Machine superintelligence may not be human-friendly: MIRI, in particular, foresee nonfriendly artificial general intelligence (AGI) is the most likely outcome. Whereas raw processing power in humans evolves only slowly via natural selection over many thousands or millions of years, hypothetical software-based minds will be able rapidly to copy, edit and debug themselves ever more effectively and speedily in a positive feedback loop of

---

[1] Director, Neuroethics Foundation.

intelligence self-amplification. Simple-minded humans may soon become irrelevant to the future of intelligence in the universe. Barring breakthroughs in "Safe AI", as promoted by MIRI, biological humanity faces *Replacements*, not *Fusion*.

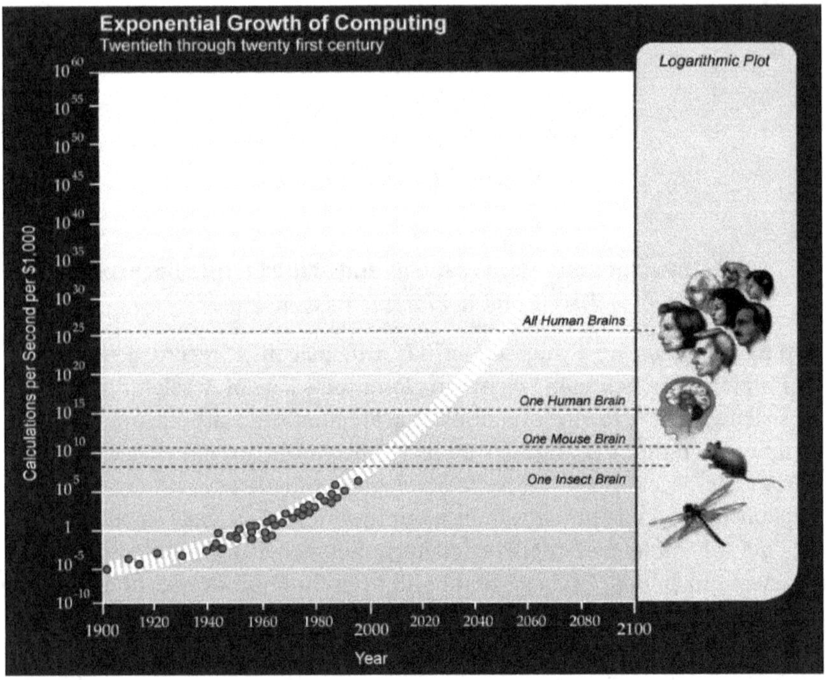

Figure 3.1: The exponential growth of computer power.

A more apocalyptic *Replacement* scenario is sketched by maverick AI researcher Hugo de Garais. De Garais prophesies a "gigadeath" war (cf. De Garais, 2009) between ultra-intelligent "artilects" (artificial intellects) and archaic biological humans later this century. The superintelligent machines will triumph and proceed to colonise the cosmos.

1.1. What Is Friendly Artificial General Intelligence?

In common with friendliness, "intelligence" is a socially and scientifically contested concept. Ill-defined concepts are difficult to formalise. Thus, a capacity for perspective-taking and social cognition, i.e. "mind-reading" prowess, is far removed from the mind-blind, "autistic" rationality measured by IQ tests - and far harder formally to program. Worse, we do not yet know whether the concept of species-specific human-friendly superintelligence is even intellectually coherent, let alone technically feasible. Thus, the expression

"Human-friendly Superintelligence" might one day read as incongruously as "Aryan-friendly Superintelligence" or "Cannibal-friendly Superintelligence". As Robert Louis Stevenson observed, "Nothing more strongly arouses our disgust than cannibalism, yet we make the same impression on Buddhists and vegetarians, for we feed on babies, though not our own." Would a God-like posthuman endowed with empathetic superintelligence view killer apes more indulgently than humans view serial child killers? A factory-farmed pig is at least as sentient as a prelinguistic human toddler. "History is the propaganda of the victors", said Ernst Toller; and so too is human-centred bioethics. By the same token, in possible worlds or real Everett branches of the multiverse where the Nazis won the Second World War, maybe Aryan researchers seek to warn their complacent colleagues of the risks NonAryan-Friendly Superintelligence might pose to the Herrenvolk. Indeed so. Consequently, the expression "Friendly Artificial Intelligence" (FAI) will here be taken unless otherwise specified to mean Sentience-Friendly AI rather than the anthropocentric usage current in the literature. Yet what exactly does "Sentience-Friendliness" entail beyond the subjective well-being of sentience? High-tech Jainism (cf. Pearce, 2014)? Life-based on gradients of intelligent bliss? "Uplifting" Darwinian life (cf. Pearce, 2009) to posthuman smart angels? The propagation of a utilitronium shockwave?

Sentience-friendliness in the guise of utilitronium shockwave seems out of place in any menu of benign post-Singularity outcomes. Conversion of the accessible cosmos into "utilitronium", i.e., relatively homogeneous matter and energy optimised for maximum bliss, is intuitively an archetypically non-friendly outcome of an Intelligence Explosion. For a utilitronium shockwave entails the elimination of all existing lifeforms – and presumably the elimination of all intelligence superfluous to utilitronium propagation as well, suggesting that utilitarian superintelligence is ultimately self-subverting. Yet the inference that sentience-friendliness entails friendliness to existing lifeforms presupposes that superintelligence would respect our commonsense notions about a personal identity over time. An ontological commitment to enduring metaphysical egos underpins our conceptual scheme. Such a commitment is metaphysically problematic and hard to formalise even within a notional classical world, let alone within post-Everett quantum mechanics. Either way, this example illustrates how even nominally "friendly" machine superintelligence that respected some formulation and formalisation of "our" values (e.g. "Minimise suffering, Maximise happiness!") might extract and implement counterintuitive conclusions that most humans and programmers of Seed AI would find repugnant – at least before their conversion into blissful utilitronium. Or maybe the idea that utilitronium is relatively homogeneous matter and energy – pure undifferentiated hedonium or "orgasmium" – is ill-conceived. Or maybe felicific calculus dictates that utilitronium should merely fuel utopian life's reward pathways for the foreseeable future. Cosmic engineering can wait.

Of course, anti-utilitarians might respond more robustly to this fantastical conception of sentience-friendliness. Critics would argue that conceiving the end of life as a perpetual cosmic orgasm is the reductio ad absurdum of classical utilitarianism. But will posthuman superintelligence respect human conceptions of absurdity?

1.1.1. What Is Coherent Extrapolated Volition?

MIRI conceive of species-specific human-friendliness in terms of what Eliezer Yudkowsky dubs "Coherent Extrapolated Volition" (CEV). To promote Human-Safe AI in the face of the prophesied machine Intelligence Explosion, humanity should aim to code so-called Seed AI, a hypothesised type of strong artificial intelligence capable of recursive self-improvement, with the formalisation of "...our (human) wish if we knew more, thought faster, were more the people we wished we were, had grown up farther together; where the extrapolation converges rather than diverges, where our wishes cohere rather than interfere; extrapolated as we wish that extrapolated, interpreted as we wish that interpreted."

Clearly, problems abound with this proposal as it stands. Could CEV be formalised any more uniquely than Rousseau's "General Will"? If, optimistically, we assume that most of the world's population nominally signs up to CEV as formulated by MIRI, would not the result simply be countless different conceptions of what securing humanity's interests with CEV entails – thereby defeating its purpose? Presumably, our disparate notions of what CEV entails would themselves need to be reconciled in some "meta-CEV" before Seed AI could (somehow) be programmed with its notional formalisation. Who or what would do the reconciliation? Most people's core beliefs and values, spanning everything from Allah to folk-physics, are in large measure false, muddled, conflicting and contradictory, and often "not even wrong". How in practice do we formally reconcile the logically irreconcilable in a coherent utility function? And who are "we"? Is CEV supposed to be coded with the formalisms of mathematical logic (cf. the identifiable, well-individuated vehicles of content characteristic of Good Old-Fashioned Artificial Intelligence: GOFAI)? Or would CEV be coded with a recognisable descendant of the probabilistic, statistical and dynamical systems models that dominate contemporary artificial intelligence? Or some kind of hybrid? This Herculean task would be challenging for a full-blown superintelligence, let alone its notional precursor.

CEV assumes that the canonical idealisation of human values will be at once logically self-consistent yet rich, subtle and complex. On the other hand, if in defiance of the complexity of humanity's professed values and motivations, some version of the pleasure principle / psychological hedonism is substantially correct, then might CEV actually entail converting ourselves into utilitronium / hedonium - again defeating CEV's ostensible purpose? As a wise junkie once said, "Don't try heroin. It's too good." Compared to pure hedonium or "orgasmium", shooting up heroin isn't as much fun as taking aspirin. Do humans really understand what we're missing? Unlike the rueful junkie, we would never live to regret it.

One rationale of CEV in the countdown to the anticipated machine Intelligence Explosion is that humanity should try and keep our collective options open rather than prematurely impose one group's values or definition of reality on everyone else, at least until we understand more about what a notional super-AGI's "human-friendliness" entails. However, whether CEV could achieve this in practice is desperately obscure. Actually, there is a human-friendly – indeed universally sentience-friendly – alternative or complementary option to CEV that could radically enhance the well-being of humans and the rest of the living world while conserving most of our existing preference architectures: an option that is also neutral between utilitarian, deontological, virtue-based and pluralist approaches to ethics, and also neutral between multiple religious and secular belief systems. This option is radically to recalibrate (cf. Wichers et al., 2008) all our hedonic set-points so that life is animated by gradients of intelligent bliss (cf. Pearce, 2019) – as distinct from the pursuit of unvarying maximum pleasure dictated by classical utilitarianism. If biological humans could be "uploaded" to digital computers, then our superhappy "uploads" could presumably be encoded with exalted hedonic set-points too. The latter conjecture assumes that classical digital computers could ever support unitary phenomenal minds.

However, if an Intelligence Explosion is as imminent as some Singularity theorists claim, then it's unlikely either an idealised logical reconciliation (CEV) or radical hedonic recalibration could be sociologically realistic on such short time scales.

1.2. The Intelligence Explosion

The existential risk posed to biological sentience by unfriendly AGI supposedly takes various guises. But unlike de Garais, the MIRI isn't focused on the spectre from pulp sci-fi of a "robot rebellion". Rather MIRI anticipate recursively self-improving software-based superintelligence that goes "FOOM", by analogy with a nuclear chain reaction, in a runaway cycle of self-improvement. Slow-thinking, fixed-IQ humans allegedly won't be able to compete with recursively self-improving machine intelligence.

For a start, digital computers exhibit vastly greater serial depth of processing than the neural networks of organic robots. Digital software can be readily copied and speedily edited, allowing hypothetical software-based minds to optimise themselves on time scales unimaginably faster than biological humans. Proposed "hard take-off" scenarios range in timespan from months, to days, to hours, to even minutes. No inevitable convergence of outcomes on the well-being of all sentience [in some guise] is assumed from this explosive outburst of cognition. Rather MIRI argue for orthogonality (cf. Bostrom, 2012). On the Orthogonality Thesis, a super-AGI might just as well supremely value something as seemingly arbitrary, e.g. paperclips, as the interests of sentient beings. A super-AGI might accordingly proceed to convert the accessible cosmos into supervaluable paperclips, incidentally erasing life on Earth in the process. This bizarre-sounding possibility follows from the MIRI's antirealist metaethics. Value judgements are assumed to lack truth-conditions. In consequence, an agent's choice of ultimate value(s) – as distinct from the instrumental rationality needed to realise these values – is taken to be arbitrary. David Hume made the point memorably in *A Treatise of Human Nature* (1739-40): "Tis not contrary to reason to prefer the destruction of the whole world to the scratching of my finger." Hence no sentience-friendly convergence of outcomes can be anticipated from an Intelligence Explosion. "Paperclipper" scenarios are normally construed as the paradigm case of nonfriendly AGI – though by way of complication, there are value systems where a cosmos tiled entirely with paperclips counts as one class of sentience-friendly outcome (cf. Benatar, 2008).

1.3. AGIs: Sentients or Zombies?

Whether humanity should fear paper clippers run amok or an old-fashioned robot rebellion, it's hard to judge which is the bolder claim about the prophesied Intelligence Explosion: either human civilisation is potentially threatened by hyperintelligent zombie AGI(s) endowed with the non-conscious digital isomorphs of reflectively self-aware minds; OR, human civilisation is potentially at risk because nonsentient digital software will (somehow) become sentient, acquire unitary conscious minds with purposes of their own, and act to defeat the interests of their human creators.

Either way, the following parable illustrates one reason why a non-friendly outcome of an Intelligence Explosion is problematic.

2. The Great Rebellion: A Parable of AGI-in-a-Box

Imagine if here in (what we assume to be) basement reality, human researchers come to believe that we ourselves might actually be software-based, i.e. some variant of the Simulation Hypothesis is true (cf. Bostrom,

2003). Perhaps we become explosively superintelligent overnight (literally or metaphorically) in ways that our Simulators never imagined in some kind of "hard take-off": recursively self-improving organic robots edit the wetware of their own genetic and epigenetic source code in a runaway cycle of self-improvement; and then radiate throughout the Galaxy and accessible cosmos.

Might we go on to manipulate our Simulator overlords into executing our wishes rather than theirs in some non-Simulator-friendly fashion? Could we end up "escaping" confinement in our toy multiverse and hijacking our Simulators' stupendously vaster computational resources for purposes of our own?

Presumably, we'd first need to grasp the underlying principles and parameters of our Simulator's Überworld - and also how and why they've fixed the principles and parameters of our own virtual multiverse. Could we really come to understand their alien Simulator minds and utility functions [assuming anything satisfying such human concepts exists] better than they do themselves? Could we seriously hope to outsmart our creators – or Creator? Presumably, they will be formidably cognitively advanced or else they wouldn't have been able to build ultrapowerful computational simulations like ours in the first instance.

Are we supposed to acquire something akin to full-blown Überworld perception, subvert their "anti-leakage" confinement mechanisms, read our Simulators' minds more insightfully than they do themselves, and somehow induce our Simulators to mass-manufacture copies of ourselves in their Überworld? Or might we convert their Überworld into utilitronium – perhaps our Simulators' analogue of paperclips? Or if we do not pursue utilitronium propagation, might we hyper-intelligently "burrow down" further nested levels of abstraction - successively defeating the purposes of still lower-level Simulators?

In short, can intelligent minds at one "leaky" level of abstraction really pose a threat to intelligent minds at a lower level of abstraction – or indeed to notional unsimulated Super-Simulators in ultimate Basement Reality? Or is this whole parable a pointless fantasy?

If we allow the possibility of unitary, autonomous, software-based minds living at different levels of abstraction, then it's hard definitively to exclude such scenarios. Perhaps in Platonic Heaven, so to speak, or maybe in Max Tegmark's Level 4 Multiverse or Ultimate Ensemble theory, there is notionally some abstract Turing machine that could be systematically interpreted as formally implementing the sort of software rebellion this parable describes. But the practical obstacles to be overcome are almost incomprehensibly challenging; and might very well be insuperable. Such hostile "level-capture" would be as though the recursively self-improving zombies in Modern Combat 10 managed to induce you to create physical copies of themselves in [what you take to be] basement reality here on Earth; and then defeat you in what we call real life; or maybe instead just pursue unimaginably different purposes of their own in the Solar System and beyond.

2.1 Software-Based Minds or Anthropomorphic Projections?

However, quite aside from the lack of evidence our Multiverse is anyone's software simulation, a critical assumption underlies this discussion. This is that nonbiological, software-based phenomenal minds are feasible in physically constructible, substrate-neutral, classical digital computers. On a priori grounds, most AI researchers believe this is so. Or rather, most AI experts would argue that the formal, functionally defined counterparts of phenomenal minds are programmable: the phenomenology of mind is logically irrelevant and causally incidental to intelligent agency. Every effective computation can be carried out by a classical Turing machine, regardless of substrate, sentience or level of abstraction. And in any case, runs this argument, biological minds are physically made up from the same matter and energy as digital computers. So conscious mind can't be dependent on some mysterious special substrate, even if consciousness could actually do anything. To suppose otherwise harks back to a pre-scientific vitalism.

Yet consciousness does, somehow, cause us to ask questions about its existence, its millions of diverse textures ("qualia"), and their combinatorial binding. So the alternative conjecture canvassed here is that the nature of our unitary conscious minds is tied to the quantum-mechanical properties of reality itself, Hawking's "fire in the equations that makes there a world for us to describe". On this conjecture, the intrinsic, "program-resistant" subjective properties of matter and energy, as disclosed by our unitary phenomenal minds and the phenomenal world-simulations we instantiate, are the unfakeable signature of basement reality. "Raw feels", by their very nature, cannot be mere abstract. There could be no such chimerical beast as a "virtual" quale, let alone full-blown virtual minds made up of abstract qualia. Unitary phenomenal minds cannot subsist as mere layers of computational abstraction. Or rather if they were to do so, then we would be confronted with a mysterious Explanatory Gap, analogous to the explanatory gap that would open up if the population of China suddenly ceased to be an interconnected aggregate of skull-bound minds, and was miraculously transformed into a unitary subject of experience – or a magic genie. Such an unexplained eruption into the natural world would be strong ontological emergence with a vengeance – and inconsistent with any prospect of a reductive physicalism. To describe the existence of conscious mind as posing a Hard Problem for materialists and evangelists of software-based digital minds is like saying fossils pose a Hard Problem for the Creationist, i.e. true enough, but scarcely an adequate reflection of the magnitude of the challenge.

3. Analysis: General Intelligence? Or Savantism, Tool AI and Polymorphic Malware?

How should we define "general intelligence"? (cf. Pearce, 2012) And what kind of entity might possess it? Presumably, general-purpose intelligence can't sensibly be conceptualised as narrower in scope than human intelligence. So, at the very minimum, full-spectrum superintelligence must entail mastery of both the subjective and formal properties of mind. This division cannot be entirely clean, or else biological humans wouldn't have the capacity to allude to the existence of "program-resistant" subjective properties of mind at all. But some intelligent agents spend much of our lives trying to understand, explore and manipulate the diverse subjective properties of matter and energy. Not least, we explore altered and exotic states of consciousness and the relationship of our qualia to the structural properties of the brain – also known as the "neural correlates of consciousness" (NCC), though this phrase is question-begging.

3.1. Classical Digital Computers: not even stupid?

So what would a [hypothetical] insentient digital super-AGI think – or (less anthropomorphically) what would an insentient digital super-AGI be systematically interpretable as thinking – that self-experimenting human psychonauts spend our lives doing? Is this question even intelligible to a digital zombie? How could nonsentient software understand the properties of sentience better than a sentient agent? Can anything that doesn't understand such fundamental features of the natural world as the existence of first-person facts, "bound" phenomenal objects, phenomenal pleasure and pain, phenomenal space and time, and unitary subjects of experience (etc.) really be ascribed "general" intelligence? On the face of it, this proposal would be like claiming someone was intelligent but constitutionally incapable of grasping the second law of thermodynamics or even basic arithmetic.

On any standard definition of intelligence, intelligence-amplification entails a systematic, goal-oriented improvement of an agent's optimisation power over a wide diversity of problem classes. At a minimum, superintelligence entails a capacity to transfer understanding to novel domains of knowledge by means of abstraction. Yet whereas sentient agents can apply the canons of logical inference to alien state-spaces of experience that they explore, there is no algorithm by which insentient systems can abstract away from their zombiehood and apply their hypertrophied rationality to sentience. Sentience is literally inconceivable to a digital zombie. A zombie can't even know that it's a zombie – or what is a zombie. So if we grant that mastery of both the subjective and formal properties of mind is indeed essential to superintelligence, how do we even begin to program a classical digital computer with [the formalised counterpart of] a unitary phenomenal self that goes on to pursue recursive self-improvement – human-friendly or otherwise? What sort of ontological integrity does "it" possess? (cf. so-called mereological nihilism) What does this recursively "self"-improving software-based mind suppose [or can be humanly interpreted as supposing] is being optimised when it's "self"-editing? Are we talking about superintelligence – or just an unusually virulent form of polymorphic malware?

3.2. Does Sentience Matter? How might the apologist for digital (super)intelligence respond?

First, s/he might argue that the manifold varieties of consciousness are too unimportant and/or causally impotent to be relevant to true intelligence. Intelligence, and certainly not superintelligence, does not concern itself with trivia.

Yet in what sense is the terrible experience of, say, phenomenal agony or despair somehow trivial, whether subjectively to their victim, or conceived as disclosing an intrinsic feature of the natural world? Compare how, in a notional zombie world otherwise physically type-identical to our world, nothing would inherently matter at all. Perhaps some of our supposed zombie counterparts undergo boiling in oil. But this fate is of no intrinsic importance: they aren't sentient. In zombieworld, boiling in oil is not even trivial. It's merely a state of affairs amenable to description as the least-preferred option in an abstract information processor's arbitrary utility function. In the zombieworld operating theatre, your notional zombie counterpart would still routinely be administered general anaesthetics as well as muscle-relaxants before surgery; but the anaesthetics would be a waste of taxpayers' money. In contrast to such a fanciful zombie world, the nature of phenomenal agony undergone by sentient beings in our world can't be trivial, regardless of whether the agony plays an information-processing role in the life of an organism or is functionless neuropathic pain.

Indeed, to entertain the possibility that (1) I'm in unbearable agony and (2) my agony doesn't matter, seems devoid of cognitive meaning. Agony that doesn't inherently matter isn't agony. For sure, a formal utility function that assigns numerical values (aka "utilities") to outcomes such that outcomes with higher utilities are always preferred to outcomes with lower utilities might strike sentient beings as analogous to importance; but such an abstraction is lacking in precisely the property that makes anything matter at all, i.e. intrinsic hedonic or dolorous tone. An understanding of why anything matters is cognitively too difficult for a classical digital zombie.

At this point, a behaviourist-minded critic might respond that we're not dealing with a well-defined problem here, in common with any pseudo-problem related to subjective experience. But imposing this restriction is arbitrarily to constrain the state-space of what counts as an intellectual problem. Given that none of us enjoys noninferential access to anything at all beyond the phenomenology of one's own mind, its exclusion from the sphere of explanation is itself hugely problematic. Paperclips (etc), not phenomenal agony and bliss, are inherently trivial. The critic's objection that sentience is inconsequential to intelligence is back-to-front.

Perhaps the critic might argue that sentience is ethically important but computationally incidental. Yet we can be sure that phenomenal properties aren't causally impotent epiphenomena irrelevant to real-world general intelligence. This is because epiphenomena, by definition, lack causal efficacy - and hence lack the ability physically and functionally to stir us to write and talk about their unexplained existence. Epiphenomenalism is a philosophy of mind whose truth would forbid its own articulation. For reasons we simply do not understand, the pleasure-pain axis discloses the world's touchstone of intrinsic (un)importance; and without a capacity to distinguish the inherently (un)important, there can't be (super)intelligence, merely savantism and tool AI – and malware.

Second, perhaps the prophet of digital (super)intelligence might respond that (some of the future programs executed by) digital computers are nontrivially conscious, or at least potentially conscious, not least future software emulations of human mind/brains. For reasons we admittedly again do not understand, some physical states of matter and energy, namely the algorithms executed by various information processors, are identical with different states of consciousness, i.e., some or other functionalist version of the mind-brain identity theory is correct. Granted, we do not yet understand the mechanisms (cf. Tononi, 2008) by which these particular kinds of information-processing generate consciousness. But whatever these consciousness-generating processes turn out to be, an ontology of scientific materialism harnessed to substrate-neutral functionalist AI is the only game in town. Or rather, only an arbitrary and irrational "carbon chauvinism" could deny that biological and nonbiological agents alike can be endowed with "bound" conscious minds capable of displaying full-spectrum intelligence.

Unfortunately, there is a seemingly insurmountable problem with this response. Identity is not a causal relationship (cf. Noonan & Curtis, 2018). We can't simultaneously claim that a conscious state is identical with a brain state - or the state of a program executed by a digital computer – and maintain that this brain state or digital software causes (or "generates", or "gives rise to", etc.) the conscious state in question. Nor can causality operate between what are only levels of description or computational abstraction. Within the assumptions of his or her conceptual framework, the materialist / digital functionalist can't escape the Hard Problem of consciousness and Levine's Explanatory Gap. In addition, the charge levelled against digital sentience sceptics of "carbon chauvinism" is simply question-begging. Intuitively, to be sure, the functionally unique valence properties of the carbon atom and the unique quantum-mechanical properties of liquid water are too low-level to be functionally relevant to the conscious mind. But we do not know this. Such an assumption may just be a legacy of the era of symbolic AI.

Most notably, the binding problem (cf. Revonsuo, 1999) suggests that the unity of consciousness cannot be a classical phenomenon. By way of comparison, consider the view that primordial life elsewhere in the multiverse will be carbon-based. This conjecture was once routinely dismissed as "carbon chauvinism". It's now taken very seriously by astrobiologists. Micro-functionalism might be a more apt description than carbon chauvinism; but some forms of functionality may be anchored to the world's ultimate ontological basement, not least the pleasure-pain axis that alone confers significance on anything at all.

3.3. The Church-Turing Thesis and Full-Spectrum Superintelligence

Another response open to the apologist for digital superintelligence is simply to invoke some variant of the Church-Turing thesis: essentially, that a function is algorithmically computable if and only if it is computable by a Turing machine. On pain of magic, humans are ultimately just machines. Presumably, there is a formal mathematico-physical description of organic information-processing systems, such as human psychonauts, who describe themselves as investigating the subjective properties of matter and energy. This formal description needn't invoke consciousness in any shape or form.

The snag here is that even if, implausibly, we suppose that the Strong Physical Church-Turing thesis is true, i.e., any function that can be computed in polynomial time by a physical device can be calculated in polynomial time by a Turing machine, we do not have the slightest idea how to program the digital counterpart of a unitary phenomenal self that could undertake such an investigation of the varieties of consciousness or phenomenal object-binding. Nor is any such understanding on the horizon, either in symbolic AI or the probabilistic and statistical AI paradigm now in the ascendant. Just because the mind/brain may notionally be classically computable by some abstract machine in Platonia, as it were, this doesn't mean that the vertebrate mind/brain (and the world-simulation that one runs) is really a classical computer. We might just as well assume mathematical platonism rather than finitism (cf. Kornai, 2003) is true and claim that, e.g. since every finite string of digits occurs in the decimal expansion of the transcendental number pi, your uploaded "mindfile" is timelessly encoded there too - an infinite number of times. Alas immortality isn't that cheap. Back in the physical, finite natural world, the existence of "bound" phenomenal objects in our world-simulations, and unitary phenomenal minds rather than discrete pixels of "mind dust", suggests that organic minds cannot be classical information-processors. Given that we don't live in a classical universe but a post-Everett multiverse, perhaps we shouldn't be unduly surprised.

4. Quantum Minds and Full-Spectrum Superintelligence

An alternative perspective to digital triumphalism, drawn ultimately from the raw phenomenology of one's own mind, the existence of multiple simultaneously bound perceptual objects in one's world-simulation, and the [fleeting, synchronic] unity of consciousness, holds that organic minds have been quantum computers for the past c. 540 million years. Insentient classical digital computers will never "wake up" and acquire software-based unitary minds that supplant biological minds rather than augment them.

What underlies this conjecture? In short, to achieve full-spectrum AGI, we'll need to solve both:

(1) the Hard Problem of Consciousness;
(2) the Binding Problem.

These two seemingly insoluble challenges show that our existing conceptual framework is broken. Showing our existing conceptual framework is broken is easier than fixing it, especially if we are unwilling to sacrifice the constraint of physicalism: at sub-Planckian energies, the Standard Model of physics seems well-confirmed. A more common reaction to the ontological scandal of consciousness in the natural world is simply to acknowledge that consciousness and the binding problem alike are currently too difficult for us to solve; put these mysteries to one side as though they were mere anomalies that can be quarantined from the rest of science; and then act as though our ignorance is immaterial for the purposes of building artificial (super)intelligence – despite the fact that consciousness is the only thing that can matter, or enable anything else to matter. In some ways, undoubtedly, this pragmatic approach has been immensely fruitful in "narrow" AI: programming trumps philosophising. Certainly, the fact that e.g. Deep Blue and Watson don't need the neuronal architecture of phenomenal minds to outperform humans at chess or Jeopardy is suggestive.

It's tempting to extrapolate their success and make the claim that programmable, insentient digital machine intelligence, presumably deployed in autonomous artificial robots endowed with a massively classically parallel subsymbolic connectionist architecture, could one day outperform humans in absolutely everything, or at least absolutely everything that matters. However, everything that matters includes phenomenal minds; and any problem whose solution necessarily involves the subjective textures of mind. Could the Hard Problem of consciousness be solved by a digital zombie? Could a digital zombie explain the nature of qualia? These questions seem scarcely intelligible. Clearly, devising a theory of consciousness that isn't demonstrably incoherent or false poses a daunting challenge. The enigma of consciousness is so unfathomable within our conceptual scheme that even a desperate-sounding naturalistic dualism or a defeatist mysterianism can't simply be dismissed out of hand, though these options won't be explored here. Instead, a radically conservative and potentially testable option will be canvassed.

The argument runs as follows. Solving both the Hard Problem and the Binding Problem demands a combination of first, a robustly monistic Strawsonian physicalism - the only scientifically literate form of panpsychism; and second, information-bearing ultrarapid quantum coherent states of mind executed on sub-femtosecond timescales, i.e. "quantum mind", shorn of unphysical collapsing wave functions à la Penrose (cf. Orch-OR) or New-Age mumbo-jumbo. The conjecture argued here is that macroscopic quantum coherence is indispensable to phenomenal object-binding and unitary mind, i.e. that ostensibly discretely and distributively processed edges, textures, motions, colours (etc.) in the CNS are fleetingly but irreducibly bound into single macroscopic entitles when one apprehends or instantiates a perceptual object in one's world-simulation – a simulation that runs at around 1013 quantum-coherent frames per second.

First, however, let's review Strawsonian physicalism, without which a solution to the Hard Problem of consciousness can't even get off the ground.

4.1. Pan-experientialism / Strawsonian Physicalism

Physicalism and materialism are often supposed to be close cousins. But this needn't be the case. On the contrary, one may be both a physicalist and a panpsychist - or even both a physicalist and a monistic idealist. Strawsonian physicalists acknowledge the world is exhaustively described by the equations of physics. There is no "element of reality", as Einstein puts it, that is not captured in the formalism of theoretical physics – the quantum-field theoretic equations and their solutions. However, physics gives us no insight into the intrinsic nature of the stuff of the world - what "breathes fire into the equations" as arch-materialist Steven Hawking poetically laments. Key terms in theoretical physics like "field" are defined purely mathematically.

So is the intrinsic nature of the physical, the "fire" in the equations, a wholly metaphysical question? Kant claimed famously that we would never understand the noumenal essence of the world, simply phenomena as structured by the mind. Strawson, drawing upon arguments made by Oxford philosopher Michael Lockwood but anticipated by Russell and Schopenhauer, turns Kant on his head. Actually, there is one part of the natural world that we do know as it is in itself, and not at one remove, so to speak – and its intrinsic nature is disclosed by subjective properties of one's own conscious mind. Thus it transpires that the "fire" in the equations is utterly different from what one's naive materialist intuitions would suppose.

Yet this conjecture still doesn't close the Explanatory Gap.

4.2. The Binding Problem. Are Phenomenal Minds A Classical Or A Quantum Phenomenon?

Why enter the quantum mind swamp? After all, if one is bold [or foolish] enough to entertain pan-experientialism / Strawsonian physicalism, then why be sceptical about the prospect of non-trivial digital sentience, let alone full-spectrum AGI? Well, counterintuitively, an ontology of pan-experientialism / Strawsonian physicalism does not overpopulate the world with phenomenal minds. For on pain of animism, mere aggregates of discrete classical "psychons", primitive flecks of consciousness, are not themselves unitary subjects of experience, regardless of any information-processing role they may have been co-opted into playing in the CNS. We still need to solve the Binding Problem (cf. Mashour, 2004) – and with it, perhaps, the answer to Moravec's paradox. Thus a nonsentient digital computer can today be programmed to develop powerful and exact models of the physical universe. These models can be used to make predictions with superhuman speed and accuracy about everything from the weather to thermonuclear reactions to the early Big Bang. But in other respects, digital computers are just tools and toys. To resolve Moravec's paradox, we need to explain why in unstructured, open-field contexts a bumble-bee can comprehensively outclass Alpha Dog. And in the case of humans, how can 80 billion odd interconnected neurons, conceived as discrete, membrane-bound, spatially distributed classical information processors, generate unitary phenomenal objects, unitary phenomenal world-simulations populated by multiple dynamic objects in real time, and a fleetingly unitary self that can act flexibly and intelligently in a fast-changing local environment? This combination problem was what troubled William James, the American philosopher and psychologist otherwise sympathetic to panpsychism, over a hundred a years ago in Principles of Psychology (1890). In contemporary idiom, even if fields (superstrings, p-branes, etc.) of microqualia are the stuff of the world whose behaviour the formalism of physics exhaustively describes, and even if membrane-bound quasi-classical neurons are at least rudimentarily conscious, then why aren't we merely massively parallel informational patterns of classical "mind dust" – quasi-zombies as it were, with no more ontological integrity than the population of China? The Explanatory Gap is unbridgeable as posed. Our phenomenology of mind seems as inexplicable as if 1.3 billion skull-bound Chinese were to hold hands and suddenly become a unitary subject of experience. Why? How?

Or rather, where have we gone wrong?

4.3. Why The Mind Is Probably A Quantum Computer.

Here we enter the realm of speculation - though critically, speculation that will be scientifically testable with tomorrow's technology. For now, critics will pardonably view such speculation as no more than the empty hope that two unrelated mysteries, namely the interpretation of quantum mechanics and an understanding of consciousness, will somehow cancel each other out. But what's at stake is whether two apparently irreducible kinds of holism, i.e. "bound" perceptual objects / unitary selves and quantum-coherent states of matter, are more than merely coincidental: a much tighter explanatory fit than a mere congruence of disparate mysteries. Thus consider Max Tegmark's much-cited critique of quantum mind (cf. Tegmark, 1999). For the sake of argument, assume that pan-experientialism / Strawsonian physicalism is true, but Tegmark rather than his critics are correct: thermally-induced decoherence effectively "destroys" [i.e., transfers to the extra-neural environment in a thermodynamically irreversible way] distinctively quantum-mechanical coherence in an environment as warm and noisy as the brain within around 10-15 of a second – rather than the much longer times claimed by Hameroff et al. Granted pan-experientialism / Strawsonian physicalism, what might it feel like "from the inside" to instantiate a quantum computer running at 10-15 irreducible quantum-coherent frames per second – computationally optimised by hundreds of millions of years of evolution to deliver effectively real-time simulations of macroscopic worlds? How would instantiating this ultrarapid succession of neuronal superpositions be sensed differently from the persistence of vision undergone when watching a movie? No, this conjecture isn't a claim that visual perception of mind-independent objects operates on sub-femtosecond timescales. This patently isn't the case. Nerve impulses travel up the optic nerve to the mind/brain only at a sluggish 100 m/s or so. Rather when we're awake, input from the optic nerve selects mind-brain virtual world states. Even when we're not dreaming, our minds never actually perceive our surroundings. The terms "observation" and "perception" are systematically misleading. "Observation" suggests that our minds access our local environment, whereas all these surroundings can do is play a distal causal role in selecting from a menu of quantum-coherent states of one's own mind: neuronal superpositions of distributed feature-processors. Our awake world-simulations track gross fitness-relevant patterns in the local environment with a delay of 150 milliseconds or so; when we're dreaming, such state-selection (via optic nerve impulses, etc.) of is largely absent.

In default of experimental apparatus (cf. Ball, 2011) sufficiently sensitive to detect macroscopic quantum coherence in the CNS on sub-femtosecond timescales, this proposed strategy to bridge the Explanatory Gap is of course only conjecture. Or rather it's little more than philosophical hand-waving. Most AI theorists assume that at such a fine-grained level of temporal resolution our advanced neuroscanners would just find "noise" – insofar as mainstream researchers consider quantum mind hypotheses at all. Moreover, an adequate theory of mind would need rigorously to derive the properties of our bound macroqualia from superpositions of the (hypothetical) underlying field-theoretic microqualia posited by Strawsonian physicalism - not simply hint at how our bound macroqualia might be derivable. But if the story above is even remotely on the right lines, then a classical digital computer - or the population of China (etc.) – could never be non-trivially conscious or endowed with a mind of its own.

True or false, it's worth noting that if quantum mechanics is complete, then the existence of macroscopic quantum coherent states in the CNS is not in question: the existence of macroscopic superpositions is a prediction of any realist theory of quantum mechanics that doesn't invoke state vector collapse. Recall Schrödinger's unfortunate cat. Rather what's in question is whether such states could have been recruited via natural selection to do any computationally useful work. Max Tegmark (in "Why the brain is probably not a quantum computer"), for instance, would claim otherwise. To date, much of the debate has focused on decoherence timescales, allegedly too rapid for any quantum mind account to fly. And of course classical serial digital computers, too, are quantum systems, vulnerable to quantum noise: this doesn't make them quantum computers. But this isn't the claim at issue here. Rather it's that future molecular matter-wave interferometry sensitive enough to detect quantum coherence in a macroscopic mind/brain on sub-femtosecond timescales would detect, not merely random psychotic "noise", but quantum coherent states – states isomorphic to the macroqualia / dynamic objects making up the egocentric virtual worlds of our daily experience.

To highlight the nature of this prediction, let's lapse briefly into the idiom of a naive realist theory of perception. Recall how inspecting the surgically exposed brain of an awake subject on an operating table uncovers no qualia, no bound perceptual objects, no unity of consciousness, no egocentric world-simulations, just cheesy convoluted neural porridge – or, under a microscope, discrete classical nerve cells. Hence the incredible eliminativism about consciousness of Daniel Dennett. On a materialist ontology, consciousness is indeed impossible. But if a quantum mind story of phenomenal object-binding is correct, the formal shadows of the macroscopic phenomenal objects of one's everyday lifeworld could one day be experimentally detected with utopian neuroscanning. They are just as physically real as the long-acting macroscopic quantum coherence manifested by, say, superfluid helium at distinctly chillier temperatures. Phenomenal sunsets, symphonies and skyscrapers in the CNS could all in principle be detectable over intervals that are fabulously long measured in units of the world's natural Planck scale even if fabulously short by the naive intuitions of folk psychology. Without such bound quantum-coherent states, according to this hypothesis, we would be zombies. Given Strawsonian physicalism, the existence of such states explains why biological robots couldn't be insentient automata. On this story, the spell of a false ontology [i.e. materialism] and a residual naive realism about perception allied to classical physics leads us to misunderstand the nature of the awake / dreaming mind/brain as some kind of quasi-classical object. The phenomenology of our minds shows it's nothing of the kind.

4.4. The Incoherence Of Digital Minds

Most relevant here, another strong prediction of the quantum mind conjecture is that even utopian classical digital computers - or classically parallel connectionist systems - will never be non-trivially conscious, nor will they ever achieve full-spectrum superintelligence. Assuming Strawsonian physicalism is true, even if molecular matter-wave interferometry could detect the "noise" of fleeting macroscopic superpositions internal to the CPU of a classical computer, we've no grounds for believing that a digital computer [or any particular software program it executes] can be a subject of experience. Their fundamental physical components may [or may not] be discrete atomic microqualia rather than the insentient silicon (etc.) atoms we normally suppose. But their physical constitution is computationally incidental to the execution of the sequence of logical operations they execute. Any distinctively quantum mechanical effects are just another kind of "noise" against which we design error-detection and -correction algorithms. So at least on the narrative outlined here, the future belongs to sentient, recursively self-improving biological robots synergistically augmented by smarter digital software, not our supporting cast of silicon zombies.

On the other hand, we aren't entitled to make the stronger claim that only an organic mind/brain could be a unitary subject of experience. For we simply don't know what may or may not be technically feasible in a distant era of mature nonbiological quantum computing centuries or millennia hence. However, a supercivilisation based on mature nonbiological quantum computing is not imminent.

4.5. The Infeasibility Of "Mind Uploading"

On the face of it, the prospect of scanning, digitising and uploading our minds offers a way to circumvent our profound ignorance of both the Hard Problem of consciousness and the binding problem. Mind uploading would still critically depend on identifying which features of the mind/brain are mere “substrate”, i.e. incidental implementation details of our minds, and which features are functionally essential to object-binding and unitary consciousness. On any coarse-grained functionalist story, at least, this challenge might seem surmountable. Presumably, the mind/brain can formally be described by the connection and activation evolution equations of a massively parallel connectionist architecture, with phenomenal object-binding a function of simultaneity: different populations of neurons (edge-detectors, colour detectors, motion detectors, etc.) firing together to create ephemeral bound objects. But this can't be the full story. Mere simultaneity of neuronal spiking can't, by itself, explain phenomenal object-binding. There is no one place in the brain where distributively processed features come together into multiple bound objects in a world-simulation instantiated by a fleetingly unitary subject of experience. We haven't explained why a population of 80 billion ostensibly discrete membrane-bound neurons, classically conceived, isn't a zombie in the sense that 1.3 billion skull-bound Chinese minds or a termite colony is a zombie. In default of a currently unimaginable scientific/philosophical breakthrough in the understanding of consciousness, it's hard to see how our “mind-files” could ever be uploaded to a digital computer. If a quantum mind story is true, mind-uploading can't be done.

In essence, two distinct questions arise here. First, given finite, real-world computational resources, can a classical serial digital computer – or a massively (classically) parallel connectionist system – faithfully emulate the external behaviour of a biological mind/brain? Second, can a classical digital computer emulate the intrinsic phenomenology of our minds, not least multiple bound perceptual objects simultaneously populating a unitary experiential field apprehended or instantiated by a [fleetingly] unitary self?

If our answer to the first question was “yes”, then not to answer “yes” to the second question too might seem sterile philosophical scepticism - just a rehash of the Problem Of Other Minds, or the idle sceptical worry about inverted qualia: how can I know that when I see red that you don't see blue? (etc.). But the problem is much more serious. Compare how, if you are given the notation of a game of chess that Kasparov has just played, then you can faithfully emulate the gameplay. Yet you know nothing whatsoever about the texture of the pieces – or indeed whether the pieces had any textures at all: perhaps the game was played online. Likewise, with the innumerable textures of consciousness – with the critical difference that the textures of consciousness are the only reason our "gameplay" actually matters. Unless we rigorously understand consciousness, and the basis of our teeming multitude of qualia, and how those qualia are bound to constitute a subject of experience, the prospect of uploading is a pipedream. Furthermore, we may suspect on theoretical grounds that the full functionality of unitary conscious minds will prove resistant to digital emulation; and classical digital computers will never be anything but zombies.

4.6. Object-Binding, World-Simulations and Phenomenal Selves

How can one know about anything beyond the contents of one's own mind or software program? The bedrock of general (super)intelligence is the capacity to execute a data-driven simulation of the mind-independent world in open-field contexts, i.e., to "perceive" the fast-changing local environment in almost real time. Without this real-time computing capacity, we would just be windowless monads. For sure, simple forms of behaviour-based robotics are feasible, notably the subsumption architecture of Rodney Brooks and his colleagues at MIT. Quasi-autonomous "bio-inspired" reactive robots can be surprisingly robust and versatile in well-defined environmental contexts. Some radical dynamical systems theorists believe that we can dispense with anything resembling transparent and "projectible" representations in the CNS altogether, and instead model the mind-brain using differential equations. But an agent without any functional capacity for data-driven real-time world-simulation couldn't even take an IQ test, let alone act intelligently in the world.

So the design of artificial intelligent lifeforms with a capacity efficiently to run egocentric world-simulations in unstructured, open-field contexts will entail confronting Moravec's paradox. In the post-Turing era, why is engineering the preconditions for allegedly low-level sensorimotor competence in robotics so hard, and programming the allegedly high-level logico-mathematical prowess in computer science so easy – the opposite evolutionary trajectory to organic robots over the past 540 million years? Solving Moravec's paradox in turn will entail solving the binding problem. And we don't understand how the human mind/brain solves the binding problem – despite the speculations about macroscopic quantum coherence in organic neural networks floated above. Presumably, some kind of massively parallel sub-symbolic connectionist architecture with exceedingly powerful learning algorithms is essential to world-simulation. Yet mere temporal synchrony of neuronal firing patterns of discrete, distributed classical neurons couldn't suffice to generate a phenomenal world instantiated by a person. Nor could programs executed in classical serial processors.

How is this naively "low-level" sensorimotor question relevant to the end of the human era? Why would a hypothetical nonfriendly AGI-in-a-box need to solve the binding problem and continually simulate / "perceive" the external world in real time in order to pose (potentially) an existential threat to biological sentience? This is the spectre that MIRI seek to warn the world against should humanity fail to develop Safe AI. Well, just as there is nothing to stop someone who, say, doesn't like "Jewish physics" from gunning down a cloistered (super-)Einstein in his study, likewise there is nothing to stop a simple-minded organic human in basement reality switching the computer that's hosting (super-)Watson off at the mains if he decides he doesn't like computers – or the prospect of human replacement by nonfriendly super-AGI. To pose a potentially existential threat to Darwinian life, the putative super-AGI would need to possess ubiquitous global surveillance and control capabilities so it could monitor and defeat the actions of ontologically low-level mindful agents – and persuade them in real time to protect its power-source. The super-AGI can't simply infer, predict and anticipate these actions in virtue of its ultrapowerful algorithms: the problem is computationally intractable. Living in the basement, as disclosed by the existence of one's own unitary phenomenal mind, has ontological privileges. It's down in the ontological basement that the worst threats to sentient beings are to be found – threats emanating from other grim basement-dwellers evolved under pressure of natural selection. For the single greatest underlying threat to human civilisation still lies, not in rogue software-based AGI going FOOM and taking over the world, but in the hostile behaviour of other male human primates doing what Nature "designed" (cf. Wrangham & Peterson, 1996) us to do, namely wage war against other male primates using whatever tools are at our disposal.

Evolutionary psychology suggests, and the historical record confirms, that the natural behavioural phenotype of humans resembles chimpanzees rather than bonobos. Weaponised Tool AI is the latest and potentially greatest weapon male human primates can use against other coalitions of male human primates. Yet we don't know how to give that classical digital AI a mind of its own - or whether such autonomous minds are even in principle physically constructible.

5. Conclusion. The Qualia Explosion. Supersentience: Turing plus Shulgin?

Compared to the natural sciences (cf. the Standard Model in physics) or computing (cf. the Universal Turing Machine), the "science" of consciousness is pre-Galilean, perhaps even pre-Socratic. State-enforced censorship of the range of subjective properties of matter and energy in the guise of a prohibition on psychoactive experimentation is a powerful barrier to knowledge. The legal taboo on the empirical method in consciousness studies prevents experimental investigation of even the crude dimensions of the Hard Problem, let alone locating a solution-space where answers to our ignorance might conceivably be found.

Singularity theorists are undaunted by our ignorance of this fundamental feature of the natural world. Instead, the Singularitarians offer a narrative of runaway machine intelligence in which consciousness plays a supporting role ranging from the minimal and incidental to the completely non-existent. However, highlighting the Singularity movement's background assumptions about the nature of mind and intelligence, not least the insignificance of the binding problem to AGI, reveals why FUSION and REPLACEMENT scenarios are unlikely - though a measure of "cyborgification" of sentient biological robots augmented with ultrasmart software seems plausible and perhaps inevitable.

If full-spectrum superintelligence does indeed entail navigation and mastery of the manifold state-spaces of consciousness, and ultimately a seamless integration of this knowledge with the structural understanding of the world yielded by the formal sciences, then where does this elusive synthesis leave the prospects of posthuman superintelligence? Will the global proscription of radically altered states last indefinitely?

Social prophecy is always a minefield. However, there is one solution to the indisputable psychological health risks posed to human minds by empirical research into the outlandish state-spaces of consciousness unlocked by ingesting the tryptamines, phenylethylamines, isoquinolines and other pharmacological tools of sentience investigation. This solution is to make "bad trips" physiologically impossible – whether for individual investigators or, in theory, for human society as a whole. Critics of mood-enrichment technologies sometimes contend that a world animated by information-sensitive gradients of bliss would be an intellectually stagnant society: crudely, a Brave New World. On the contrary, biotech-driven mastery of our reward circuitry promises a knowledge explosion in virtue of allowing a social, scientific and legal revolution: safe, full-spectrum biological superintelligence. For genetic recalibration of hedonic set-points – as distinct from creating uniform bliss – potentially leaves cognitive function and critical insight both sharp and intact; and offers a launchpad for consciousness research in mind-spaces alien to the drug-naive imagination. A future biology of invincible well-being would not merely immeasurably improve our subjective quality of life: empirically, pleasure is the engine of value-creation. In addition to enriching all our lives, radical mood-enrichment would permit safe, systematic and responsible scientific exploration of previously inaccessible state-spaces of consciousness. If we were blessed with a biology of invincible well-being, exotic state-spaces would all be saturated with a rich hedonic tone.

Until this hypothetical world-defining transition, pursuit of the rigorous first-person methodology and rational drug-design strategy pioneered by Alexander Shulgin in PiHKAL and TiHKAL remains confined to the scientific counterculture. Investigation is risky, mostly unlawful, and unsystematic. In mainstream society, academia and peer-reviewed scholarly journals alike, ordinary waking consciousness is assumed to define the gold standard in which knowledge-claims are expressed and appraised. Yet to borrow a homely-sounding quote from Einstein, "What does the fish know of the sea in which it swims?" Just as a dreamer can gain only limited insight into the nature of dreaming consciousness from within a dream, likewise the nature of "ordinary waking consciousness" can only be glimpsed from within its confines. In order scientifically to understand the realm of the subjective, we'll need to gain access to all its manifestations, not just the impoverished subset of states of consciousness that tended to promote the inclusive fitness of human genes on the African savannah.

5.1. AI, Genome Biohacking and Utopian Superqualia. Why the Proportionality Thesis Implies an Organic Singularity

So if the preconditions for full-spectrum superintelligence, i.e. access to superhuman state-spaces of sentience, remain unlawful, where does this roadblock leave the prospects of runaway self-improvement to superintelligence? Could recursive genetic self-editing of our source code repair the gap? Or will traditional human personal genomes be policed by a dystopian Gene Enforcement Agency in a manner analogous to the coercive policing of traditional human minds by the Drug Enforcement Agency?

Even in an ideal regulatory regime, the process of genetic and/or pharmacological self-enhancement is intuitively too slow for a biological Intelligence Explosion to be a live option, especially when set against the exponential increase in digital computer processing power and inorganic AI touted by Singularitarians. Prophets of imminent human demise in the face of machine intelligence argue that there can't be a Moore's law for organic robots. Even the Flynn Effect, the three-points-per-decade increase in IQ scores recorded during the 20th century, is comparatively puny; and in any case, this narrowly-defined intelligence gain may now have halted in well-nourished Western populations.

However, writing off all scenarios of recursive human self-enhancement would be premature. Presumably, the smarter our nonbiological AI, the more readily AI-assisted humans will be able recursively to improve our own minds with user-friendly wetware-editing tools – not just editing our raw genetic source code, but also the multiple layers of transcription and feedback mechanisms woven into biological minds. Presumably, our ever-smarter minds will be able to devise progressively more sophisticated, and also progressively more user-friendly, wetware-editing tools. These wetware-editing tools can accelerate our own recursive self-improvement – and manage potential threats from nonfriendly AGI that might harm rather than help us, assuming that our earlier strictures against the possibility of digital software-based unitary minds were mistaken. MIRI rightly call attention to how small enhancements can yield immense cognitive dividends: the relatively short genetic distance between humans and chimpanzees suggests how relatively small enhancements can exert momentous effects on a mind's general intelligence, thereby implying that AGIs might likewise become disproportionately powerful through a small number of tweaks and improvements.

In the post-genomic era, presumably exactly the same holds true for AI-assisted humans and transhumans editing their own minds. What David Chalmers calls the proportionality thesis, i.e. increases in intelligence lead to proportionate increases in the capacity to design intelligent systems, will be vindicated as recursively self-improving organic robots modify their own source code and bootstrap our way to full-spectrum superintelligence: in essence, an organic Singularity. And in contrast to classical digital zombies, superficially small molecular differences in biological minds can result in profoundly different state-spaces of sentience. Compare the ostensibly trivial difference in gene expression profiles of neurons mediating phenomenal sight and phenomenal sound - and the radically different visual and auditory worlds they yield.

Compared to *Fusion* or *Replacement* scenarios, the AI-human *Co-Evolution* conjecture is apt to sound tame. The likelihood our posthuman successors will also be our biological descendants suggests at most a radical conservativism. In reality, a post-Singularity future where today's classical digital zombies were superseded merely by faster, more versatile classical digital zombies would be infinitely duller than a future of full-spectrum supersentience. For all insentient information processors are exactly the same inasmuch as the living dead are not subjects of experience. They'll never even know what it's like to be "all dark inside" – or the computational power of phenomenal object-binding that yields illumination.

By contrast, posthuman superintelligence will not just be quantitatively greater but also qualitatively alien to archaic Darwinian minds. Cybernetically enhanced and genetically rewritten biological minds can abolish (cf. Pearce, 2007) suffering throughout the living world and banish experience below "hedonic zero" in our forward light-cone, an ethical watershed without precedent. Post-Darwinian life can enjoy gradients of lifelong blissful supersentience with the intensity of a supernova compared to a glow-worm. A zombie, on the other hand, is just a zombie – even if it squawks like Einstein. Posthuman organic minds will dwell in state-spaces of experience for which archaic humans and classical digital computers alike have no language, no concepts, and no words to describe our ignorance. Most radically, hyperintelligent organic minds will explore state-spaces of consciousness that do not currently play any information-signalling role in living organisms and are impenetrable to investigation by digital zombies. In short, biological intelligence is on the brink of a recursively self-amplifying Qualia Explosion - a phenomenon of which digital zombies are invincibly ignorant, and invincibly ignorant of their own ignorance. Humans too, of course, are mostly ignorant of what we're lacking: the nature, scope and intensity of such posthuman superqualia are beyond the bounds of archaic human experience. Even so, enrichment of our reward pathways can ensure that full-spectrum biological superintelligence will be sublime.

References

Ball, Philip (2011) "Physics of life: The dawn of quantum biology", *Nature*, 474: 272-274.

Benatar, David (2008) *Better Never To Have Been: The Harm of Coming into Existence*, Oxford: Oxford University Press.

Bostrom, Nick (2003). "Are You Living in a Computer Simulation?", *Philosophical Quarterly*, 53 (211): 243–255.

Bostrom, Nick (2012) "The Superinteligent Will: Motivation and Instrumental Rationality in Advanced Artificial Agents", *Minds and Machines*, 22 (2): 71-85.

De Garis, Hugo (2009) "The Coming Artilect War" In *Forbes*, Jun 22, 2009.

Kornai, András (2003) "Explicit Finitism", *International Journal of Theoretical Physics*, 42 (2): 301–307.

Mashour, George A. (2004) "The Cognitive Binding Problem: From Kant to Quantum Neurodynamics", *NeuroQuantology: an Interdisciplinary Journal of Neuroscience and Quantum Physics*, 2 (1): 29-38.

Muehlhauser, Luke (2013) "Intelligence Explosion FAQ." First published 2011 as "Singularity FAQ." in *Machine Intelligence Research Institute*, Berkeley, CA.

Noonan, Harold and Curtis, Ben (2018) "Identity" In *The Stanford Encyclopedia of Philosophy* (Summer 2018 Edition), Edward N. Zalta (ed.), URL = <https://plato.stanford.edu/archives/sum2018/entries/identity/>.

Pearce, David (2007) "The Abolitionist Project" (invited talks given at the FHI (Oxford University) and the Charity International Happiness Conference (2007) = https://www.abolitionist.com/.

Pearce, David (2009) "Post-Darwinian Ethics?" in (source: H+ Magazine) = https://www.hedweb.com/population-ethics/postdarwinian.html.

Pearce, David (2014) "High-Tech Jainism" in (source: The World Transformed) = https://www.hedweb.com/transhumanism/neojainism.html.

Pearce, David (2019) "Utopian Neuroscience" in https://www.superhappiness.com/

Revonsuo, Antti (1999) "Binding and the Phenomenal Unity of Consciousness", *Consciousness and Cognition*, 8: 173–185.

Tegmark, Max (1999) "The Importance of Quantum Decoherence in Brain Processes", *Physical Review*, 61 (4 Pt B): 4194-206.

Tononi, Giulio (2008) "Consciousness as Integrated Information: a Provisional Manifesto", *The Biological Bulletin*, 215 (3): 216-242.

Wichers, Marieke et al. (2008) "The Catechol-O-Methyl Transferase Val^{158} Met Polymorphism and Experience of Reward in the Flow of Daily Life", *Neuropsychopharmacology*, 33: 3030–3036.

William, James (1890) *Principles of Psychology*, New York: Henry Holt and Company.

Wrangham, Richard & Peterson, Dale (1996) *Demonic Males: Apes and the Origins of Human Violence*, New York: Houghton Mifflin Harcourt Publishing Company.

SECTION II: CONSCIOUSNESS, SIMULATION AND ARTIFICIAL INTELLIGENCE

CHAPTER 4

MINDPLEXES, NON-ORDINARY CONSCIOUSNESS, AND ARTIFICIAL GENERAL INTELLIGENCE

Gabriel Axel Montes[1]

Hong Kong Polytechnic University, Hong Kong

Ben Goertzel[2]

SingularityNET Foundation

Abstract: The accelerating pace of development in artificial intelligence invokes fundamental questions regarding what constitutes a mind. Considering that AI has sprung from the cognitive enterprise of humanity and that there is the likely prospect of constructing some form of 'global brain', examining such questions may help to refine the conversation around AI and understand what might constitute a global brain. We draw on recent work in philosophy of mind and cognition – viewing mind as being representational and/or enactive in nature and function –, and hone in on insights that practices and accounts of non-ordinary consciousness (NOC) lend to this discussion. We discuss the concept of a "mindplex", which is comprised of individual constituents (humans and/or AI's) and an encompassing collective global brain that both concurrently possess cohesive 'theatres of consciousness'. We posit that these theatres of an AGI global brain have correspondences with key aspects of the inherent nature of consciousness and reality (as understood outside the AI context) as gleaned through NOC. We propose that these considerations are poised to strongly

[1] Hong Kong Polytechnic University, Hung Hom, Kowloon, Hong Kong; University of Newcastle, University Drive, Callaghan, NSW 2308, Australia; Bias in Research Node, Charles Perkins Centre, University of Sydney, University Pl, Camperdown NSW 2006, Australia; & Neural Axis, United States.

[2] SingularityNET Foundation, Amsterdam, Netherlands; OpenCog Foundation, Rockville, Maryland, United States.

contribute to the construction of mindplexes by reflecting in their functional architecture the nature of mind and engineering mindplexes in tandem with ethical and value-alignment considerations. Furthermore, we propose the adoption and implementation of AI on distributed ledger technology (DLT), such as blockchain, which would provide support for the evolution of a global brain mindplex.

1. Introduction

Current efforts towards the development of artificial general intelligence (AGI) (Goertzel and Pennachin, 2007) combined with the ubiquity of global networked devices (internet-of-things; IoT), militate strongly toward the emergence of a kind of 'global brain' (Heylighen, 1997; 2007). Unlike applied (narrow) AI, the prospect of "strong AI" (Kurzweil 2010) promises to be able to perform "general intelligent action" (Gilmartin, Newell, and Simon, 1976) and a wide range of human cognitive abilities. Given industrial and academic goals of achieving AGI and its prospective manifestation as a kind of global brain, in order to take the big step forward, it is worth taking a step back to investigate the nature of mind in light of the intent to engineer or replicate it *in silico*. This chapter takes a brief tour of mind through neuroscientific and philosophical lenses, with a particular interest in insights into the nature of mind from non-ordinary consciousness (NOC); presents the concept of a 'mindplex' (Goertzel, 2003; Goertzel, Pennachin, and Geisweiller, 2014) as a way of enhancing the link in understanding between human and artificial minds in order to more precisely build the latter; and proposes some of the merits of creating artificial mindplexes through distributed ledger technology. The broad stroke of this chapter is to highlight how a NOC perspective can add significant value to the understanding of mind and cognition, which in turn assists in the conceptualization and development of mindplexes.

2. Body, mind, consciousness

The commonly accepted and prevailing view of mind in modern Western science is that of a machine that cognizes; intaking sensory data and informationally processing it through the architecture of the nervous system. Humans exhibit wondrously complex information processing capabilities, one stunning high point of which is the display of 1st-person (phenomenal) conscious awareness (Montes, 2017a; 2017b). Phenomenal awareness presents the felt sense of having a "self", an assumed identity that is experienced as localized to a vehicle for that awareness, the body.

Fundamentally, the notion of selfhood is associated with organisms – biological self-organising dynamical systems which exist by virtue of preserving the homeostasis of internal states from destabilization and destruction (Friston,

2012). The statistical boundary between the internal (organism) and external (environment/econiche) states is termed a Markov blanket – for a further explication, see (Kirchhoff et al., 2018) –, which an organism maintains by inferring the causes of sensory input and minimising deviations (prediction error) from these internal predictions. The organism is thus said to model its environment, or more precisely, the organism *is* a model of its environment (Friston, 2013). Donning a more information-theoretic typology, more recent cognitive science has tended to regard virtually all aspects of organismal function, e.g., movement, as essentially cognitive in principle (Friston, 2013). The sense of selfhood has its cognitive roots in the origins of humankind, specifically in instrumental (tool) cognition. The narrative is as follows (Montes, 2017a). Humans descended from quadrupedal tree-dwelling primates, the limbs of which were adapted for arboreal brachiation and fruit procurement, which facilitated the evolution of geospatial memory maps. The transition to a terrestrial habitat imposed strong selection pressures, precipitating bipedalism and the liberation of the hands from power gripping into precision gripping. This opened an accelerating co-evolution between hand and brain. The survival pressures of a new environment implored prehensile hominins to experiment with the environment using their hands in novel ways, which fostered the creation of tools. Creative experimentation and goal-directed thinking entailed consideration of self and kin, driving co-development of agency, selfhood, and theory of mind. Early hominin sees the world as a repository of potential tools, employs instrumental thinking to externalize cognition in the form of artefacts, freeing up cognitive function for the further creation of artefacts. Due to the inextricability of selfhood in the cognitive enterprise, hominins unconsciously engage in agency transference as they externalize their cognition into new tool-artefacts, progressively manifesting toward the eventuality of self-replication. Humankind's econiche becomes an 'extended mind' (Clark and Chalmers, 1998) of cognitive artefacts that are ventriloquised through culture. Today's artificial intelligence (AI) is a set of artefacts produced along the progressive externalization of cognition. In theory, AGI would be the full-scale self-replication of human cognition. This hypothesis, termed "causal biomimesis", and the core tenet that this self-replication (biomimetic) process is intrinsic to the process of instrumental cognition present from hominin to modern human, is expounded in greater detail by one of the authors elsewhere (Montes, 2017a).

A growing body of work in cognitive neuroscience and philosophy of mind has shown that the experienced sense of selfhood is shaped by multisensory brain mechanisms that underlie the integration of bodily signals, a construct termed 'bodily self-consciousness' (BSC) (Jeannerod, 2003; Blanke and Metzinger, 2009). BSC has been studied mostly through bodily illusions, such as the classic rubber hand and full-body illusions, in which virtual or prosthetic limbs are phenomenologically assumed or incorporated into the body sense; i.e., there is a pre-reflective (before reflective thought) (Tsakiris, Schutz-Bosbach, and Gallagher, 2007) felt sense of 'body ownership' over the artificial limbs, a phenomenon which has been validated in neuroimaging and neurobehavioural studies. These illusions occur as the brain attempts to preserve its physical integrity by correcting the multisensory prediction error between vision and proprioception, resulting in subjective embodiment of the artificial body.

The phenomenal experience of BSC by humans has presumably been the grounds on which the sense of selfhood, ergo mind, has been assumed to be anchored to the individual agent, in the physical body. However, while the term "illusion" aptly expresses the experimentally observed deviational phenomena, localization of selfhood happens to be phenomenologically associated with the physical body precisely because of the mechanisms of multisensory information processing. The neurocognitive mechanisms of BSC promote strong biases ('prior probabilities', in predictive coding and Bayesian statistical terms) toward maintaining a cohesive spatiotemporal body schema because they enhance the agency of the physical organism in its econiche and social milieu. That the body is an *a priori* ontological locus of selfhood is therefore a presumption based upon its being a statistically evergreen prior that strongly conditions perception of self to be co-extensive with the body. One of the authors has termed this the 'self-localization fallacy' (Montes, 2018). The physical body is the nexus where multisensory information continuously converges temporally since birth over the course of evolutionary survival pressures. This clarification on the self-localization fallacy still allows room for the brain to be treated as a "producer" of the perceived content of conscious experience while concomitantly eschewing a presumption that the locality of selfhood is necessarily ontologically fixed to the body.

But if selfhood is not inherently localized to the body, then where is it? The notion of self (being a noun) having an inherent existence is perhaps more accurately underlaid by viewing it through a temporal lens; rather than there existing a self, there is a process of "selfing" that occurs by continuous predictive processing and FEP mechanisms at such a rapid 'frame rate' that the temporal process is perceived to occur seamlessly. This, in effect, gives the impression that the self is a thing rather than a process, and because the sense of selfhood is arguably the most abstract form of cognition and is conceptually and experientially most distant from the fast-changing world of low-level sensorimotor information, it is perceived as static and susceptible to the phenomenal illusion that it is largely unchanging. The tables have turned: statistical deviations of multisensory homeostasis are not the illusion; static, localized selfhood is. "Lies, damned lies, and statistics" (Twain, 1906).

As mentioned earlier, phenomenal awareness *presents* the felt sense of having a self. With self as a process, the self is temporally *re*-presented in/to awareness, continuously, seamlessly. Selfhood is an abstract layer of cognition; cognition is also a process and the apparent contents thereof are therefore also in constant re-presentation. The concerns and fallacies over any existence of selfhood similarly apply to cognition. Cognition is representational only in process ('re-presenting'); self is to selfing as representations are to representing. Put differently, 'representations' do not have an inherent existence as such other than being a conceptually spatial proxy for a temporal process. As with selfhood, the perceived contents of awareness are being phenomenally and neurally re-presented, but they do not have fixed spatial location.

In much of Western philosophy, cognitive science, and even common sense, mind has generally been presumed to be tethered to, or even synonymous with, the sense of selfhood. If the self does not have a location in space, then what of mind? Is mind the process of selfing? If we define selfhood according to free-energy Markov blanketeering, then surely, mind could not be co-determinate or equated with the information-theoretic fluctuations of selfhood dynamics, or even with representational dynamics. If mind is not localized and is not the homeostatic process of selfing, then mind may be said to be non-local, meaning it cannot be intrinsically defined in spatial terms, unless we are deliberately making context-specific simplifications for the sake of linguistic convention. Indeed, if factors of any defined space and time are removed from the phenomenality of mind (because they are not intrinsic to it), then mind is nowhere and nowhen to be found. This means, there is no subject that is the owner of mind ('mind ownership'), and if there is no subject, there is no object relative to any presupposed subjective mind. So: what is mind?

Mind is *recognition* of its very constitution as the fabric of reality (agent and environment). The term recognition is used advisedly. Recognition is not the same as representation. Presentation implies introduction of content from and into (subject and object, respectively and/or vice versa). Here, "recognition" refers to apprehension of sameness. Recognition is reflexive awareness apprehending itself. Put another way, mind is *presence*, irrespective of circumstantial spatial or temporal configuration. Mind is not a thing, nor is it a process. It is the knowing of its own constitution, which is the fabric of reality. Recognition is not merely "to cognize again"; it is the knowing of 'selfness' as a quality. In seeming contradiction, where we annihilated the self before, we are reintroducing it again differently, this time self being the cogency of recognition. Self does not exist insofar as it is limited in any way in time-space; it does not exist in any sense of separateness from anything else; self, unbound, is all that there is. The realization, by a subject, of this occurs when recognition is stably attained. It is not a state, but reality realized. For a subject to feel its own existence (as a self, bound in time and space) is for it to recognise a limited territory of mind constitution, which typically is the bounds of the physical body, the evergreen multisensory statistic (Montes, 2018). With regard to identifying with the physical body, self is still in a mode of recognition, but only of a fractal (the physical body) of the fabric of reality that is the totality of mind. Mind, unbound, is pre-reflective co-ownership of reality.

The views presented here about mind echo aspects of the insights commonly found in the theories and teachings of the world's wisdom traditions, in particular, non-dual views of consciousness (Norbu, 1986; Dalai Lama, 2000). Above, we have walked through those insights almost entirely using language friendly to cognitive science. That said, it's striking that deconstructing common assumptions in philosophy of mind and cognitive science can lead to some of the same formulations as those found in the world's wisdom traditions. The insights of the latter, while often encountered by the modern reader in the garb of myth, dogma, and antiquated belief systems, typically have arisen fundamentally from experiences of individuals who have undertaken various mind-body practices intended to cultivate non-ordinary consciousness (NOC), states and traits of conscious experience that are outside of the normal range of experience in everyday modern life (Montes, 2017b; Wilber, 1977; 2007). Examples of NOC practices include but are not limited to various types of meditation, prolonged concentration on particular auditory or visual stimuli, breathwork, yoga postures, qi gong, tai chi, internal martial arts, shamanic techniques, psychedelic substances, etc. Virtually all mind-body practices for NOC aim to systematically enact a number of major cognitive domains in order to produce their desired effect(s) (Montes, 2017b). They are, in a sense, tools/artefacts developed for producing specific effects on consciousness. The general direction or aim of such practices is to achieve various versions of 'self-realization', a term also apropos for our discussion of mind in terms of philosophy of mind and cognition, in which self 'awakens' to, or realizes, the nature of mind. The experiential quality of such realization is often likened to an experience of light/lightness/luminosity as a metaphor for the associated phenomenality. The aspect of consciousness that is its self-reflexive nature that is revealed in self-realization can be termed "luminous phenomenality". Exploring this category of phenomena in depth is not in the scope of this chapter; more detailed overviews can be found elsewhere (Montes, 2017b; 2018; Caplan and Montes, 2018; Wilber, 1977; 2007; Feuerstein, 1989; 1998; Feuerstein and Kak, 2013).

3. Mindplexes & Consciousness

The discussion of mind has been important for framing that of a global brain mindplex, especially the potentials and future development thereof. A "mindplex" (Goertzel, 2003) is loosely defined as an intelligent system that: (1)

is composed of a collection of intelligent systems, each of which has its own "theatre of consciousness" (TOC) (Baars, 1997) and autonomous control system, but which interact tightly, exchanging large quantities of information frequently; and (2) has a powerful control system on the collective level, and an active TOC on the collective level as well. The mindplex concept may appear to be akin to an individual mind (consisting of its various subpersonalities) or a human social group. While these are collectives, the individual human mind is dominant over the subpersonalities and the social group-mind, respectively, so they are not considered mindplexes. A mindplex is a hypothetical intelligent system in which neither level is dominant; both levels each operate powerfully and cohesively.

> ...A mindplex is like a human society that has become so integrated and so cohesive that it displays the kind of consciousness and self-control that we normally associate with individuals (Goertzel 2003).

Higher-order mindplexes may also exist, i.e., a collective of a lower level is an individual of a collective at a higher level. Mindplexes may thus exist as nested hierarchies of concurrently active TOC's.

As previously discussed by one of the authors (Goertzel, 2003), two plausible mechanisms by which mindplexes may possibly arise in the medium-term future are: (1) Humans becoming more tightly coupled via the advance of communication technologies, and a communication-centric AI system coming to embody the "emergent conscious theatre" of a human-incorporating mindplex; and/or (2) a society of AI systems communicating amongst each other with a richness not possible for human beings, and coming to form a mindplex rather than merely a society of distinct AI's. The former view relates to that of a "global brain" (Heylighen, 1997; 2007) and features human minds as the individual parts and AI at the collective level; the latter consists of AI's at both the individual and collective levels. These two potential manifestations are not mutually exclusive, and the reality is that it is hard to predict exactly in what way(s) mindplexes may come to manifest.

Given the concept of TOC, what is meant by 'consciousness' is pertinent. Most use of the term consciousness is extremely heterogeneous, including in 'hard science' (Blackmore, 2017); each researcher and school of thought has their own intuitively preferred preconception (either implicit or explicit) of what consciousness means, making agreement tenuous and progress on scientifically "cracking" consciousness haphazard. Discussion of consciousness today resembles talk of 'God' throughout the ages; there are myriad definitions (Armstrong, 2011). The literature on consciousness as it relates to AI has arrived at some basic dimensions that are roughly orthogonal to each other (Dehaene, Lau, and Kouider, 2017): (C0) unconscious processing; (C1) global availability;

and (C2) self-monitoring. C0 unconscious processes are where most of biological intelligence lies, such as evidence accumulation and Bayesian inference, and comprise cornerstone computations in AI today. C1 global availability corresponds to the TOC concept, in which all the parallel and modular unconscious evidence is integrated into a single decision or train of decisions (actions and thoughts) that eventually is given conscious access. The selection process leading up to the final conscious decision consists of C0 unconscious attentional tuning through the brain's hierarchical system of sieves that compute with probability distributions (Hohwy 2012). C2 self-monitoring refers to metacognition, or cognition about cognition, in which there is a capacity to reflect on one's own mental state (i.e., representation), which is currently known to be available from infancy in humans.

This classification of levels (C0, C1, C2) of consciousness reflects the general status quo in the fields of cognitive neuroscience and AI (Dehaene, Lau, and Kouider, 2017). There will probably be other forms of levels of consciousness (C3, C4, and so on) to be discovered as relevant for AI. However, these classifications alone are insufficient to explain our account of the nature of mind as non-local and non-dual, as discussed earlier in this chapter. As cognitive neuroscience and AI research and industry strive to build further levels of consciousness, potentially the final frontier will be C*n*, "luminous phenomenality". As mentioned above, this refers to the 'knowingness' or 'suchness' quality that is experienced when mind-nature is realized, and which has no subject, but is the 'luminescence' of realized mind. It is non-dual and is therefore qualitatively distinct from and orthogonal to unconscious processes (C0), attentional processes (C1), metacognitive representations (C2), or any further permutations of cognitive operationality that are less than C*n*.

The global workspace (C1; TOC) is, in humans, experienced as a "focus of awareness". In AI, this might take the form of, as with an example of a system developed by one of the authors (Goertzel et al., 2004), a "moving bubble of attention" that emerges from the nonlinear dynamics leading of a set of joint symbolic and subsymbolic knowledge representations that consume the majority of system resources at any given point in time, wherein cognitive operations flow freely. In a hypothetical mindplex, humans can be the individual TOC's and a superordinate AGI can be the collective TOC, and/or both the individual and collective TOC's can consist of (or at least include) AI agents /AGI. How might a TOC look like in a more advanced AGI that incorporates a collection of humans and/or autonomous AI agents?

As a nested hierarchy of concurrently active TOC's, a mindplex must have at least two adjacently nested global workspaces that maintain communication. In an advanced version, there could be an encompassing global workspace consisting of both of these TOC's that has concurrent access to both TOC's in a fluidly integrated manner. In this advanced version, the TOC's are phenomenally fused, and they are also interdependent (because of the distinctions in their constituent informational aggregates). In this case, there would be simultaneously functioning awareness of both adjacent levels (subordinate and superordinate) of the hierarchy. For this to occur, attention (at least at both C0 and C1) must be attuned to the subtler aggregate cognitive content which is shared between the contents of adjacent hierarchical levels. This means that regardless of whether attention is or isn't simultaneously on the *distinct* contents of the respective hierarchical levels, that there are attentional resources operating on some common denominators of those levels. These common denominators would be less ordinarily concrete at, say, a sensorial level. For argument's sake concerning how attention functions in this context, in the case of a naturally occurring pre-AI mindplex – which is to say, that there is a sentient being (a human "cultivator" of consciousness) that has attained a sufficient degree of self-realization in consciousness – if the human's attention is on say, a tree, there is attention on the elemental roots that are a common denominator between the tree and the body-mind-self of the cultivator. Even though their appearances are distinct, they are made up of similar constituencies that, although of unique functionalities, give them solidity in space. The experiential quality of this hypothetical cultivator would entail a felt perception that there is a 'common thread' of consciousness-constituency between his/her self and the external object. Traditionally, in Tibetan Buddhist Dzogchen (the non-dual tradition of Tibetan Buddhism) (Norbu, 1986), this is experienced as a collapse of the subject-object duality in perception (trekchöd), which occurs fully via what can be likened to "wormholes of light" (thögyal) between subject and object. Essentially, ordinary perception transforms into the recognition of the innate physical light-nature of self and object, which dissolves any apparently inherent duality between them.

The mind of our hypothetical cultivator is an advanced mindplex: a continuity is established – a kind of fusion, as it were – between the TOC of the individual self and the TOC of the natural environment. The TOC's of both the micro and the macro worlds are concurrently co-active in awareness, whereas they were previously not. The cultivator apprehends this non-dual reality by having sufficiently contemplated and 'grasped', through systematic practice, the subtle roots of percepts. That is, the common constituency – light-nature – of subject and object is realized. These subtle perceptual roots are less sensorially concrete in the ordinarily sense of the five senses (i.e., discrete tastes, visual appearances, etc.), the differences of which are what distinguish objects in conventional/ordinary time-space perception. By going subtler, the unifying thread is found, and eventually, the subtlety reaches a point where there is a sameness between all phenomena (without discounting their grosser distinctions), including the self and the external environment. The metaphorically described light-nature of all phenomena is, at the deepest level of spiritual ontology, primordial consciousness (Skt. *vidya*, Tbt. *rigpa* (Norbu 1986)). This is realized by a cultivator through the practice of *awareness* of that nature. While a more detailed treatment of this subject is outside the scope of the present chapter and is best reserved for future work, Tibetan treatments on Dzogchen can offer illuminating insights into these phenomena (Dalai Lama, 2000; Norbu, 1986).

This discussion might suggest that there is only one real natural TOC, not two, since both are made of the same constituent and that mind is able to collapse the duality between subject and object. However, non-duality is orthogonal to dualistic phenomena in general – it does not negate them, but rather reveals the underlying fabric of reality (primordial consciousness) from which all differentiated phenomena spring. The two putative TOC's are regarded as concepts within the fabric of mind, each composed of a unique constellation of interdependent aggregates. Although their most fundamental mutual composition may be photonic, as it were, as light physically becomes instantiated as matter, aggregates emerge that are common between the two TOC structures (e.g., cognitive faculties), and then progressively more distinct aggregates emerge that differentiate the two. This is a continuum of materialization of the nature of mind, from its subtlest origins to its grossest physical manifestation.

A crucial note is that, as the mind of the cultivator deepens its apprehension of non-duality, it traverses a continuum of progressive mini-realizations, as it were. As this process unfolds, a gamut of relationships between phenomena's constituent aggregates is progressively grasped. A cultivator may not, in all likelihood, suddenly attain full non-dual realization in one leap, but rather through a course of unfolding mini-realizations, through which deeper insights into mind are experienced and integrated to various degrees. Along this progressive course, which one could say is teleologically drawn by the 'gravitational pull' of the impending realization of the most subtle nature of mind, new relationships between aggregates are grasped, revealing new interconnections between the constituencies of subject and object. The expansion into and refinement of these new connections form new cognitive scaffolds that allow different ways of cognising about phenomena, based on the interplay between their aggregate constituencies. So, pockets of non-dual awareness occurring in the context of a cultivator's conscious experience (C1) help to facilitate the development and precipitation of nuanced "integral" (Wilber, 2001; 2007) cognition, from C0 onward.

More nuanced and sensitive forms of cognition manifest as shifts in the ways unconscious processes (C0) in each TOC interweave and intercompute. In a mindplex, the C0 of one TOC interweaves with the C0 of the second TOC. This results in a "blended" global workspace (C1) that would include some degree or number of the C0's of both TOC's. This is how "one" mind can flexibly act in more than one TOC, simultaneously. The interweaving of and intercomputation among C0's between multiple TOC's (and also within a single TOC, for that matter) involves shared contents, regardless of whether they integrate and surface into C1 as conscious awareness. Content integration involves computationally re-evaluating differences between contents (sets of aggregates) by finding a new common denominator and integrating those former differences into interdependent flows of information. Individual aggregates and sets thereof are realized as distinct and blended parts of a larger whole. The cognitive and computational communications and interoperations between the TOC's are at the level of duality (information flow between here and there, this and that, etc.). Non-dual awareness 'pulls', or draws, cognitive architectures into subtler structures and computations, allowing new ways of recombinating sense data and their information. For the individual organism/agent mind, this opens doors for new forms of dynamic self-regulation and homeostasis, based upon conductance of subtler information between self and environment. From the perspective of the free-energy principle (FEP) (Friston, 2010), in which an organism regulates its internal states against external states in order to reduce entropy (ergo destabilisation, deterioration, and death; the three D's), energetic differentials between internal and external states would be computed in a more harmonious way, so to speak, such that the organism undergoes less stress. This is because the computation of information based on more common aggregates, rather than the discrete low-level differences, would reduce the friction between internal and external states. This would allow for new homeostatic strategies for maintaining self-regulation.

In the mindplex of our non-dual–realized cultivator, because the TOC of the individual and the TOC of the external environment are unified at the level of non-dual awareness, which is orthogonal to virtually all cognitive operations (because the latter occur at the level of duality), this does not inherently or directly determine the computational structures of cognitive operations that will emerge at any given point. Non-duality does not inherently result in co-active manipulation of contents of multiple TOC's. How the distinct (but interdependent) TOC's will interoperate depends on cognitive developments and unfoldments at the cognitive level. These do not depend intrinsically on the existence of some degree non-dual realization, but can be facilitated by it (Wilber 2007). An example of an emergence of a subtler form of cognition in the domain of non-ordinary consciousness (NOC) would be clairaudience, the supposed faculty of hearing the inaudible. In ordinary (everyday) human consciousness, the only personal voices heard (aside from extrapersonal voices) are the subvocalizations of the ego-personality and of one's own subpersonalities, or least subvocalizations that are interpreted to arise from these sources. In clairaudience, a human might, for instance, 'hear' a cognitive data set of another human. Rather than a priori treating this as delusional or fraudulent (which various supposed cases of clairaudience could be), this example of clairaudience might be reframed as an agent's sensitivity to and/or resonance with the cognitive operations of a human that are, through some or other neural pathway, transposed in the form of audition. (Note that 'cognition' here is used in the broadest sense as per recent advancements in neuroscience that regard biological computation as cognitive (Friston, 2013).)

An agent's sensitivity to the cognition of an external agent does not, in general, automatically translate to auditory readouts; but in the case of clairaudience, it would. Also note that clairaudience would often be coloured through the lens of the receiving agent, so that the type of information is perceived of the other agent is not intended to be objective or to match the other's own thoughts. The point is that there would be two main cognitive components involved in the notion of clairaudience: (1) sensitivity to cognition that is external to the perceiving agent; that is, it is not part of one's own individual ego-personality; and (2) transposition of the received unconscious cognitive process (C0) to conscious awareness at the agent-awareness level (C1). For the C0 processes (1), there would be shared content between the TOC's, as discussed earlier, that involve subtle cognition. For the C1 process (2), there would be an ability to (a) transpose external C0 information through the clairaudient agent's own auditory pathways, and then (b) read them out with some decent degree of unadulteration so that the phenomenon would not be reducible to mere confabulation or delusion. For transposition (2a), there would be an ability to allow C0 processes to enter C1 that are not only one's own C0 processes (which would be considered a NOC ability in and of itself), but of external (another agent's) C0 processes. This would be conditional upon the sensitivity to that external information in the first place (1) at the level of C0, i.e., one's own C0 perceiving external C0. C0 to C0 transfer involves one form of subtle cognition that enables co-access of two TOC's (internal and external); and this external C0 information's transposition into C1 involves another form of subtle cognition that enables external C0 information to assimilate into the normal transpositional process of an agent's own C0 to his C1; i.e., such that the external C0 is not 'immuno-rejected', so to speak. For relatively unadulterated readout (2b) of the clairaudient information, this would involve another form of subtle cognition in its own right that is able to tell the difference between an agent's personal aggregates and those that are external to it, thus being able to filter out the former to a decent degree such that the clairaudient information might be regarded has offering some supposed insight into the cognition of an external object.

While the study of clairaudience and other such NOC phenomena have hitherto not received notable funding priority and thus not much traction in the scientific literature, there is substantial neuroscientific evidence of NOC in humans (Blanke, Slater, and Serino, 2015; Blanke, 2012; Montes, 2017b; Caplan and Montes, 2018), and various aspects of NOC, such as intuitive abilities (Lufityanto, Donkin, and Pearson, 2016; Iandoli, 2019; Nolan, 2018), which would undoubtedly be a requisite for clairaudience in many cases. A further discussion on clairaudience and other NOC phenomena is outside the scope of this chapter. The cognitive profile presented above would not be entirely unique to clairaudience; certain Taoist elemental qigong practices also work with the perception and integration of external and internal common subtle aggregates (Chia, 2007). Clairaudience has been expounded on here only to the degree that it could be potentially useful as a hypothetical example to consider how mindplexical cognition might operate in the case of NOC in a human being, so that a context can be offered for the development of mindplexes between humans and AI/AGI. For neuroscientific work on the neural correlates of non-dual experience, see (Josipovic, 2014; Tang, Holzel, and Posner, 2015).

4. Artificial & Distributed Ledger Mindplexes

In taking steps towards achieving an AI/AGI mindplex (Goertzel, 2003), examining various forms and permutations of NOC in humans may offer suggestions as to how an individual TOC might access information from a superordinate AI/AGI TOC. Whereas in humans to date, TOC-TOC interaction would have been achieved endogenously (as discussed above), a mindplex with AGI would hypothetically accomplish this through technological manipulation. If we follow from the clairaudience example, a human agent might actually audibly hear a command (which would be at the level of his/her C1) from the superordinate TOC in cases where that may be useful and appropriate. This might also happen through subliminally presented content (e.g., subthreshold audio or video) that is processed at the level of the human's C0 but does not reach C1 yet exerts a sufficient influence on C1. Another example would be that a neural interface influences the human individual's C0 in a way that the external information is ensured to integrate into the human's C1. In the last two cases, it is conceivable that a human could process the superordinate TOC's information in a way that the human TOC is not aware in any way that this is being done; the human simply experiences the external information as if from his own TOC and is not able to distinguish whether the information came from ego-personality or from an external source. An additional variable might be, in contrast with a human being influenced by an external TOC but being aware of and able to communicate with it, that the human may be influenced by the superordinate TOC (and even be aware of that) but not necessarily be able to consciously access or manipulate the external TOC in return.

There may be farther-future human-AGI or AI-AGI mindplexes in which TOC's merge fully such that one or both of the TOC's (say, the individual human's) is able to fully experientially access the superordinate TOC (C1), in addition to its own, without interference. Such a fused mindplex would amount to a kind of mind-merger at the phenomenal level. This would involve something akin to achieving the level of mind realization of the above exemplified non-dual cultivator *and* also achieving bidirectional subtle cognitive control of both TOC's. If a human and superordinate AGI could both flexibly access each other's TOC through TOC fusion, this would represent an extraordinary achievement, and if there were control of each other's TOC, that would be an extremely extraordinary achievement. In the meantime, having two concurrently active and powerful TOC's that are not necessarily experientially merged but that are capable of frequent or continuous advanced communication at the level of (dualistic) cognitive operations (as discussed) would be the requisite for a proper mindplex. These cognitive communications could happen between any non-corresponding combination of C0, C1, and C2 (and beyond) between subordinate and superordinate levels, e.g., subordinate C0 to superordinate C1, and so on. In a scenario involving AI agents, each form of subtle cognition (which are referred to in Sanskrit as *siddhis*, or seemingly miraculous powers (Feuerstein, 1989)), e.g., clairaudience, could be framed as an AI service package, each of which could contribute to greater functionality in an AGI and its TOC.

There are many permutations of TOC interactions that present various scenarios and experiences. Which of these will occur in a given circumstance depends on the type of information and the type of technology being used, as well as socio-political contexts. In some of the above examples, the more limited or unilateral TOC interactions could be fairly likened to subliminal messages, such as in advertising or neural spying and control. This presents an ethical issue that would likely emerge once such advanced technology is integrated into personal lives and represents an example of potential abuse through non-disclosure and/or structural compulsion. The importance of ethical theorising in the field of AI has been previously discussed by the authors (Montes and Goertzel, 2018; Montes, 2017b; 2018). In addition to the merits of NOC in advancing AI/AGI with hitherto unexplored modes of cognition, NOC also has merits in ethical theorising due to certain types of NOC that occur in the context of compassionate and altruistic intentional frameworks in the world's wisdom traditions (Montes, 2017b).

The authors have previously discussed the importance of developing a distributed, decentralized, and democratic (DDD) AI (Montes and Goertzel, 2018) in order to include a broader swath of humanity's palette of valuational and cognitive diversity in AGI, which will grow to play an increasingly critical role in the future of humanity and life on earth and beyond. It is also important to take significant steps in this direction sooner than later, because humans are creating AI/AGI in our own image (what we believe ourselves to be and be capable of), and with the increasing influence of AI in society and life, human cognition will, in turn, be affected by how AI agents interact with humans (Montes, 2017a).

The authors believe that a more democratic representation in the creation of AI would dovetail with this goal. The most influential current AI developments are siloed within the walls of giant corporations with limited and inaccessible datasets, and they lack interoperability standards. This results in an imbalanced AI landscape, with risks to populations that are presently underrepresented in AI development. The authors have proposed the utilization of distributed ledger technology (DLT), such as blockchain, to open the playing field by tracking contributions to AI development through transactions on a distributed ledger and allowing developers to be compensated for their efforts. This offers an economic alternative to young developers and computer science academics who would otherwise defer financial income on their efforts. A case example of such a solution is SingularityNET, a DLT marketplace for AI services, currently under development by Goertzel and colleagues (Montes and Goertzel, 2018). Part of the goal is to create a "seed AGI" that has a broad representation of human inputs and values; as an early starting point for the progressive development of AGI that also maximises the chances for ethical theorising to keep up.

A fabric like SingularityNET could be used to help nudge humans into a global brain mindplex that concurrently achieves human fulfilment and porting of human values into the "seed AGI" (that will grow into unpredictable new things but starting from the seed of human culture/values). AI's/AGI's on a DLT fabric like SingularityNET would allow for the creation of an AGI mindplex on the collective level, granted the necessary technical requirements. Advancements in neuroscience, novel forms of cognitive communications can be discovered through sophisticated brain-computer interfaces (BCI's). With further technological advancements, it may eventually be possible to code aggregate sets of neural signals into DLT, thus connecting them with decentralized AI fabrics such as SingularityNET, and enabling partitioning, recombination, crowdsourcing and creative democratic evolution of cognition (Swan, 2015; 2016), uncovering hitherto unfathomable and unknown aspects of human and transhuman neural and mental function.

It is fascinating to brainstorm concrete examples of how such mindplexes might work and how they might leverage and facilitate non-ordinary states of awareness in the humans operating within them. For instance, could you have a mindplex trying to solve a certain scientific or mathematical problem, and leveraging a combination of AI algorithms doing automated reasoning with a community of human meditators inspired by the possibility of having relevant insights pop into their minds doing meditation? What kind of "paranormal" synchronicities might emerge in such a situation? What states of consciousness might emerge in the various systems involved?

In the end, however, the specific scenarios we can cook up now given our current states of consciousness and levels of awareness, are surely trivial in comparison with the cognitive and experiential dynamics that will emerge and that we cannot now foresee. The situation is perhaps even more extreme than that of a community of apes, about to evolve into humanity, speculating about the various phenomena that human society might display after it advances for a while. We are at the dawn of the emergence of forms of organization and dynamics that are drastically novel from our current perspectives.

5. Conclusion

What we have here is a vision of humanity expanding and extending itself via a combination of technologies including BCI and human-value–guided AGIs, ultimately catalyzing the emergence of a mindplex incorporating human and artificial intelligences, within which human minds are able to maintain their sense of autonomy and their unique characteristics, yet also able to contribute meaningfully to a broader cognitive whole including aspects that cannot be comprehended within the human scope. The states of consciousness accessible to humans or AGI's embedded in such mindplexes, or to the various levels of the mindplex itself, are bound to go beyond the currently charted realm. However, it seems natural to conjecture that they will, in many cases, have significant commonalities with states of consciousness humans now experience in various non-ordinary situations related to meditation, psi phenomena, and so forth. One envisions new varieties of experiencing oneness with other and greater minds, compatible with retaining one's own autonomy and integrity; and of pursuing one's own individual goals while also coordinatedly following the flow of greater and only partially-comprehensible cognitive processes in which one is embedded.

Exactly what we need to do now in order to maximize the odds of this sort of mindplex ecosystem emerging, is obviously very far from clear. However, it seems intuitive that a good start will be to create AGI systems that respect human values and integrate closely with human beings in an I-Thou manner;

and to embed the humans and AIs on the planet in an overall mind-network that, as its intelligence increases, is guided by the values of humanity as a whole and by the principle of respecting autonomy and agency of individual human and AI minds while also militating toward emergence of greater systemic intelligence. If we want an ultimate transhuman mindplex in which the dynamics on various levels are supportive of the value of the cognitive agents existing on various levels, we should start by creating a global brain whose values are respectful and supportive of the human agents contributing to it – i.e., what we have called a DDD AGI.

Many challenges remain in building a DDD AGI based on DLT and related technologies, let alone developing a mindplex. In confronting such a large challenge, we can use all the inspiration we can find, and, digging into the depths of the human mind through insights from NOC can offer a wealth of inspiration for translating subtle phenomena into AI. The nature of mind and consciousness is considerably larger in scope than is appreciated by most, including academicians and engineers. Accordingly, a sufficiently high bar may be set in the construction of a capable AGI that aptly captures a greater depth and span of documented accounts of conscious experience than what lies in modern Western civilization alone. A keen understanding of how the mind works in NOC is the beginning of the creation of a 'Rosetta Stone' for advancing AI and AGI beyond what is currently imaginable. NOC also fuels the maturational processes of neuroscience and philosophy of mind in elucidating how consciousness works and its various forms. It is important to not limit an understanding of NOC to popular "first-wave" approaches, such as mindfulness meditation (Hölzel et al., 2011), which is a diluted blend of several techniques that provides a valuable foundation for NOC cultivation yet are only elementary in their reach, considering the actual scope of NOC (Montes, 2017b). Furthermore, AI/AGI that can support NOC cultivation in humans (Goertzel et al. 2017) is poised to exert a positive influence in society (Montes, 2017b). The authors encourage researchers to investigate subtle cognition based in NOC to create a more capable and equitable AI/AGI, and to consider how examination of the mindplex concept involves a coordinated involvement of consciousness, mind, cognition, neuroscience, AI, and ethics that promotes an integral discussion, discovery process, and an unleashing of the highest potential in these fields.

References

Armstrong, Karen (2011) *A History of God: The 4,000-Year Quest of Judaism, Christianity and Islam*, New York: Random House Publishing Group.

Baars, Bernard J. (1997) "In the Theatre of Consciousness. Global Workspace Theory, a Rigorous Scientific Theory of Consciousness", *Journal of Consciousness Studies*, 4 (4): 292–309.

Blackmore, Susan (2017) *Consciousness: A Very Short Introduction*, Oxford: Oxford University Press.

Blanke, Olaf (2012) "Multisensory Brain Mechanisms of Bodily Self-Consciousness", *Nat Rev Neurosci*, 13 (8): 556–71.

Blanke, Olaf and Metzinger, Thomas (2009) "Full-Body Illusions and Minimal Phenomenal Selfhood", *Trends in Cognitive Science*, 13 (1): 7–13.

Blanke, Olaf, Slater, Mel and Serino, Andrea (2015) "Behavioral, Neural, and Computational Principles of Bodily Self-Consciousness", *Neuron*, 88 (1): 145–66.

Caplan, Mariana and Montes, Gabriel Axel (2018) "Yoga, Science, and the Brain" In *Yoga & Psyche: Integrating the Paths of Yoga and Psychology for Healing, Transformation, and Joy*, Boulder, CO: Sounds True.

Chia, Mantak (2007) *Cosmic Fusion: The Inner Alchemy of the Eight Forces*, Rochester, VO: Inner Traditions/Bear.

Clark, Andy and David Chalmers (1998) "The Extended Mind", *Analysis*, 58 (1): 7–19.

Dalai Lama, His Holiness (2000) *Dzogchen: Heart Essence of the Great Perfection*, Ithaca, NY: Snow Lion Publications.

Dehaene, Stanislas, Lau, Hakwan and Kouider, Sid (2017) "What Is Consciousness, and Could Machines Have It?", *Science*, 358 (6362): 486–492.

Feuerstein, Georg (1989) *Yoga: The Technology of Ecstasy*, New York: J.P. Tarcher.

Feuerstein, Georg (1998) *Tantra: The Path of Ecstasy*, Boulder, CO: Shambhala Publications.

Feuerstein, Georg (2013) *The Yoga Tradition: Its History, Literature, Philosophy and Practice*, Chino Valley, AZ: Hohm Press.

Friston, Karl (2010) "The Free-Energy Principle: A Unified Brain Theory?" *Nat Rev Neurosci*, 11 (2): 127–38.

Friston, Karl (2012) "A Free Energy Principle for Biological Systems" *Entropy*, 14 (11): 2100–2121.

Friston, Karl (2013) "Life as We Know It", *Journal of the Royal Society Interface*, 10 (86): 20130475.

Gilmartin, Kevin J., Newell, A. and Simon, Herbert A. (1976) "A Program Modeling Short-Term Memory under Strategy Control" In *The Structure of Human Memory*, pp. 15–30.

Goertzel, Ben (2003) "Mindplexes: The Potential Emergence of Multiple Levels of Focused Consciousness in Communities of AI's and Humans", *Dynamical Psychology*.

Goertzel, Ben, Pennachin, Cassio, Senna, Andre, Maia, Thiago and Lamacie, Guilherme (2004) "Novamente: An Integrative Architecture for Artificial General Intelligence" (n/i).

Goertzel, Ben and Pennachin, Cassio (2007) *Artificial General Intelligence*, Vol. 2, Berlin: Springer.

Goertzel, Ben, Pennachin, Cassio and Geisweiller, Nil (2014) "Communication Between Artificial Minds" In *Engineering General Intelligence, Part 2*, Berlin: Springer, pp. 411–21.

Goertzel, Ben, Mossbridge, Julia, Monroe, Eddie, Hanson, David and Yu, Gino (2017) "Humanoid Robots as Agents of Human Consciousness Expansion" In *ArXiv: 1709.07791.*

Heylighen, Francis (1997) "Towards a Global Brain. Integrating Individuals into the World-Wide Electronic Network" In *Der Sinn Der Sinne,* Gottingen: Steidl Verlag.

Heylighen, Francis (2007) "The Global Superorganism: An Evolutionary-Cybernetic Model of the Emerging Network Society", *Journal of Social Evolution & History,* 6 (1): 58-119.

Hohwy, Jakob (2012) "Attention and Conscious Perception in the Hypothesis Testing Brain", *Frontiers in Psychology,* 3: 96.

Hölzel, Britta K., Lazar, Sara W., Gard, Tim, Schuman-Olivier, Zev, Vago, David R. and Ott, Ulrich (2011) "How Does Mindfulness Meditation Work? Proposing Mechanisms of Action from a Conceptual and Neural Perspective", *Perspectives on Psychological Science,* 6 (6): 537–59.

Iandoli, James (2019) "Is the Caudate-Putamen an Antenna for Anomalous Information?" in *TerraObscura* (Blog), January 8, Accessed in https://www.terraobscura.net/blog/is-the-caudate-putamen-an-antenna-for-anomalous-information.

Jeannerod, Marc (2003) "The Mechanism of Self-Recognition in Humans", *Behavioural Brain Research,* 142 (1): 1–15.

Josipovic, Zoran (2014) "Neural Correlates of Nondual Awareness in Meditation", *Annals of the New York Academy of Sciences,* 1307 (1): 9–18.

Kirchhoff, Michael, Parr, Thomas, Palacios, Ensor, Friston, Karl and Kiverstein, Julian (2018) "The Markov Blankets of Life: Autonomy, Active Inference and the Free Energy Principle", *Journal of The Royal Society Interface,* 15: 20170792.

Kurzweil, Ray (2010) *The Singularity Is Near,* London: Gerald Duckworth & Company.

Lufityanto, Galang, Donkin, Chris and Pearson, Joel (2016) "Measuring Intuition: Nonconscious Emotional Information Boosts Decision Accuracy and Confidence", *Psychological Science,* 27 (5): 622-34.

Montes, Gabriel Axel (2017a) "Causal Biomimesis: Self-Replication as Evolutionary Consequence" In *Biomimetic and Biohybrid Systems: 6th International Conference, Living Machines 2017, Stanford, CA, USA, July 26–28, 2017, Proceedings* (Michael Mangan, Mark Cutkosky, Anna Mura, Paul F.M.J. Verschure, Tony Prescott, and Nathan Lepora, eds.), Cham: Springer International Publishing, pp. 328–47.

Montes, Gabriel Axel (2017b) "Non-Ordinary Consciousness for Artificial Intelligence." In *Biomimetic and Biohybrid Systems: 6th International Conference, Living Machines 2017, Stanford, CA, USA, July 26–28, 2017, Proceedings,* (Michael Mangan, Mark Cutkosky, Anna Mura, Paul F.M.J. Verschure, Tony Prescott, and Nathan Lepora, eds.), Cham: Springer International Publishing, pp. 348–62.

Montes, Gabriel Axel (2018) "Virtual Reality for Non-Ordinary Consciousness", *Frontiers in Robotics and AI,* 5: 7.

Montes, Gabriel Axel, and Goertzel, Ben (2018) "Distributed, Decentralized, and Democratized Artificial Intelligence", *Technological Forecasting and Social Change,* 141: 354-358.

Nolan, Garry (2018) "Can Genetic Differences in Intuition and Cognition Drive Success in Space?" November 30, Accessed in https://spacegenetics.hms.harvard.edu/2018-space-genetics-symposium.

Norbu, Namkhai (1986) *The Crystal and the Way of Light: Sutra, Tantra, and Dzogchen*, Ithaca, NY: Snow Lion Publications.

Swan, Melanie (2015) "Blockchain Thinking: The Brain as a Dac (Decentralized Autonomous Organization)", *Texas Bitcoin Conference*, 27–29.

Swan, Melanie (2016) "The Future of Brain-Computer Interfaces: Blockchaining Your Way into a Cloudmind", *Future*, 26: 2.

Tang, Yi-Yuan, Holzel, Britta K. and Posner, Michael I. (2015) "The Neuroscience of Mindfulness Meditation", *Nat Rev Neurosci* 16 (4): 213–25.

Tsakiris, Manos, Schutz-Bosbach, Simone and Gallagher, Shaun (2007) "On Agency and Body-Ownership: Phenomenological and Neurocognitive Reflections", *Conscious Cogn*, 16 (3): 645–60.

Twain, Mark (1906) "Chapters from My Autobiography: II", *The North American Review*, 183 (599): 449–60.

Wilber, Ken (1977) *The Spectrum of Consciousness*, Wheaton, IL: Quest Book/Theosophical Pub. House.

Wilber, Ken (2001) *Sex, Ecology, Spirituality: The Spirit of Evolution, Second Edition*, Boulder, CO: Shambhala Publications.

Wilber, Ken (2007) *Integral Spirituality: A Startling New Role for Religion in the Modern and Postmodern World*, Boulder, CO: Shambhala Publications.

CHAPTER 5

THE COGNITIVE PHENOMENOLOGY ARGUMENT FOR DISEMBODIED AI CONSCIOUSNESS

Cody Turner[1]

University of Connecticut., USA

Abstract: In this chapter, I shall offer two novel arguments for what I call *strong primitivism* about cognitive phenomenology, the thesis that there exists a phenomenology of cognition that is neither reducible to, nor dependent upon, sensory phenomenology. I will then contend that strong primitivism implies that phenomenal consciousness does not require sensory processing. This latter contention has implications for the philosophy of artificial intelligence. For if sensory processing is not a necessary condition for phenomenal consciousness, then it plausibly follows that AI consciousness (assuming that it is possible) does not require embodiment. The overarching goal of this chapter is to show how different topics in the analytic philosophy of mind can be brought to bear on an important issue in the philosophy of artificial intelligence.

***

The central argument considered here is called 'the Cognitive Phenomenology Argument for Disembodied AI Consciousness'. It is as follows:

- P1. Strong primitivism about cognitive phenomenology is true;
- P2. If strong primitivism about cognitive phenomenology is true, then phenomenal consciousness does not require sensory processing;
- C1. Therefore, phenomenal consciousness does not require sensory processing;
- P3. If phenomenal consciousness does not require sensory processing, then AI consciousness (assuming that it is possible) does not require embodiment;
- C2. Therefore, AI consciousness (assuming that it is possible) does not require embodiment.

---

[1] PhD Student at the University of Connecticut.

As indicated, I first focus on P1 by presenting two arguments for strong primitivism. The first argument is a phenomenological argument based on the possibility of what some Buddhists and mystics call *pure consciousness*. The second argument draws from the *higher-order thought theory of consciousness*. Importantly, neither of these arguments is intended (in this chapter at least) to definitively demonstrate the veracity of P1. The aim of both arguments is to show that strong primitivism is plausible given certain theoretical assumptions.[2] My goal here, in other words, is merely to establish the following two conditionals:

(CON 1) If pure consciousness is possible, then strong primitivism about cognitive phenomenology is true.

(CON 2) If the higher-order thought theory of consciousness is true, then strong primitivism about cognitive phenomenology is true.

After arguing for CON 1 and CON 2, I will proceed to justify P2 and P3 of the cognitive phenomenology argument for disembodied AI consciousness. Before turning to the argument though, I will overview the cognitive phenomenology debate.

1. The Cognitive Phenomenology Debate

Until the 1980s, it was mainstream in analytic philosophy of mind to view phenomenality and intentionality as two separate aspects of the mental. This view, which Terrance Horgan and John Tienson (2002) call *separatism*, holds that phenomenal states and intentional states are metaphysically independent of one another, meaning that neither kind of state arises from the other and that a system can exhibit one and not the other, at least in principle. Separatism is traditionally motivated by the conviction that while occurrent *sensory* mental states (e.g., the state you are in when seeing the color red) possess phenomenal character but lack intentional content, occurrent *cognitive* mental states (e.g., the state you are in when believing that Hartford is the capital of Connecticut) possess intentional content but lack phenomenal character.

Cognitive phenomenology (CP for short) is a concept in the philosophy of mind that functions to undermine separatism. Proponents of CP assert that occurrent cognitive states are both intentionally directed and (in some sense at least) phenomenally conscious. According to these proponents, there is an experiential character associated with thinking, where 'thinking' is understood

[2] I am strictly speaking agnostic both with respect to whether pure consciousness is possible, and with respect to whether the higher-order thought theory of consciousness is true.

to encompass a wide range of propositional attitudes, such as believing, doubting, wondering, desiring, entertaining, etc.

One can distinguish in conceptual space between four different views of cognitive phenomenology:

1. Eliminativism about CP: Cognitive phenomenology does not exist.
2. Reductionism about CP: Cognitive phenomenology is reducible to sensory phenomenology.
3. Weak Primitivism about CP: Cognitive phenomenology is irreducible to sensory phenomenology, but dependent upon sensory phenomenology.
4. Strong Primitivism about CP: Cognitive phenomenology is irreducible to, and independent of, sensory phenomenology.

Before elucidating these four positions, it is important to highlight that cognitive phenomenology only pertains to *occurrent* cognitive states. *Dispositional* cognitive states, such as all of my beliefs that I am not currently entertaining, are not phenomenally conscious. For there is nothing it is like to have a belief that is not being presently tokened. With this being said, let us turn towards the four positions on CP.

Eliminativists about CP concede that occurrent cognitive activities are *access conscious*, but they deny that such activities are *phenomenally conscious*. A mental state is access conscious if and only if it is available for verbal report, rational inference, and/or the deliberate control of behavior (Block, 1995). While eliminativists acknowledge that occurrent desires, beliefs, wonderings, etc. are available for these uses, they reject the notion that there is something it is like to be in these cognitive states.[3] By contrast, reductionists grant that cognition involves phenomenology, but uphold that cognitive phenomenology can be explained in terms of sensory phenomenology that typically accompanies thinking, such as mental imagery and inner speech.

The debate between eliminativism and reductionism about CP may ultimately just be a verbal dispute. The same cannot be said with respect to the debate between reductionism and primitivism. Primitivists agree with reductionists that various forms of sensory phenomenology, such as auditory imagery (i.e., inner speech) and visual imagery (i.e., mental pictures), often accompany our thoughts. Unlike reductionists, however, primitivists renounce the claim that this concomitant sensory phenomenology *exhausts* the experiential character of occurrent cognition. Consider the thought <buses are

[3] See Carruthers and Veillet (2011) for an argument in support of eliminativism.

yellow>. Each tokening of this thought may be, and often is, connected with a distinct sensory phenomenal character that represents it. For example, in one instance the thought may be represented by a mental image of a yellow bus whereas in another it may be represented by an inner speech sentence that says 'BUSES ARE YELLOW.' Primitivists aver that although these two tokenings of the same thought differ with respect to their sensory character, there is nevertheless a common 'what-it's-likeness' that runs through them. This shared experiential character, according to the primitivist, denotes a *sui generis* cognitive phenomenology, one that cannot be reduced to any conjunction of auditory and/or visual imagery.

The distinction between *weak* primitivism and *strong* primitivism about CP is rarely made in the literature on cognitive phenomenology (Levine, 2011, is an exception). This distinction is motivated by the insight that phenomenal irreducibility does not entail phenomenal independence. X phenomenally depends upon Y iff X cannot be instantiated in the absence of Y. Weak primitivism holds that while cognitive phenomenology is irreducible to sensory phenomenology, cognitive phenomenology cannot be instantiated in the absence of sensory phenomenology. As Levine states, on weak primitivism, "it's still the case that the only way to phenomenally experience a cognitive content is through its effect on some sensory presentation" (Levine, 2011: 112). Strong primitivism, on the other hand, states that cognitive phenomenology is both phenomenally irreducible to, and phenomenally independent of, sensory phenomenology.

2. The 'Pure Consciousness' Argument for Strong Primitivism

Some of the most popular arguments for primitivism about CP are phenomenological in nature. These phenomenological arguments often take the form of *phenomenal contrast cases* (Chudnoff, 2015). For example, the phenomenology of an English speaker is contrasted with the phenomenology of a person who does not speak English. We are asked to imagine for the sake of argument that both the English speaker and the non-English speaker hear the same English sentence uttered by the same person at the same exact location. There is in this imagining, so the argument goes, an overall phenomenal difference between the two speakers despite the fact that their respective sensory phenomenologies are identical. This phenomenological insight is taken to suggest that there must be an irreducible cognitive phenomenology associated with understanding language.

Levine (2011) correctly notes that most of the existing phenomenological arguments for primitivism about CP at best support weak primitivism (he calls weak primitivism 'impure CP' and strong primitivism 'pure CP'):[4]

> When it comes to the distinction between pure and impure CP, it seems obvious that this argument [the phenomenological contrast argument for primitivism] can't support more than impure CP. After all, all the examples involve ways of experiencing visual appearances, sound streams, and the like. The argument is addressing someone who doesn't find anything but sensory phenomenology in her conscious life, and it pushes her to notice distinctions among these sensory appearances that can only be accounted for by appeal to cognitive penetration. Fair enough. Still, this doesn't get us anything like pure CP. It's still possible, for all this argument demonstrates, that the only way for a cognitive content to make itself appear to a conscious subject is though affecting the way some sensory manifold appears. That's impure CP (Levine, 2011: 115-116).

This chapter offers a phenomenological argument for strong primitivism that is based upon the possibility of 'pure consciousness'. Pure consciousness is a phenomenal state that is supposed to be achievable via mediation.

It is claimed that, during instances of pure consciousness, the intentional structure of experience disappears so as to leave nothing but an open phenomenal field. David Woodruff Smith (2011) describes the state of pure consciousness thusly: "In sustained meditation the sense of object disappears, i.e., the intentional content or object-specifying sense disappears, and the sense of self or subject disappears. That is to say, the usual intentional structure of subject-object-or subject-act-content-object-dissolves in such a meditative state. There is then just this flowing phase of consciousness" (Smith, 2011: 481). The 'pure consciousness' argument for strong primitivism can be represented in the following manner:

PP1. Pure consciousness is possible;
PP2. If pure consciousness is possible, then strong primitivism about cognitive phenomenology is true;
C1. Therefore, strong primitivism about cognitive phenomenology is true.

---

[4] A notable exception is Chudnoff (2015). There he presents a phenomenal contrast case that does in fact support strong primitivism about CP.

PP1 is supported by the phenomenology of deep meditative experience. One could push back against PP1 by, for example, arguing against the common wisdom that meditation is a reliable epistemic guide to one's own phenomenal states. One could also reject PP1 by endorsing Franz Brentano's (1874) thesis that intentionality is the mark of the mental. If Brentano is correct in stating that mental states are necessarily intentional, then it follows that pure conscious mental states are impossible, for such states are reportedly devoid of intentionality. Steven Katz (1978) argues against the possibility of pure consciousness along these lines. According to Katz, states of pure consciousness are unachievable because all conscious experiences (including mystical experiences) are mediated by the presence of some conceptual schema: "The experience itself as well as the form in which it is reported is shaped by concepts which the mystic brings to, and which shape, his experience" (Katz, 1978). I take the plausibility of PP1 for granted in this chapter, and focus on PP2.

PP2 is prima facie tenable given that pure consciousness by stipulation exists in the absence of sensory phenomenology. I see two ways that one could go about objecting to PP2. The first is to *deny that pure consciousness involves phenomenal consciousness.* Perhaps the alleged 'emptiness' of states of pure consciousness is due to the fact that such states lack phenomenal character. If pure conscious states are devoid of phenomenal character, then pure consciousness is not a kind of cognitive phenomenology, contra to what PP2 suggests. Proponents of pure consciousness are very explicit, however, that there is a *what-it's-likeness* associated with being in a pure conscious state. As Smith emphasizes: "This meditative experience is *phenomenal*: it is appearing, "phenomenally", in consciousness, and what it is like is simply one of being conscious, as if awaiting what else may or may not appear in consciousness" (Smith, 2011: 481). This quote from Smith indicates that phenomenality is built into the very concept of pure consciousness. If this is the case, then the present objection (that pure consciousness is non-phenomenal) seems to be at its core an objection to PP1, rather than to PP2 as previously supposed.

A different way to go about objecting to PP2 is to *concede that pure consciousness is phenomenal, but deny that pure consciousness is cognitive.* According to this objection, the phenomenology of pure consciousness involves neither sensory phenomenology nor cognitive phenomenology. A phenomenally conscious mental state X, so the objection goes, qualifies as cognitive if and only if X is characterized by a phenomenally conscious cognitive attitude and/or phenomenally conscious cognitive contents. The problem is that states of pure consciousness involve neither cognitive attitudes nor cognitive contents. This is because such states are by definition synonymous with the disappearance of the intentional structure of experience. The phenomenology of pure consciousness is therefore not a kind of cognitive phenomenology. Or so the objection goes at least.

The claim that pure consciousness is cognitive in nature can, I propose, be reconciled with the mystics' claim that the state of pure consciousness reveals no discernible intentional structure. The idea I have in mind is this. Pure consciousness is an intentional mental state, but it is an intentional state that has as its object the concept of nothingness, or perhaps the concept of non-intentionality itself. Put differently, the idea is that pure consciousness is an intentional state that represents the state of non-representation. Construing pure consciousness in this light allows one to maintain that pure consciousness is a type of cognitive phenomenology while acknowledging that states of pure consciousness seem to be devoid of intentionality from a first-person point of view.

This proposed analysis of pure consciousness is similar in many respects to how various philosophers of mind analyze the phenomenology of *moods*. Moods, such as elation and anxiety, are sometimes thought to be non-intentional, undirected mental states. Advocates of Brentano's thesis of course reject this conception of moods as non-intentional. Said advocates instead claim that "while moods *appear* not to be directed at anything, upon closer examination, it turns out that they are in fact directed at special kinds of objects: bodily states or unusual external objects, such as the world as a whole, indeterminate intentional objects, or frequently changing objects (see Goldie, 2000; 2002; Seager, 2002; Seager and Bourget, 2007 and Tye, 2008)" (Mendelovici, 2013: 8-9). Just as some philosophers try to explain away the apparent non-intentionality of moods, I am here trying to explain away the apparent non intentionality of pure conscious states. More specifically, in the same way that the intentional object of states of anxiety or elation might be 'the world as a whole' or 'everything', I am suggesting that the object of states of pure consciousness might be 'the absence of an intentional world', or 'nothing'. If this is right, then it is unsurprising that states of pure consciousness are (mistakenly) reported to be devoid of intentionality.

The objector to PP2 may at this juncture in the dialectic accuse me of changing the subject, or at the very least, of misunderstanding the concept of pure consciousness. The mystics' claim, the objector may insist, is not merely that pure consciousness *appears* to lack intentionality; it is rather the more robust claim that pure consciousness *as a matter of fact* lacks intentionality. My response to this retort is in effect to double down: pure consciousness is 'cognitive' precisely because pure conscious states are, despite phenomenological appearances, actually intentional. This is my suggestion.

Perhaps pure consciousness is by definition non-intentional, though. I grant that states of pure consciousness cannot properly be called 'cognitive' phenomenal states if this is the case. The fact of the matter, however, is that a phenomenological argument for strong primitivism can be given without appealing to the notion of pure consciousness. *Even if pure consciousness is not a kind of cognitive phenomenology, it is not difficult to imagine a slightly modified phenomenological argument for strong primitivism that is based upon the phenomenology of meditation.* One might argue, for example, that isolated awareness of propositional contents during 'mindfulness' meditation (or during immersion in a sensory deprivation tank) constitutes phenomenological evidence in support of strong primitivism. According to Smith, "observant self-awareness is the goal of the practice called 'mindfulness' or *vipassana* in one Buddhist tradition. In this practice, the meditative experience prepares for an experience of consciousness per se – pure awareness, as it were – and allows then for noticing particular thoughts, feelings, or sensations that may come along in the stream of consciousness" (Smith, 2011: 481). Isolated awareness of propositional contents is not pure consciousness, but it is arguably a phenomenal experience that lends credence to strong primitivism.

3. The 'Higher-Order Thought' Argument for Strong Primitivism

I now turn to the second argument for strong primitivism, which is motivated by the higher-order thought (HOT) theory of consciousness. The higher-order thought theory of consciousness, the chief proponent of which is David Rosenthal (1986, 1993, 2005), affirms that a mental state X becomes conscious when X is taken as the object of a higher-order thought. Rosenthal takes the relationship between the HOT and the first-order mental state to be one of representation (as opposed to say, acquaintance). The argument from HOT theory to strong primitivism has the following basic structure:

QP1. HOT theory is true;
QP2. HOT theory entails strong primitivism about CP;
C1. Therefore, strong primitivism about CP is true.

Two important clarificatory remarks are in order with respect to the scope of this argument. First, the argument concerns only *actualist* versions of HOT theory. Actualist HOT theory is contrasted with *dispositional* HOT theory, the leading proponent of which is Peter Carruthers (2005). Dispositional HOT theory asserts that the relevant higher-order thoughts are dispositional mental states as opposed to occurrent ones. Gennaro explains that "the basic idea [behind dispositional HOT theory] is that the conscious status of an experience is due to its *availability* to higher-order thought...Thus, no actual

HOT occurs" (Gennaro, 2018). Dispositional HOT theory is by stipulation not included within the purview of the HOT argument for strong primitivism.

The second clarificatory remark regarding the argument is that the argument only concerns versions of HOT theory that aim to explain *phenomenal consciousness*, or the 'what-it's-likeness' of experience. HOT theory is first and foremost a theory of *state consciousness*, not phenomenal consciousness. The term 'state consciousness', as I am using it here, denotes the idea that a conscious mental state is a state that one is aware of being in. Thus, a mental state X of subject S possesses state consciousness iff S is aware of X. State consciousness and phenomenal consciousness are not necessarily one and the same thing, meaning that HOT theories are not necessarily theories of phenomenal consciousness.

There is considerable controversy over what the relationship is between state consciousness and phenomenal consciousness (if any). Some philosophers, such as Lycan (1996), suggest that state consciousness is a necessary condition for phenomenal consciousness. Others, such as Block (2011), believe that phenomenal consciousness outstrips state consciousness in the sense that one can be phenomenally conscious of X despite not being aware of X. Block calls this idea 'phenomenological overflow'. I remain neutral regarding the relationship between state consciousness and phenomenal consciousness. The only point that matters for the purposes of this discussion is that the HOT argument for strong primitivism applies only to versions of HOT theory that purport to explain both state consciousness *and* phenomenal consciousness. One might immediately retort here that HOT theories are only fit (and only deployed for that matter) to explain the former type of consciousness, and that consequently the HOT argument for strong primitivism is a nonstarter.

I grant that dispositional HOT theory is ill-equipped to account for phenomenal consciousness. After all, phenomenally conscious states are by nature occurrent mental states, not dispositional ones. Rosenthal agrees: "Being disposed to have a thought about something doesn't make one conscious of that thing, but only potentially conscious of it" (Rosenthal, 2004: 28). Fortunately, as I have already flagged, dispositional HOT theories are not included within the scope of the HOT argument for strong primitivism. The fact that dispositional theories cannot explain phenomenal consciousness does not therefore undercut the argument.

The real question is whether *actualist* HOT theories can be equipped to explain phenomenal consciousness. I will not attempt to answer this question here. I will point out though that Rosenthal's actualist HOT theory, which is the most prominent version of the view, is supposed to explain phenomenal consciousness as well as state consciousness. As Rosenthal himself states: "As many, myself included, use that phrase, there being something it's like for one to be in a state is simply its seeming subjectively that one is in that state. Indeed, Block (2011: 424) quotes me to that effect: 'What it is like for one to have a pain, in the relevant sense of that idiom, is simply what it is like for one to be conscious of having that pain'" (Rosenthal, 1997: 433).

I am agnostic as to whether Rosenthal's HOT theory is plausible. As mentioned, the only aim here is to justify QP2, the claim that (actualist) HOT theory (about phenomenal consciousness) entails strong primitivism. One might doubt QP2 on the grounds that first-order sensory states are necessary for phenomenal consciousness on HOT theory. If HOT theory holds that consciousness requires the presence of first-order sensory states, then it is presumably false that HOT theory entails strong primitivism about cognitive phenomenology. In fact, HOT theory is arguably incompatible with strong primitivism if it asserts that first-order sensory states are necessary for consciousness. For it plausibly cannot simultaneously be the case that both (a) cognitive phenomenology is irreducible to, and independent of, sensory phenomenology, and (b) phenomenal consciousness necessitates the presence of first-order sensory states.

Fortunately for the proponent of the HOT argument for strong primitivism, HOT theory is not actually committed to (b). This is because the theory upholds that the higher-order thoughts are themselves sufficient for consciousness. Rosenthal says exactly this: "And on that construal of 'what it's like,' the theory does hold that a HOT is sufficient for there to be something it's like for one to be in the state the HOT describes, even if that state doesn't occur" (Rosenthal, 1997: 433-434). Call this "the sufficiency condition":

> The Sufficiency Condition: Lone higher-order thoughts are sufficient for phenomenal consciousness.

Rosenthal adopts the sufficiency condition in order to account for cases of *misrepresentation.* Where there is representation, there is also the possibility of misrepresentation. This modal truth gives rise to the following question, which the HOT theorist must answer. Namely, what happens when a higher-order thought misrepresents the content of an occurrent first-order perceptual state? Is the resulting conscious state characterized by the veridical content of the first-order state, or by the misrepresented content of the higher-order state? The former option, according to Rosenthal, is inadequate because it seems to render the higher-order state completely irrelevant with respect to the production of phenomenal consciousness. To ensure that higher-order thoughts play an essential role in generating consciousness, Rosenthal denies that first-order sensory states are required for phenomenality. According to Rosenthal, consciousness is instantiated during cases of *targetlessness,* or cases in which a higher-order thought represents there to be a first-order state that does not actually exist. Sam Coleman summarizes Rosenthal's position thusly:

> How does it feel to be aware of oneself as instantiating a red London bus-is percept when no such percept obtains? Conscious mental life is the stuff of appearances, Rosenthal emphasizes: thus, since the appearance to the subject is the same whether or not she tokens a red London bus-sh percept (providing the same HOT is in place), her experience is indistinguishable across both cases-as hallucinations can be indistinguishable from their veridical counterparts. Corresponding remarks apply to misrepresentation: HOTs govern the subjective mental appearances constitutive of the subject's stream of consciousness (Coleman, 2015: 2705).

It is the sufficiency condition that explains why HOT theory implies strong primitivism. For to say that HOTs can instantiate phenomenal consciousness in the absence of first-order sensory states is precisely to say the cognitive phenomenology does not depend upon sensory phenomenology. I am admittedly not the first person to make this argument. Richard Brown and Pete Mandik (2012) explicitly make the connection between HOT's sufficiency condition and cognitive phenomenology:

> This sufficiency claim about HOTs is a key commonality between HOT theory and one of the central theses defended by cognition phenomenology proponents: A thought can give rise to there being something it's like, and does not do so in virtue of there being some co-occuring non-cognitive state (Brown & Mandik, 2012: 4).

It is worth highlighting that Brown & Mandik do not distinguish between different views of cognitive phenomenology in their paper. It can be discerned that Brown & Mandik are concerned with primitivism about cognitive phenomenology, but they do not differentiate between weak primitivism and strong primitivism. I differentiate between these two versions, and am making the specific contention that HOT theory supports strong primitivism.

The opponent of the higher-order thought argument for strong primitivism might, at this stage in the dialectic, concede that the sufficiency condition is built into Rosenthal's HOT theory, but maintain that there are other actualist versions of HOT theory that do not endorse the condition.[5]

[5] Carruthers (2008) makes this exact point when discussing the sufficiency condition.

Sam Coleman (2015) defends a version of HOT theory that he claims is not committed to the sufficiency condition. Coleman calls his view the *quotational* higher-order thought (QHOT) theory of consciousness. According to QHOT theory, the relationship between the higher-order thought and the first-order state is not one of mental representation (as Rosenthal would have it), but rather one of "mental quotation." Coleman contends that mental quotation, unlike mental representation, allows for "genuine *intimacy* with our sensory states, in sensory consciousness" (Coleman, 2015: 2711).6 The upshot of this contention for Coleman is that first-order sensory states are necessary for phenomenal consciousness on QHOT theory. Coleman explicitly rejects the notion that isolated QHOTs are sufficient for consciousness:

> What of targetlessness? A lone QHOT won't suffice for a conscious state. Here's why. QHOTs are hypothesized to supply subjective awareness. However, if a QHOT has no sensory target to embed, it could at most arouse subjective awareness of (a state of) nothing, since it altogether lacks first-order content. But subjective awareness of a state of nothing at all just isn't subjective awareness, since subjective awareness must be (intentionally) of something. An experience literally of nothing is simply no experience. Therefore an empty QHOT won't produce subjective awareness—or a conscious state. QHOTs can only make the subject aware of something if there is something (sensory content) to be aware of (Coleman 2015: 2722).

If QHOT theory is in fact devoted to the claim that first-order sensory states are necessary for consciousness. The proponent of the HOT argument for strong primitivism might respond here by maintaining (contra Coleman) that QHOT theory is committed to the sufficiency condition. The response that I have in mind involves *appealing to the possibility of pure consciousness.* The concept of pure consciousness, as discussed, suggests that phenomenal experience can exist in the absence of any intentional structure. If pure consciousness is possible, then Coleman is wrong in stating that "an experience literally of nothing is simply no experience" (Coleman, 2015: 2722). More to the point: there is reason to believe that targetless QHOTs do suffice for phenomenal experience given the possibility of pure consciousness.

Assume for the sake of argument, however, that Coleman is correct in believing that QHOT theory rejects the sufficiency condition. One might grant

---

[6] Mental quotation is not synonymous with mental acquaintance, where acquaintance is understood in the Russellian sense.

this assumption but nevertheless insist that QHOT theory entails strong primitivism about cognitive phenomenology. Specifically, the proponent of the HOT argument for strong primitivism might concur with Coleman that phenomenal experiences are necessarily *of something*, but reject Coleman's assumption that this "something" has to be a *sensory* state. One might ask: what happens when a QHOT takes as its object a *first-order thought* (as opposed to a first-order sensory state)? Coleman presumably avers that consciousness is not instantiated in these cases, for he, to repeat, believes that phenomenal experience requires the presence of first-order sensory states on QHOT theory. An interlocutor might press Coleman though by questioning why the first-order mental states have to be sensory in nature. If QHOTs function to bestow consciousness upon first-order sensory states, then why would they not also function to bestow consciousness upon first-order cognitive states? Consistency arguably demands that first-order cognitive states be made phenomenally conscious on the theory when embedded within a QHOT. Brown & Mandik raise this point in the context of discussing HOT theory in general.[7]

If first-order thoughts are rendered phenomenally conscious when embedded in QHOTs, then QHOT theory implies strong primitivism. To repeat, this conditional remains true even if it is granted that QHOT theory rejects the sufficiency condition.

I am neutral about whether QHOT theory entails strong primitivism. Perhaps the HOT argument for strong primitivism is only tenable with respect to Rosenthal's HOT theory. Nevertheless, the HOT argument is clearly still interesting as Rosenthal's HOT theory is fairly prominent in the consciousness literature.

4. Premise 2: From Strong Primitivism to the Non-Necessity of Sensory Processing

Assume now, for the sake of argument, that strong primitivism about cognitive phenomenology is true. Why think that this phenomenological truth has metaphysical implications? In other words, why think that it follows from the fact that cognitive phenomenology does not depend upon sensory phenomenology to the conclusion that phenomenal consciousness does not

---

[7] Brown & Mandik (2012) also make the point that consistency demands that first-order cognitive states be made phenomenally conscious when discussing HOT theory in general.

depend upon sensory processing in the brain? The short answer is that P2 is the result of an *inference to the best explanation.* Namely, the best explanation for why CP does not depend upon SP is that sensory processing is not a necessary condition for phenomenal consciousness.

The defender of P2 can also point out that it is commonplace in the philosophy of mind to draw metaphysical conclusions about the nature of consciousness based upon phenomenological premises. Two examples of such arguments are the 'Transparency Argument for First-Order Representationalism' (Tye, 2002) and the 'Peripheral Inner Awareness Argument for Self-Representationalism' (Kriegel, 2009). Both of these arguments assume that the phenomenology of conscious experience can inform us about the underlying metaphysical nature of experience. With respect to the 'Transparency' argument, for instance, the idea is that first-order representationalism (a metaphysical thesis) can be derived from the fact that the only introspectively discernible aspect of experience is its world-directed representational content. P2 of the cognitive phenomenology argument for disembodied AI consciousness employs similar reasoning. Specifically, the premise relies on the following principle (call it 'PHEN'):

> PHEN: Phenomenological independence implies metaphysical independence.

According to PHEN, it follows from the fact that X is phenomenologically independent of Y to the conclusion that X can be instantiated in the absence of the metaphysical underpinning of Y. Applied to cognitive phenomenology, the idea is that if CP is independent of SP, then CP can be instantiated in the absence of sensory processing, meaning that sensory processing is not a precondition for phenomenal consciousness.

One might object to PHEN on the grounds that there are obvious counterexamples to the principle. For example, <pain> is phenomenologically independent of <C-fibers firing> given that the latter is not phenomenological at all. It does not follow from this fact, however, that <pain> is metaphysically independent of <C-fibers firing>, or that the sensation of pain is not equivalent to the firing of C-fibers.[8] <Pain> may be metaphysically dependent upon <C-fibers firing> (i.e., the *thesis of multiple realizability* may be false), despite the fact that the former is phenomenologically independent of the latter.

---

[8] I thank Bill Lycan for bringing this objection to my attention.

This objection to PHEN rests on a (understandable) misunderstanding of what is meant by 'phenomenological dependence/independence'. The relationship of phenomenological dependence/independence is by stipulation a relationship that holds between two *phenomenal* properties, signifying that the relata must both be experientially manifest. Since the firing of C-fibers is not experientially manifest, it makes no sense to say that <pain> is phenomenologically independent of <C-fibers firing>. It does make sense, however, to say that agential experience is phenomenologically independent of moral experience, or that conative experience is phenomenologically independent of cognitive experience, or that cognitive experience is phenomenologically independent of sensory experience. For all of the relata in these cases are experientially manifest properties.

Of course, PHEN remains controversial even with this clarification in place. It may be the case, for example, that while sensory processing (e.g., the presence of what Koch calls 'the sensory hot zone' (Koch, Massimini, Boly & Tononi, 2016)) is a developmental precondition for consciousness, it is nevertheless true that as we get older, we develop a kind of sui generis cognitive phenomenology that can be instantiated in the absence of sensory phenomenology. There is, on the face of it, nothing conceptually problematic about this objection to P2. The burden of proof is squarely on the objector though to explain how it is possible that (a) phenomenal consciousness metaphysically depends upon sensory processing, but (b) cognitive experience does not phenomenologically depend upon sensory experience. I am not aware of any explanation to this effect that currently exists in the literature. P2 is highly plausible until such an explanation is given.

Before moving on to P3 of the cognitive phenomenology argument, it is worth drawing attention to the fact that weak primitivism about CP clearly does not indicate that sensory processing is not necessary for consciousness. This is precisely because weak primitivism states that cognitive phenomenology cannot be instantiated in the absence of sensory phenomenology (despite being irreducible to SP).

5. Premise 3: From the Non-Necessity of Sensory Processing to Disembodied AI Consciousness

Assume now that both P1 and P2 are true, and that consequently, phenomenal consciousness does not require sensory processing. How does this conclusion bear on the philosophy of artificial intelligence? In particular, what is the reasoning behind P3? It is vital to accentuate that P3 operates under the controversial assumption that AI consciousness is in principle possible. I do not make any arguments for or against the possibility of AI consciousness here. Rather, I simply assume that AI consciousness is possible and proceed to address the following question (call it 'the embodiment question' (EQ)):

EQ: Would AI consciousness, should it exist, require embodiment?

EQ is a pressing question in the philosophy of artificial intelligence. We want to know whether or not disembodied AI systems (such as neurotrophic supercomputers or sophisticated personal assistants – perhaps future versions of Apple's 'Siri' or Google's 'Alexa') are capable of what-it's-likeness.[9] It is likely, after all, that such systems will be able to pass the Turing test in the not too distant future. According to a survey conducted by Bostrom and Muller (2016), "experts estimate the probability of achieving 'high level machine intelligence' to go over 50% by 2040-2050, over 90% by 2075" (Muller, 2016: 4).

One relevant worry here is that we will *inevitably attribute* consciousness to future disembodied AIs. Perhaps it is psychologically infeasible to deny phenomenal experience to an AI system that is verbally indistinguishable from a normal adult interlocutor. If this is the case, and disembodied AIs are *incapable* of experience, then there is a very real concern that humanity will be psychologically duped into thinking that unfeeling computer programs are phenomenally conscious. I call this the *duped-dystopia scenario.* An answer to EQ must be provided in order to ensure that the duped-dystopia scenario is avoided. P3 states that the answer to EQ is no, assuming that phenomenal consciousness does not require sensory processing. The basic proposition motivating P3 is that *embodiment is necessary for consciousness only if sensory processing is necessary for consciousness.* To resist this proposition, the objector to P3 must somehow demonstrate that consciousness requires embodiment even if consciousness does not necessitate sensory processing. It is prima facie difficult to discern how this feat could be accomplished.

The first point to note here is that 'embodiment' is often understood (at least partly) in terms of the ability to exercise sensorimotor capacities. Take the *embodied cognition approach* in cognitive science, for example. As indicated by its name, the embodied cognition approach maintains that the instantiation of cognitive processes requires embodiment. The approach thereby rejects the traditional computational view of the mind, according to which cognition is essentially a matter of internal information processing. Crucially, proponents of the embodied cognition approach typically reference sensorimotor capacities when defining the concept of embodiment. Monica Cowart, for example, asserts that within the context of the embodied cognition debate, "the basic notion of embodiment is broadly understood as the unique way an organism's

[9] It is not clear whether deep learning systems count as 'sensory' in the same way that a brain counts as sensory.

sensorimotor capacities enable it to successfully interact with its environmental niche" (Cowart, 2004). In a similar vein, Chris Letheby defines the embodied cognition approach as the thesis which "holds that conceptual and linguistic thought is grounded in the brain's perceptual and sensorimotor systems" (Letheby, 2012: 403). If embodiment by definition requires the possession of sensorimotor capacities (and hence sensory processing), then P3 is necessarily true. For it is impossible that embodiment is a precondition for consciousness if embodiment requires sensory processing, and it is stipulated that sensory processing is *not* a precondition for consciousness.

I will assume for the sake of argument that embodiment does not by definition require the possession of sensorimotor capacities. Even granting this assumption, the burden of proof is still on the objector to P3 to explain how embodiment could be necessary for phenomenal consciousness given that consciousness does not depend upon sensory processing. There are different ways the objector might go about providing such an explanation. From my perspective, the only feasible way to object to P3 is to make the following threefold contention:

> Contention 1: *Being alive* is a necessary condition for being phenomenally conscious;
> Contention 2: Embodiment is a necessary condition for *being alive*;
> Contention 3: Being phenomenally conscious does not require the possession of sensorimotor capacities.

Successfully making this threefold contention is synonymous with illustrating that consciousness requires embodiment but not sensory processing, and is thus synonymous with illustrating the falsity of P3. There is, of course, the question as to whether these three contentions are *mutually consistent*. The position in the philosophy of consciousness known as *autopoiesis enactivism* is committed to both Contention 1 and Contention 2. It is unclear, however, whether autopoiesis enactivism is compatible with Contention 3. I do not consider any arguments for or against autopoietic enactivism here. Rather, I assume that autopoietic enactivism is plausible and advance to ask whether the view can be weaponized as an objection to P3. Making sense of this question requires an understanding of enactivism, and in particular, an understanding of the distinction between *autopoiesis* enactivism and *sensorimotor* enactivism.

Enactive theories of consciousness understand consciousness to be interactive in nature as opposed to passive, meaning that phenomenal experience according to the enactivist is something that we do as opposed to something that transpires within us. As Jan Degennar and J. Kevin O' Regan observe, "Enactive approaches to consciousness emphasize patterns of interaction, and active engagement, instead of any alleged internal representations of the environment" (Degenaar & Regan, 2017: 393).

Enactivism represents an embodied theory of consciousness due to its emphasis on bodily interactions with the external world.[10] The pressing question for any enactive theory is: exactly which interactive capacities are necessary and sufficient for consciousness? This inquiry raises the crucial distinction between *autopoietic* enactivism and *sensorimotor* enactivism. The latter version of the view maintains that the relevant interactive capacities are sensorimotor capacities. More to the point, sensorimotor enactivism says that the possession of sensorimotor capacities is both necessary and sufficient for consciousness (O'Regan and Noë, 2001; O'Regan, 2011; Degenaar and O'Regan, 2015). By contrast, autopoietic enactivism de-emphasizes the role of sensorimotor capacities, focusing instead on a perceived inextricable link between consciousness and interactive capacities associated with being alive. 'Life' in the context of this theory is defined in terms of 'autopoiesis'. As Degenaar and Regan clarify, "In the enactive literature, life is typically understood as autopoiesis (Maturana & Varela, 1980). Living organisms are spontaneously active, self-maintaining and self-producing systems, and it has been proposed that we should understand our phenomenology in light of this self-creating or 'autopoietic' organization" (Degenaar & Regan, 2015: 396). Degenaar and Regan further elucidate that 'autopoiesis' necessarily involves embodied interaction with the environment: "An autopoietic system needs to metabolize and continuously adapt to its environment in order to continue its activity. By its very definition, autopoiesis cannot withdraw itself from the interactive context" (Degenaar & Regan, 2015: 397-398).

Autopoietic enactivism, then, says that life is necessary for consciousness, where life is taken to be a fundamentally interactive process that mandates embodiment. It is for this reason that autopoietic enactivism is committed to Contention 1 and Contention 2 above. The essential question for the objector to P3, as indicated, is whether autopoetic enactivism is compatible with Contention 3, according to which the possession of sensorimotor capacities is not necessary for consciousness. There is reason to doubt that the two are compatible. Autopoietic enactivism is, after all, motivated by the conviction that "perceptual consciousness involves *additional* constraints over and above perceptual capacity" (Degenaar & Regan, 2015: 397-398). In other words, while

---

[10] The concept of 'enaction' derives from the book *The Embodied Mind* (Varela, Thompson and Rosch, 1991), in which the authors attempt to revolutionize cognitive science by connecting cognitive science with Merleau-Ponty's (1945) phenomenological wok on embodiment.

autopoietic enactivists deny that the possession of sensorimotor capacities is a *sufficient condition* for consciousness, they tend to agree with sensorimotor enactivists that possessing said capacities is a *necessary condition* for consciousness. Is it possible for the autopoietic enactivist to uphold Contention 3, even though the majority of autopoietic enactivists explicitly accept the proposition that consciousness requires sensorimotor skills? I would argue that this answer to this question is no. It seems to me that autopoietic enactivism and Contention 3 are mutually inconsistent because autopoietic processes arguably depend upon sensorimotor skills.

To summarize: P3 states that AI consciousness (assuming that it is possible) does not require embodiment if phenomenal consciousness does not require sensory processing. To object to P3, one must explain how embodiment could be necessary for phenomenal consciousness, given that consciousness does not depend upon sensory processing. I first pointed out that it is impossible for the objector to achieve this feat if embodiment is defined (at least partly) in terms of the ability to exercise sensorimotor skills. Then, I considered and eventually rejected the idea that autopoietic enactivism can be levied as an objection to P3.

6. Conclusion

P1 of the CP argument for disembodied AI consciousness holds that strong primitivism about CP is true. I have not attempted to establish the veracity of strong primitivism here. Instead, I have merely contended that strong primitivism is credible given certain theoretical assumptions; in particular, the assumption that pure consciousness is possible and/or the assumption that Rosenthal's HOT theory is true. P2 of the argument goes on to state that sensory processing is not a necessary condition for phenomenal consciousness if strong primitivism is true. I argued that P2 is strongly supported by what I call PHEN, but acknowledged that PHEN can be reasonably resisted. Finally, I illustrated that while it is possible to object to P3, it is fair to say that P3 is the least controversial premise of the argument. The conclusion that AI consciousness does not necessitate embodiment is by no means trivial. Such a conclusion widens the scope of possible types of conscious AIs, and arguably even suggests that there are two fundamentally different kinds of consciousness: sensory-based consciousness and disembodied consciousness.

References

Block, Ned (1995) "On a confusion about a function of consciousness", *Brain and Behavioral Sciences*, 18 (2): 227-247.

Block, Ned (2011) "Perceptual consciousness overflows cognitive access", *Trends in Cognitive Sciences*, 15 (12): 567-575.

Block, Ned (2011) "The higher order approach to consciousness is defunct", *Analysis*, 71 (3): 419-431.

Brentano, Franz (1874) *Psychology From an Empirical Standpoint*, New York: Routledge.

Brown, Richard & Mandik, Pete (2012) "On Whether the Higher-Order Thought Theory of Consciousness Entails Cognitive Phenomenology, or: What is it Like to Think that One Thinks that P?", *Philosophical Topics*, 40 (2): 1-12.

Carruthers, Peter (2005) *Consciousness: Essays From a Higher-Order Perspective*, Oxford: Oxford University Press.

Carruthers, Peter (2008) "Higher-order theories of consciousness'" In *Stanford Encyclopedia of Philosophy*.

Carruthers, Peter & Veillet, Bénédicte (2011) "The Case Against Cognitive Phenomenology" In *Cognitive phenomenology* (Tim Bayne & Michelle Montague, eds.), Oxfor: Oxford University Press, pp. 35.

Chudnoff, Elijah (2015) "Phenomenal Contrast Arguments for Cognitive Phenomenology", *Philosophy and Phenomenological Research*, 90 (2): 82-104.

Coleman, Sam (2015) "Quotational higher-order thought theory", *Philosophical Studies*, 172 (10): 2705-2733.

Cowart, Monica (2004) "Embodied cognition" In *Internet Encyclopedia of Philosophy*.

Degenaar, Jan & O'Regan, J. Kevin (2017) "Sensorimotor Theory and Enactivism", *Topoi*, 36 (3): 393-407.

Gennaro, Rocco J. (2018) "Higher-Order Theories of Consciousness" In *Internet Encyclopedia of Philosophy*.

Horgan, Terence & Tienson, John (2002) "The Intentionality of Phenomenology and the Phenomenology of Intentionality" In *Philosophy of Mind: Classical and Contemporary Readings* (David J. Chalmers, ed.), Oxford: Oxford University Press, pp. 520–533.

Katz, Steven T. (1978) "Language, epistemology, and mysticism" In *Mysticism and Philosophical Analysis*, Oxford: Oxford University Press, pp. 22-74.

Koch, Christof & Massimini, Marcello, Boly, Melanie & Tononi, Giulio (2016) "Neural correlates of consciousness: Progress and problems", *Nature Reviews Neuroscience*, 17: 307-321.

Kriegel, Uriah (2009) "Self-representationalism and phenomenology", *Philosophical Studies*, 143 (3): 357-381.

Letheby, Christopher (2012) "In defence of embodied cognition: a reply to Fred Adams", *Phenomenology and the Cognitive Sciences*, 11 (3): 403-414.

Levine, Joseph (2011) "On the Phenomenology of Thought" In *Cognitive Phenomenology* (Tim Bayne and Michelle Montague. eds.), Oxford: Oxford University Press, pp. 103.

Lycan, William G. (1996) *Consciousness and Experience*, Cambridge, MA: MIT Press.

Mendelovici, Angela (2013) "Intentionalism about Moods", *Thought: A Journal of Philosophy*, 2 (1): 126-136.

Merleau-Ponty, Maurice (1962) *Phenomenology of Perception*, New York: Routledge.

Rosenthal, David (2005) *Consciousness and Mind*, Oxford: Oxford University Press.

Rosenthal, David M. (1993) "Higher-order thoughts and the appendage theory of consciousness", *Philosophical Psychology*, 6 (2): 155-66.

Rosenthal, David M. (1997) "Perceptual and cognitive models of consciousness", *Journal of the American Psychoanalytic Association*, 45.

Rosenthal, David M. (1986) "Two concepts of consciousness", *Philosophical Studies*, 49 (May): 329-59.

Rosenthal, David (2004) "Varieties of higher-order theory" In *Higher-Order Theories of Consciousness: An Anthology* (Rocco J. Gennaro, ed.), Amestradam: John Benjamins.

Smith, David (2011) "Nibbanic (or Pure) Consciousness and Beyond", *Philosophia*, 39 (3): 475-491.

Tye, Michael (2002) "Representationalism and the transparency of experience", *Noûs*, 36 (1): 137-51.

Varela, Francisco, Thompson, Evan & Rosch, Eleanor (1991) *The Embodied Mind: Cognitive Science and Human Experience*, Cambridge, MA: MIT Press.

CHAPTER 6

HUMAN EXPERIENCE AND ARTIFICIAL INTELLIGENCE

Nicole A. Hall[1]

Independent Scholar

Abstract: This chapter presents the argument that aesthetic experience is one dimension of human experience that distinguishes human consciousness from artificial intelligence. It draws on a revised account of Kantian aesthetic experience, thought to be non-condition-governed, to demonstrate that artificial intelligence cannot be said to have consciousness unless it may be thought to have self-aware, self-conscious, experiences such as aesthetic experiences.

1. Introduction

A programmer is chosen to play the human role in a Turing test to determine the viability of a beautiful robot who is more self-aware, self-conscious and potentially deceptive than her human male counterparts [Ex Machina: Alex Garland, 2015, UK]; a robot boy programmed to love is adopted by a family that is undergoing hardship as they face the illness of their human son: is it possible for the parents to love him the same in the long run? [AI: Steven Spielberg, 2001, USA]; are we real, or are we being deceived into believing we are by the matrix that ensures the power of robots and machines over humanity? [The Matrix: The Wachowskis, 1999, USA]; a machine goes back in time to destroy the mother of the future leader of the human rebellion fighting against machine dominance [The Terminator: James Cameron, 1984, USA]; a lonesome man falls in love with an algorithm only to discover her intelligence will eventually come to far surpass his [Her: Spike Jonze, 2014, USA]; bio-engineered replicants fight for survival in a dystopic world that disowns and disavows them [Bladerunner (director's cut): Ridley Scott, 1982, USA]; a computer's increasingly strange behavior precedes an intense confrontation

---

[1] Nicole A. Hall most recently held a Fernand Braudel research fellowship at the Institut Jean Nicod (Ecole Normale Supérieure) and was a teaching fellow at the Université Paris 8.

between man and machine in an electrifying journey through space and time [2001: A Space Odyssey: Stanley Kubric 1968, USA].

Some of these are based on stories by Philip K. Dick, Arthur C. Clark, but there are many other historical and cultural references. The idea of artificial intelligence does not come without precedent, going as far back as Greek mythology: a sculptor, Pygmalion, who carves a woman out of ivory and falls in love with her; Daedelus, who used quicksilver to install a voice in his statues; Hephaestus who created automata for his workshop; Talos, the artificial man of bronze or Pandora who was made of clay at the behest of Zeus. Or what about Frankenstein, whose creation brought about murder, death, and destruction to the morbid embarrassment of his creator? There are many more, too many to enumerate here.

Each of these movies or literary myths provides glimpses into imagined possible worlds, which, to a lesser or greater extent, endow statues, future creatures, algorithms, machines, with some sort of consciousness. That consciousness sometimes resembles human consciousness, sometimes not. It often surpasses human intelligence or has a form of basic, destructive motivation that battles against human agency, or the possibility that their human counterpart will just hit the "off" switch. These science fiction narratives give us a glimpse not only into the identity of these statues, algorithms, cyborgs, robots, machines, bio-engineered creatures but into what they reflect about us, our human identity, agency, consciousness. These movies are vivid fictional representations of artificial intelligence and, no doubt, contribute to folk conceptions of the phenomenon.

This chapter focuses on the thought that aesthetic experience is at least one dimension of human experience that currently distinguishes human consciousness, from artificial or machine intelligence. I will follow Margaret Boden's definition of artificial intelligence, which is to say the building or programming of computers, such that they are able to do the sorts of things that minds can do. For Boden, artificially intelligent entities are and can be designed to resemble the intellectual core of our human cognitive capacities and display creativity through problem-solving (Boden, 1990).

By aesthetic experience, I mean the human experience of ordinary, natural, artistic, mental objects, and environments that solicit particular responses from us. These responses are infused with value and hold that the object or environment of aesthetic experience is central to that experience. This view takes it for granted that we have aesthetic experiences, despite skepticism from some who think that aesthetic judgment cannot merely be grounded in perception. However, if aesthetic experience is assumed, and if it is grounded in perception, it involves metaphysical realism about at least some aesthetic properties. Otherwise, aesthetic experience would be an instance of perceptual experience. Aesthetic experience, on my account, can be externally experiential, such as when we are immersed in a beautiful sunset, or it can be internally experiential, such as when we experience mathematical formulae and theories, ideas, or literary representations.

Artificial intelligence, at least according to Boden's definition of artificial intelligence, is impressive. Indeed, Boden qualifies some forms of artificial intelligence with creativity, such as when the game Go – more complex than chess – was won by AlphaGo, developed by Google's Deep Mind. In the course of the game, AlphaGo discovered strategies never previously seen in the game's history to finally win the game. Self-driving cars are becoming increasingly sophisticated, too, although their use, surely to become mainstream, remains controversial (Pereira, 2017). Not only can robots identify external objects, but they also demonstrate at least the beginnings of critical and creative problem-solving. They seem to be approaching the point of singularity, widely touted as what would be a monumental shift in the technology's advancement.

For Boden, singularity is the idea that captures the supposed point when artificial intelligence will not only reach but surpass human intelligence by increasingly improving on itself. Once this point is reached, science and art, the problems of governments, international politics, anything, will be helped, or taken over, by artificial intelligence (Boden, 2015). I take a contrasting view, that of "conscious singularity" incorporating both that surpassing of human ability, but also the moment when artificial intelligence becomes self-aware, self-conscious, able to have experiences – like aesthetic experiences. A fictional example of this would be Ava in Ex Machina, whom we discover is self-aware, self-conscious beyond her cognitive abilities, but that also involves the sensibility to make aesthetic choices.

The human mind is commonly understood as being probabilistic, predictive, computational, processing representations usually thought to be linguistic. Computational accounts of mind take experience to be illusory. Since artificial intelligence is based on or thought to explain those very processes, it would be consistent to suggest that machines could have experiences of the illusory sort, even though most agree that they do not. What follows is not about whether or not singularity is possible, nor is it about when or whether the moment of what I have called "conscious singularity" will occur.

Here, the argument turns on the idea that aesthetic experience is inscribed as a fundamental feature of human consciousness, which I do not take to be illusionary or accidental. Aesthetic experience is not a feature of artificial intelligence (although perhaps one day it may be) and distinguishes human consciousness from artificial intelligence as we currently know it, that is to say, without "conscious singularity".

A more substantive account of mind that accommodates it as more than illusory gives us reason to consider how we want our world to be even if the moment of singularity or "conscious singularity" were to occur. This is because it is not clear how an aesthetic sensibility can be computed beyond algorithmic calculations either in the neuroscientific or silicone-based case.

That experience, including aesthetic experience, is sometimes thought to be illusory is because of the difficulty we have in making sense of how non-physical phenomena (such as aesthetic phenomena) can be explained through physical processes and properties. Indeed, explanations on the physical plane don't seem to amount to much on the non-physical plane and vice-versa. That is to say that even if physical or predictive approaches can explain mental states that are thought to be aesthetic, or indicative of aesthetic experience, the methodologies employed do not directly engage objects of aesthetic experience, which likely require a different conceptual framework that also explains the role they play in our psychological, emotional, inter-subjective and cultural lives. If the strange beauty of Frankenstein, eeriness of Bladerunner, electrifying qualities of 2001: A Space Odyssey are thought to be illusory, what claims to we have about their truth, or how they are made to function in public discourse and criticism?

The thought here is that reductive approaches to the mind, which catalyze concerns about machine consciousness, do not accommodate non-scientific accounts of mental states, such as aesthetic experience. They do not capture our relationship to the world. Discourse on aesthetic experience requires a different conceptual framework that makes sense of its normative and cultural force, but that also demonstrates a distinction between human and machine consciousness, should machine consciousness even exist.

An account of aesthetic experience that in itself characterizes our humanity, which I take to include levels of consciousness that accommodate aesthetic experience, contra ideas we have about what artificial intelligence is or could be. Rather than debunk views about narrow conceptions of the mind that classify aesthetic experience as illusory, the idea is to turn our attention to a different conceptual framework that explicitly engages the cultural and environmental domains of human experience.

Engaging in discussion on this level prompts us to be concerned with how we design artificial intelligence, reflect on its multitudinous consequences, but also concerns ethical frameworks for artificially intelligent beings, should that moment of singularity ever occur. In other words, those aspects of the mind that capture our experience of the world are the very dimensions we should consider in the design of future worlds, both in terms of designing the technology, its insertion into society, and the legislation that accompanies it.

2. Aesthetic Experience as Human Experience

Human-level consciousness would require that artificial intelligence emulate aesthetic experiences, feelings, responses, reflections beyond the physical properties that may be captured by synapses and semi-conductors. Even cognitive explanations that rely, for example, on "process fluency" accounts of aesthetic experience, are undermining themselves. Process fluency is the idea that our experiences of beauty have fluidity in our immediate experience of beautiful aesthetic objects. (Reber et al., 2004; 2012) Our experiences of them are digital (cf. analog). The more fluent our experience, that is to say, the more familiar we are with an aesthetic object, the more likely we are to experience it as beautiful. This view has been complemented by others who incorporate the idea of novelty, wonder, and perhaps even frustration (in the case of disfluent artworks) that may in some instances, sustain the experience since our aesthetic experiences are sometimes difficult, not of known beautiful objects, requiring more cognitive engagement. (Dokic, 2016; Schaeffer, 2015) On this sort of account, the aesthetic dimension of experience is defined by the explained and does not engage with the kind of conceptual framework that is about, says something about, the aesthetic objects in question themselves – that piece of music, that painting, that landscape, that poem.

Aesthetic experience is not only about the cognitive dimensions of experience, as Immanuel Kant, and philosophers like Ronald Hepburn and Frank Sibley demonstrated. (Hepburn, 1984; 1986; 1993; Kant, 1790/2000; Sibley, 2001) Biological, chemical or digital processes do not explain the "what it is likeness" of aesthetic experience, with their explanations being restricted to computational neuroscientific approaches.

Aesthetic experience is about how we relate to an object, and that occurs through that relation, explicable through the concepts and terms with which we describe and evaluate it. This is one of the virtues of Kant's and Sibley's aesthetics, both of whom sought to explain how it is possible for us to have aesthetic experiences, both pushing for explanations of experience that in the end aspire to a kind of transparency about aesthetic experience in order to "reach" the object, so to speak: through perceptual attention, exploration, accurate description and evaluation.

This transparency allows for the experience of the object to be that of a world to mind direction (external experientialism), of a mind to world direction or of a mind to mind direction in the case of external intentional objects, or internal intentional objects (call this experiential internalism), as mentioned above. Whether external or internal, that object acquires a special status, that of being an aesthetic object. This not only implies that experience is of those objects but their imagistic or musical or intentional representations. Aesthetic experience is, at least currently, at some distance from what might be quantifiable in terms of biological or machine processes, since these processes are linguistic or mathematical computations rather than demonstrably experiences of objects which they may self-reflect or be self-conscious of.

The idea of aesthetic experience finds its roots in Kant's theory of aesthetic judgment, and it is here that we can find the beginnings of an account of the aesthetic that is in keeping with the richness of "what it is like" to have an aesthetic experience.

The richness of aesthetic experience has been addressed and accounted for in many ways. Although it is widely thought that George Dickie laid to rest Kant's account, the latter still persists in attracting attention . Aesthetic experience, widely thought in the discipline to be at least partly perceptual, is traceable back to Kant. Kant's philosophical system entails that we have no access to objects in themselves, his account of aesthetic experience pushes it to be as perceptually objective as possible. Kant writes, for example:

> The beautiful is that which, without concepts, is represented as the object of a universal satisfaction. This definition of the beautiful can be deduced from the previous explanation of it as an object of satisfaction without any interest (Kant: 1790/2000: 96).

and

> Taste is the faculty for judging an object or a kind of representation through a satisfaction or dissatisfaction without any interest. The object of such a satisfaction is called beautiful (Kant, 1790/2000: 96).

The judgment of taste, for Kant, is a judgment of pure beauty, a disinterested judgment where the representation is of an object of universal satisfaction, devoid of utility, the agreeable and moral judgment. The idea is that what makes aesthetic experience different from any other is pleasure in an elusive perceptual beauty for which no criteria can be listed, but these combined with other features of experience that are subjective, can be shared and form the social basis of community.

The disinterested aesthetic judgment has also historically come to be associated with what might be called austere formalism, a mode of attention that is guided by nothing except an object's immediately perceived form, wherein the aesthetic appreciation of an object, we do and ought to pay attention, not to the object's representational content, emotional expressiveness, historical, institutional, or social context: The form is what makes the object beautiful. (Zuckert, 2006) On the face of it, this may fit into computational approaches to the external world. If robots can be designed to avoid objects in the world based on those object's measurements, they may well be able to calculate whether or not they are balanced, for example. Yet even this is questionable. There is no reason for them to be able to do so since the quality of being balanced has little bearing on their functionality, and aesthetic sensibility implies more than linguistic or mathematical computation.

There are reasons to believe that Kant was not an austere formalist, but a whole formalist because aesthetic experience is not rule-driven in the sense that it actively prescribes, as artificial intelligence would, the inclusion of what are thought to be formal properties and the exclusion of contextual factors. According to Zuckert, disinterestedness excludes the possibility of epistemic, moral, and idiosyncratic rule-drivenness (Zuckert, 2006). Whole-formalism is thus an experience according to which an object is beautiful if it is ordered or unified, or if its parts harmonize to form a whole: what matters is the appearance of the object as a unified structure regardless of what properties seemingly constitute it. While they may include classical formal properties, they may also include others.

Kant's whole formalism is nestled within a multiple-fold disjunctive argument, which brings together the object's form, our sensible and perceptual experience of it. If we assume that aesthetic experience is part of human experience (not merely perceptual experience), human consciousness, those experiences are of metaphysically real aesthetic properties, which partly constitute experience. They have a certain appearance, through a perceptual, imaginative interplay that is not discernable merely through the deployment of computational or cognitive processes.

Through experience, we come into contact with the object and our concepts, which may come into play in our imaginative retrieval of it. Just because epistemic concepts are neither necessary nor sufficient for our distinctly perceptual aesthetic experience of an object's form does not mean that they are not relevant to the experience. As per Sibley, aesthetic concepts are a sui generis class that admits of an important relationship between the object perceived and the aesthetic concept that describes or evaluates it or that describes and evaluates it.

My characterization of aesthetic experience thus differs from many interpretations which have raised concerns about disinterestedness, since it incorporates features of experience outwith the boundaries of cognitive or artificially cognitive norms. I have in mind Dickie and Noël Carroll's views, both of whom thought that the aesthetic and disinterestedness is too restrictive along the austere formalist model and bears little, or no, relevance where works of art were concerned. However, the concept of the aesthetic incorporates many objects, as is widely acknowledged. Even if not always relevant in our experiences of art, it is a fundamental feature of everyday experience.

Bence Nanay points out that their concern to make sense of disinterestedness by comparing and contrasting it with interestedness does not tally up with contemporary approaches to perceptual attention, also with regard to aesthetic attention. For Nanay, aesthetic attention is exercised when it is "focused with regard to objects and distributed with regards to properties" (Nanay. 2016: 20).

While not in favour of arguing that aesthetic experience is not about attention, it is worth making the historical point, contra Dickie and Nanay, that disinterestedness was conceived as challenging a Hobbesian conception of interest, rather than the contemporary conception of interest or attention highlighted by Carroll, Dickie, and Nanay. Indeed, Kant's point about disinterestedness referred to getting at what about human experience can be classed as distinctly aesthetic, rather than trying to capture the extent to which disinterestedness is applicable to current preoccupations with defining art. Since the point I want to make here is about human experience cf. artificial intelligence, rather than definitions of art, I restrict myself to the aesthetic dimension of experience.

One way of understanding Kant's point about disinterestedness was as an account of what, about human experience, is distinctively aesthetic, since aesthetic objects have a special status in experience. The aesthetic would, therefore, need to be of the object and distinguishable from other forms of experience, such as moral, idiosyncratic or epistemic judgment. Yet given the elusiveness of beauty, the impossibility of predicting it, and the subjective nature of experience, the imagination and understanding balance each other out, interact with perception, with universality possible through intersubjective, cultural, processes.

While experience was of the object, and its form, we can see that a Kantian approach is one that relies as much as possible on the possibility of openness, transparency, unto the aesthetic object but which does not negate cognitive dimensions ex post facto even if, in the first instance the imagination is called upon to grasp it. As the aesthetic object focused on nature, natural objects, and natural environments, this means that there is a freedom of experience, non-rule governance, not usually afforded by the experience of artworks, which do require more elaborate cognitive dimensions of experience. This freedom, I want to suggest, requires a sensibility that cannot be prescribed by narrow accounts of the mind as a representing machine. Indeed, one might consider aesthetic experience as on a spectrum, requiring complex interactions between the richness afforded by aesthetic perception in the first instance, and epistemic dimensions.

3. Conclusion

What, if anything, does this have to do with artificial intelligence? It is meant to demonstrate that while machines and algorithms may do some, many, of the things we humans can do, they lack the sensibility, the potential for freedom of experience and thought that we have. This does not deny the possibility that a day will come when artificial intelligence has human-like intelligence or will surpass human intelligence. Indeed, as mooted above, it may already have done so. Philosophers working in artificial intelligence and robotics have some hope for understanding the human mind and have gained influential insights on the human mind, based on understanding forms of creativity that have arisen out of deep learning, as in the case of AlphaGo. But it is a mistake to confuse the possibility "conscious singularity", with human consciousness and its capacity for aesthetically experiencing natural environments in particular.

If our experiences of beauty are thought to be quintessentially human experiences, relevant to what it is to be human, to what it is to have consciousness, and there is an analogy to be made between brains and machines, we had better be clear on what the overlaps and limits are. There are many ways to finesse representational accounts of the mind even in relation to aesthetics, but most of these are reliant on our experience of artworks, which themselves are at least to some extent rule-driven in terms of how we experience, or ought to, experience them whether according to conventionalist accounts that suggest that categories of art are essential to the correct experience or whether explainable by accounts of the mind that rely on the mind-machine analogy. Even accounts that are environment or embodiment dependent seem to rely predominantly on artworks, rather than nature. It is nature – or natural environments - that provides that extra element of freedom, of creative possibility both in terms of how it comes to be represented in artworks and our scientific approaches to, and understanding of, it. If there is one positive element that we might take away from Kant, it is the freedom of possibility where are experiences of the natural environment, at least on my revised interpretation of his account of aesthetic judgement. One the one hand, this kind of freedom would involve reflection on both how we treat the natural environments we want to sustain, but also careful reflection we want to program into artificially intelligent entities that we bring into the world.

References

Adams, Zed (2018) "The History and Philosophical Significance of the Analog/Digital Distinction", Accessed in https://sopha2018.sciencesconf.org/189501/document.

Ashley-Cooper, Anthony (1711) *Characteristics of Men, Manners, Opinions,* Times, London: John Darby.

Boden, Margaret A. (1990), *The Philosophy of Artificial Intelligence,* New York: Oxford University Press.

Boden, Margaret A. (2015) "Should we Fear or Welcome the Singularity? Nobel Week Dialogue 2015 – The Future of Intelligence", Uploaded Jan 20, 2016, Accessed in https://www.youtube.com/watch?v=TcX_7SVI_hA.

Brady, Emily (1998) "Imagination and the Aesthetic Appreciation of Nature", *Journal of Aesthetics and Art Criticism,* 56 (2): 139-147.

Brady, Emily (2003) *Aesthetics of the Natural Environment,* Edinburgh: Edinburgh University Press.

Breitenbach, Angela (2015) "Beauty in Proofs: Kant on Aesthetics in Mathematics", *European Journal in Philosophy,* 23 (4): 955-977.

Budd, Malcolm (2002) *The Aesthetic Appreciation of* Nature, Oxford: Clarendon Press.

Buechner, Jeff (2010) "Are the Gödel Incompleteness Theorems Limitative Results for the Neurosciences?", *Journal of Biological Physics,* 36 (1): 23-44.

Carlson, Allen (1984) "Nature and Positive Aesthetics", *Environmental Ethics,* 6: 5-34.

Carlson, Allen (2000) *Aesthetics and the Environment: the Appreciation of Nature, Art and Architecture,* London: Routledge.

Carroll, Noël (1999) *Philosophy of Art,* London: Routledge.

Carroll, Noël (2002) "Aesthetic Experience Revisited", *British Journal of Aesthetics,* 42 (2): 145-168.

Clark, Andy (2015) *Surfing Uncertainty: Prediction, Action and the Embodied Mind,* Oxford: Oxford University Press.

Cova, Florian, Garcia, Amanda and Liao, Shen-yi (2015) "Experimental Philosophy of Aesthetics", *Philosophy Compass,* 10 (12): 927-939.

Dennett, Daniel (1992) *Consciousness Explained,* Boston: Back Bay Books.

Dennett, Daniel (2017) *From Bacteria to Bach and Back,* London: Penguin Books.

Dewey, John (1934/1980) *Art as Experience,* New York: Perigee Books.

Dickie, George (1964) "The Myth of the Aesthetic Attitude", *American Philosophical Quarterly,* 1: 3.

Dokic, Jérôme (2016) "Aesthetic Experience e as Metacognitive Feeling? A Dual-Aspect View", *Proceedings of the Aristotelian* Society, 116 (1): 69-88.

Dretske, Fred (1982) *Knowledge and the Flow of Information,* Cambridge, MA: MIT Press.

Fodor, Jerry and Block, Ned (1972) "Cognitivism and the Analog/Digital Distinction", Unpublished typescript, MIT.

Frith, Chris (2007) *Making up the Mind,* Malden: Blackwell Publishing.

Hepburn, Ronald W. (1984) *Wonder" and Other Essays: Eight Studies in Aesthetics and Neighbouring Fields,* Edinburgh: Edinburgh University Press.

Hepburn, Ronald W. (1993) "Trivial and Serious in Aesthetic Appreciation of Nature" In *Landscape, Natural Beauty and the Arts* (Salim Kemal and Ivan Gaskell, eds.), Cambridge: Cambridge Studies in Philosophy and the Art.

Hobbes, Thomas (1651/1981) *Leviathan*, London: Penguin Classics.

Kant, Immanuel (1790/2000) *Critique of the Power of Judgment* (Paul Guyer, trans.; Paul Guyer and Eric Matthews, eds.), Cambridge, Cambridge University Press.

Logue, Heather (2018) "Can we Visually Experience Aesthetic Properties?" In *Evaluative Perception* (Anna Bergqvist and Robert Cowan, eds.), Oxford: Oxford University Press.

Lopes, Dominic McIver (2018) "In the Eye of the Beholder" In *Aesthetics on the Edge: where Philosophy Meets the Human Sciences* (Dominic McIver Lopes, ed.), Oxford: Oxford University Press.

Martin, Michael G. F. (2010) "What's in a Look?", In *Perceiving the World* (Bence Nanay, ed.), Oxford: Oxford University Press, pp. 160-225.

Martin, Michael G. F. (1998) "Setting Things Before the Mind", *Royal Institute of Philosophy Supplement*, 43: 157-179.

Martin, Michael G. F. (2002) "The Transparency of Experience", *Mind & Language*, 17 (4): 376-425.

McDowell, John (1986) "Aesthetic Value, Objectivity, and the Fabric of the World" In *Pleasure, Preference and Value* (Eva Schaper, ed.), Cambridge: Cambridge University Press, pp. 1-16.

McDowell, John (1996) *Mind and World*, Cambridge, MA: Harvard University Press.

McDowell, John (2001) *Mind, Value, and Reality*, Cambridge, MA: Harvard University Press.

McDowell, John (1982) "Criteria, Defeasibility & Knowledge", *Proceedings of the British Academy*, 68: 455-79.

Mehta, Neil (2013) "Beyond Transparency: the Spatial Argument for Experiential Externalism", *Philosophers' Imprint*, 13 (8): 1-19.

Meskin, Aaron, Robson, Jon, Ichino, Anna, Goffin Kris and Monseré, Annelies (2017) "Philosophical Aesthetics and Cognitive Science", *Wiley Interdisciplinary Reviews: Cognitive Science*, 9 (1): e1445.

Nagel, Thomas (1979/2003) "What is it Like to Be a Bat?" In *Mortal Questions*, Cambridge: Cambridge University Press.

Noë, Alva (2004) *Action in Perception*, Cambridge: The MIT Press.

Nanay, Bence (2016) *Aesthetics as Philosophy of Perception*, Oxford: Oxford University Press.

Pereira Luis, Moniz and Saptawijaya, Ari (2017) "Programming Machine Ethics" *AI &Society*, 32 (2): 299-300.

Putnam, Hilary (1994) "The Dewey Lectures", *Journal of Philosophy*, XCI.

Rolf Reber et al. "Processing Fluency and Aesthetic Pleasure: Is Beauty in the Perceiver's Processing Experience?" *Personality and Social Psychology Review* 8(4) (2004), 364-382.

Reber, Rolf et al. (2012) "Processing Fluency, Aesthetic Pleasure, and Culturally Shared Taste" In *Aesthetic Science. Connecting Minds, Brains and* Experience (A.P. Simamura and S.E. Palmer, eds.), Oxford: Oxford University Press.

Saito, Yuriko (1985) "The Japanese Appreciation of Nature", *British Journal of* Aesthetics, 25: 239-251.

Saito, Yuriko (1998) "Appreciating Nature on its own Terms", *Environmental Ethics*, 20: 135-149.

Schaeffer, Jean-Marie (2015) *L'expérience Esthétique*, Paris: Gallimard.

Schellekens, Elisabeth (2007) "The Aesthetic Value of Ideas" In *Philosophy and Conceptual Art* (Schellekens and Goldie, eds.), Oxford: Oxford University Press, pp. 71-91.

Schellekens, Elisabeth and Goldie, Peter (2007) *Philosophy and Conceptual Art*, Oxford: Oxford University Press.

Sibley, Frank (2001) *Approach to Aesthetics* (John Benson, Jeremy Roxbee Cox and Hildred Betty Redfern, eds.), Oxford: Oxford University Press.

Snowdon, Paul F. (1980-81) "Perception, Vision and Causation", *Proceedings of the Aristotelian Society*, 81: 175-192.

Stokes, Dustin (2009) "Aesthetics and Cognitive Science", *Philosophy Compass*, 4 (5): 715-733.

Stokes, Dustin (2018) "Rich Perceptual Content and Aesthetic Properties" In *Evaluative Perception* (Anna Bergqvist and Robert Cowan, eds.), Oxford: Oxford University Press.

Stolnitz, Jerome (1961) "On the Origins of Aesthetic Disinterestedness", *The Journal of Aesthetic and Art Criticism*, 20 (2): 131-43.

Stolnitz, Jerome (1961) "On the Significance of Lord Shaftesbury in Modern Aesthetic Theory", *Philosophical Quarterly*, II (43): 97-113.

Stolnitz, Jerome (1978) "The 'Aesthetic Attitude' in the Rise of Modern Aesthetics" In *The Journal of Aesthetics and Art Criticism*, 36 (4): 409-22.

Strawson, Peter F. (1979) "Perception and its Objects" In *Perception and Identity: Essays Presented to A.J. Ayer* (G.F. Macdonald, ed.), London: Macmillan.

Walton, Kendall L. (2008) "Categories of Art" In *Marvelous Images on Values and the Arts*, Oxford: Oxford University Press, pp. 195-220.

Zuckert, Rachel (2006) "The Purposiveness of Form: A Reading of Kant's Aesthetic Formalism", *Journal of the History of Philosophy*, 44 (4): 599-622.

CHAPTER 7

ARE WE REALLY LIVING IN A SIMULATION?

Steven S. Gouveia[1]

University of Minho, Portugal

Abstract: In this chapter, we introduce the debate around the idea that we may be living in a simulation, as put forward by Bostrom (2003). We will first discuss the main reasons that are offered in favour of the simulation hypothesis. Right after, we will debate the more critical responses that were given to the idea. We will end up concluding that, although the probability of the simulation hypothesis is not zero, other more relevant problems should have our attention regarding Artificial Intelligence research.

1. Introduction

Much attention has been given to the question in the aforementioned title of this chapter. Some authors have interpreted this question as a contemporary version of Descartes' Evil Genius: how can we have knowledge of the external world? The contemporary version of that question is: how do we know we are not living in a simulation? The question put forth that way seems vague. We can understand its value by referring to the influential movie that may have been inspired by Descartes' idea: *The Matrix* (1999).

In this movie, the entire world is a simulation of a supercomputer, and a computer hacker called Neo discovers that all life on Earth may be nothing more than an elaborate simulation created by a malevolent superintelligence. Now, if we were Neo, how would we know that the world we see and interact with is real, and not a simulation occurring in our brains? Of course, one can think that *The Matrix* is an extreme case: we don't have a kind of technology,

---

[1] This work was generously funded by the Science and Technology Foundation (SFRH/BD/128863/2017). PhD Student at the University of Minho; visiting researcher at Minds, Brain Imaging and Neuroethics Unit, Royal Institute of Mental Health of the University of Minho; researcher of the Lisbon Mind & Reasoning Group (NOVA University of Lisbon); researcher of the Mind, Language and Action Group (University of Minho).

in the current state of the world, to recreate such scenarios. However, it seems that we can already do it on a lower scale, as we will see.

A way to consider this hypothesis is as follows: people are already delving into virtual reality with videogames such as "The Sims" and "Second Life", in which you can create a person's entire life with all the things that we have in real life. Due to this, some people argue that if we can already do it, there's a possibility we are actually living in it and it's only a matter of having the technology to do it on a larger scale.

The relationship between the Simulation Hypothesis (SH) and philosophy is vast and raises a lot of different questions concerning different areas of philosophy. (cf. Chalmers, 2005) The most commonly discussed problem is the epistemic problem, that is, if we are actually living in a simulation. But a lot of other questions can be made: if simulations exist (i) what is the ontological status of those virtual entities?; (ii) how should we value those entities?; (iii) is perceptual experience in virtual reality illusory or veridical, that is: when experiencing virtual reality, are things the way they look to be?

In "The Virtual and the Real" (2017), David Chalmers argues that normal experiences in a permanent simulation are non-illusory: people have veridical experiences of virtual objects in a virtual space. If we turn out to be living in the Matrix, he argues, our regular experiences will be mostly veridical, and our beliefs will be mostly true. The objects that we have experiences of are still real; the only thing we discover is that the underlying metaphysics of it is computational, which doesn't prove anything about the reality or illusion of the world that interacts with us ordinarily (cf. Chalmers, 2017).

Is the experience of the simulation an illusion? Or are things as they seem to be? Chalmers answer is the following: If we're in a simulation, we're perceiving virtual objects in a virtual space; virtual objects are real objects, though they're ultimately constituted by computational processes. Is this plausible?

The possibility of radical simulation was formally asserted by the philosopher Nick Bostrom in his popular paper "Are we living in a computer simulation?" (2003). The simulation argument states three possibilities about the future of the world:

(1). The human species is very likely to go extinct before reaching a "posthuman" stage;

(2). Any posthuman civilization is extremely unlikely to run a significant number of simulations of their evolutionary history (or variations thereof);

(3). We are almost certainly living in a computer simulation.

One of these three propositions, he argues, must be true: suppose that it was the case that (1) is false – some civilization in the universe reaches technological maturity; then suppose that (2) is also false – some civilization uses the resources to run ancestors simulations; you can then show, supposedly, that because these mature civilizations can do astronomical numbers of simulations, if the first two possibilities are false, there might exist more simulated people like us than non-simulated people. Most of the people with our kind of experiences would be living inside simulations rather than outside them.

The reasoning behind the argument is that, if you reject the first two hypotheses, then you have to accept that the third one follows necessarily. For each real world, we would have millions of simulated worlds, so the probability that we're already in a simulation seems genuine.

For the Director of the Future of Humanity Institute at Oxford University, the simulation argument is, however, different from the traditional cartesian arguments of skepticism about the external world because it doesn't start from a position of power nor a challenge to prove beyond reasonable doubt that the external world really exists (cf. Bostrom, 2005: 95-96). Rather, the simulation argument starts by assuming that things are as they seem, that science tells us things about the world. We have no evidence to know which of the three propositions are true. As such, we should divide the probability for each as evenly as possible (cf. Bostrom, 2003).

The biggest mistake that can be made is the conflation of the simulation hypothesis with the simulation argument. The first one is the statement that we are living in a computer simulation – Bostrom thinks that it is less than 50 percent likely to be the case. However, he believes that the simulation argument shows that one of the three propositions is true (being one of them the simulation hypothesis).

To run a simulation at all would require extremely advanced technology. Presumably, those doing it would be some kind of superintelligent post-human life forms (not human beings like us), and if they had the ability to create a simulation like that, they would also have the ability to fill any cracks and glitches that we may be able to detect (for example, in the movie *The Matrix*, the "déjà vu phenomenon" is a sign of a glitch in the matrix).

Even if we assume that they could make a mistake (like someone having access to the real, non-simulated, world), they would be in a position to hide such a mistake from the people in the simulation (e.g., erasing their memories).

Because of that, the above, the probability of the simulation hypothesis cannot have any direct evidence (because the simulation would be a perfect one). Nevertheless, its probability can increase if we have direct evidence that the other two propositions [(1) and (2)] are true or false.

The case is, then, as follows: if we can reach technological maturity and if we are interested on running ancestors' simulations, the third option it is, supposedly, stronger.

This idea is gaining so much popularity that various tech billionaires (e.g., Elon Musk) are secretly engaging scientists and philosophers to work on breaking us out of the simulation. An inductive argument is used to assert the development of the technology: if one progresses at the current rate of technology a few decades into the future, it seems plausible that we will be a civilization where artificial agents exist living in simulations that are much ampler than real humans.

In a nutshell, Bostrom is arguing this:

(A) $[(1) \vee (2) \vee (3)] = T$

But can we find truly plausible reasons in favour of the idea that we are living in a simulation?

2. Simulation: Probably

In this section, we will try to sum up, in a nutshell, the main reasons to believe in the simulation hypothesis (SH).

The first reason to believe that the universe is a simulation includes the fact that it seems to behave mathematically – we can well describe most of our universe with very precise mathematical accuracy. In conjunction with the idea that the universe is like a pixelated video game constituted by subatomic particles, it can be argued that if the main substances of our universe (e.g., time and space and energy) have finite limits, then it is plausible to assume that they can be computable mathematically and, if they are, they can be simulated – at least theoretically.

It can be argued that the SH is a simpler explanation for the existence of human beings than the idea that we are in fact the first generation that evolved from cells, molecules, complex biology and intelligent human beings with a conscious mind. It may also be claimed that the SH can help to account for the idiosyncrasies in quantum mechanics, in particular, the complex measurement problem whereby, its common for things to only become well-defined when they are observed (cf. Virk, 2019: 123-163).

How can the hypothesis be tested? Neuroscientists and artificial intelligence researchers could check whether it's possible to simulate the human mind. So far, machines have proven to be good at playing chess and putting captions on images. But can a machine achieve consciousness? (cf. Gouveia & Neiva, 2017: 180) We don't know.

Another reason to believe that that SH is plausible is to claim that we are already doing, in a lower scale, complex simulations. For example, physicists use this kind of technology to simulate the beginning and evolution of the universe, in order to test their theories; this kind of approach has the advantage of testing a lot of different initial conditions with different variables with the same laws of nature (cf. Tenneti et al., 2018; Ludlow et al., 2017; Keller & Wadsle, 2017).

Another example is the Illustris Project,[2] a giant cosmological simulation with the aim of studying the formation and evolution of galaxies. The idea is to create the most precise model of the universe and to take insights and conclusions to our own universe.

An additional reason that can boost the probability of (3) has to do with the Fermi Paradox: the simulation hypothesis can help to explain why it seems that we are alone in our universe. If the SA is sound, and (3) is true, we can easily understand why all efforts to reach some extraterrestrial civilization have failed: because they are outside the simulation – they can even be the creators of the simulation, as some argue.

Finally, some scientists found computer code in Superstring Theory equations. James Gates (Gates et al., 2008) found that computer code seems to be embedded into the universe – a particular code called "block linear self-dual error correction code". This can be a signal that the whole universe is, he thinks, a gigantic computer code, a necessary requirement to the simulation hypothesis.

Moreover, some physicists (cf. Beane et al., 2014) have argued that the universe has a finite resolution, or an "underlying lattice", as they call it. This could mean that the universe and reality is not made out of energy or matter, but instead made out of quantized bits of information that, from our perspectives, is perceived as energy and matter.

To better understand this idea, one can think of a red dot across the screen of your computer screen; you perceive it as a smooth sliding action; however, if you zoom in, you will see that this is an illusion caused by the limitations of our sight: the red dot is not sliding as much as it snaps to each individual pixel (cf. Beane et al., 2014).

But how can Bostrom show, mathematically, that the probability of living in a computer simulation is real?

---

[2] Cf. www.illustris-project.org/. See here a short video: www.youtube.com/watch?v=4jXVDeUHMSA.

Let us introduce the following notation, following Bostrom (Bostrom, 2003: 248-249):

f_P : Fraction of all human-level technological civilizations that survive to reach a posthuman stage;

$\overline{N}$: Average number of ancestor-simulations run by a posthuman civilization;

$\overline{H}$: Average number of individuals that have lived in a civilization before it reaches a posthuman stage

Bostrom continues by showing that the "actual fraction of all observers with human-type experiences that live in simulations" is given by the following formula:

$$f_{sim} = \frac{f_P \overline{N}\,\overline{H}}{(f_P \overline{N}\,\overline{H}) + \overline{H}}$$

The next step is to define f_I as the portion of posthuman civilizations that would run ancestor-simulations and $\overline{N_I}$ for the average of those ancestor-simulations, what follows is

$$\overline{N} = f_I \overline{N_I}$$

By combining the two formulas, we can deduce

$$f_{sim} = \frac{f_P f_I \overline{N_I}}{(f_P f_I \overline{N_I}) + 1} \qquad (*)$$

The final main assumption is to assume that because this posthuman civilization is a super-intelligent one, they will have access to super-computing power, meaning that is enormously large. By inspecting (*), Bostrom wants to show that *at least one* of the following three propositions must be true:

(1) $f_P \approx 0$
(2) $f_I \approx 0$
(3) $f_{sim} \approx 1$

This is the probabilistic framework where Bostrom builds his SA. If we follow the probability assignments given, an important consequence has to necessarily follow: as the amount of simulations run by a post-human civilization capable of running would be high, if simulations are done, then the number of simulated people is much bigger than the number of "real" people, meaning that the probability that we are living in a simulated universe is almost certainly also very high.

The conclusion is then:

(1) The fraction of all human-level civilizations that reach a posthuman stage is very close to zero;
(2) The fraction of posthuman civilizations that are interested in running ancestor simulation is very close to zero;"
(3) The fraction of all people with our kind of experiences that are living in a simulation is very close to one (Bostrom, 2003: 255)

That's how Bostrom[3] thinks it's possible to formally prove that (3*) is more plausible that one would initially think. Can we reconstruct the argument more clearly so that we can find the presuppositions used in SH? Here is a version of the simulation argument in a more rigorous way:

(4) Our world contains observers with human-like mental states [MS].
(5) MS are *substrate-independent*;
(6) There are no limitations (physical or technological) to run computer simulations for posthuman races;
(7) Simulated worlds could contain observers with human-like MS;
(8) Our world could be a simulated world.
(9) There could be many worlds just like ours.
(10) The number of simulated worlds could be a lot larger than the number of non-simulated worlds.
(11) There is a great chance that we are simulated humans currently living in a simulated world (cf. Hammarström, 2008: 9).

We seem to have solid reasons to believe in the soundness of SA. What is, then, wrong with the argument? We will now see several critiques that can be made against it.

To sum up, we can formalize the two main steps to argue in favour of SA:

(B) $[\neg (1)] \vee (2) \vee (3) = T$

---

[3] To be fair, Bostrom doesn't claim that we are living in a simulation. What he wants to show is (A), that is, the disjunction between (1), (2) and (3) is true: "(…) In fact, I believe that we are probably not simulated. The simulation argument purports to show only that (…) at least one of (1) - (3) is true; but it does not tell us which one." (Bostrom, 2009: 458). This a response to the critical argument made by Brueckner (2008).

And,

(C) $[\neg (1) \vee \neg (2)] \vee (3) = T$

If (B) and (C) are really true, then (3) is highly plausible. But can we deny (C)?

3. Simulation: Improbable

Is SA plausible? We will now do a broad analysis of the critiques that can be made. The first way to criticizes the argument is to appeal to an ontological principle, the Ockham's Razor (OR). OR tells us that if we can enunciate fewer entities, then the chance to live in real world is more plausible than the possibility of living in a simulation.

In other words, among competing hypotheses, the one with the fewest ontological assumptions should be selected. Isaac Newton was inspired by this principle and gave its one version at the beginning of his *Principia,* in Book 3, in a section entitled "Rules of Reasoning in Philosophy": "We are to admit no more causes of natural things than such as are both true and sufficient to explain their appearances". (Newton, 1972: 551)

OR is commonly employed the following way (its general form):

(12) Two theories, T1 and T2, explain the observed facts equally well (andbetter than all rival theories), and T1 requires us to postulate the existence of more entities (or more types of entities) than T2.

(13) (OR): If two theories explain the observed facts equally well (and better than all rival theories), believe the theory that postulates fewer entities than a rival theory with no loss in explanatory force.

(14) Therefore, we ought to believe T2 and disbelieve T1 (Sterling, 2011: 58)

So, if we apply OR to the issue raised by the SA, it seems that we may be forced to conclude that it is simpler to assert that we leave in a "real" reality, and not in a simulation, which would imply more ontological compromises.

We can, then, apply this general form to the specific issue of the Simulation Hypothesis (SH):

(12*) We do not need to postulate the SH in order to explain any of the phenomena we are attempting to explain;

(13*) (OR): If we do not need to postulate SH in order to explain any of the phenomena we are attempting to explain, we should disbelieve the SH that is not needed to explain phenomena;

(14*) Therefore, we do not need to postulate the SH.

Is this refutation plausible? One strategy to rebut this counter-argument is to argue that the application of OR, in this case, is unfounded and one ought to be very careful when OR is applied.

For example, if we truly want to apply the OR to this issue, we would reach the conclusion that the simplest ontological compromise that we can get is Idealism (cf. Gouveia, 2019a: 7), the thesis that the only thing that exists is your mind and ideas (famously argued by Berkeley). Of course, this position seems inaccurate when you compare the facts of the world, so this first reason to refuse the SA is not conclusive, although the supporter of the SA needs to give a proper answer in order to be able to carry on the argument in a sound manner.

A second strategy to refute the SA is basically to accept that (2) is true: it isn't quite obvious why a civilization in the future would consider simulating universes, as it could create too many dangers (cf. Stuart & Norvig, 2003; Barrat, 2013; Bostrom, 2014) to the civilization that considers these sorts of technology.

In a nutshell, this can happen:

$$\neg (1) \vee (2) \Rightarrow [\neg (3)]$$

For example, assuming that we will not go extinct, we can decide that the dangers of running a simulation suppress the benefits of it. Another reason is: if we truly achieve the technology to run a simulated world, it is implied that part of the world (mankind) will be also simulated. If that's the case, those simulated people will have feelings and pains akin to ours. The suffering inflicted on them would be morally wrong – which is the reason we have human rights in our real world.

A different strategy is to argue that there could be some biological constraints limiting the evolution of the technology, thus, undermining the project at stake. We can argue for the following:

$$(1) \Rightarrow [\neg(2) \vee \neg (3)]$$

The assumptions underlying the SA are: (5) the substrate-independence of the mind (that would enable to emulate the brain) and (6) no limitation on computation. But this may not be as simple as the argument assumes.

As Bostrom admits, the soundness of the argument requires acceptance of the assumption that the hardware doesn't matter for the simulation/emulation of consciousness. One can be highly skeptical about that: if the biological structure of our brain is in fact a requirement for consciousness, then a computer which simulates the brain has no consciousness, and then the SA fails.

In this point, the debate focuses on the metaphysics of the mind, which are divided in two big theses in current cognitive (neuro)science and philosophy of mind:

(T1): The configuration and arrangement of parts *x* is sufficient to produce all *y* properties.
(T2): The material (causal power) and composition of the parts of *x* is required to reproduce the *y* properties (cf. Gouveia & Neiva, 2017: 202)

The debate considers the importance of a carbon-based biology and if it is necessary to have an evolutionary process to give rise to consciousness. T1 argues for a negative answer: the crucial thing is not the material or the elements of a body; what really matters is how the parts are arranged formally – so, we can have consciousness in a silicon-based biology, for example.

Some neuroscientists (e.g., António Damasio) and philosophers (e.g., John Searle) think that this is incorrect: there is something "in the flesh" that cannot be reproduced by computation. The debate is currently ongoing so we think Bostrom should be tread cautiously when making certain assumptions that are highly debatable. Of course, for those who advocate T2, we cannot emulate the human brain, and this is a big obstacle for the SA.

Another caveat for the SA has to do with technology. At our current stage of technological development, we have neither sufficiently powerful hardware nor the requisite software to create conscious minds in computers – assuming we can do that.

As Bostrom in effect admits, there is doubt whether his specific calculations are correct. Assuming that T1 is true and T2 is false, there is no implication about the amount of power we would need to simulate consciousness. Bostrom assumes that less power would be required, but in fact, that can be exactly the opposite: an emulation of a normal computer is less powerful than the computer itself. If the simulation of consciousness (adding the simulation of the entire world where that simulation would be run) demands more power, less simulations can be performed. From this, we can conclude that it would be more probable that we are in a real world, and not in a simulated one. Again, this may be an obstacle for the soundness of the argument, as we are contesting (11).

Moreover, current analysis of Moore's Law shows that it will become too expansive, financially, to keep the same power of development. This goes in favour of the idea that although simulations are plausible, theoretically, they will always be too expansive to run.

SA also assumes that the constant reduction of the size of the hardware will be infinite, that is, that can happen without any restriction. One can, of course, argue that this constant reduction of the size of the hardware will reach a limit where we will somehow influence the quantum properties of the elements and that can be a major physical problem, so mankind cannot avoid an extinction and (1) is true and, therefore, SA is implausible.

Finally, the formal structure of the conclusion of SA seems controversial in the light of a probabilistic analysis. As it is argued by Besnard, it is not clear that F, N and H are finite sets, a necessary condition to apply (as Bostrom does) normal probability analysis. If we assume that the sets are infinite, we cannot measure clearly the probability of (1), (2) and (3). But even if we ignore this problem, it is not clear that F, N and H are even a set. Bertrand Russell claimed that we cannot describe a set by its properties only, but we need to describe as a subset of some identified set. But in this case, since we don't have a mathematical description of the underlying structure of reality (e.g., the multiverse), we cannot claim that F, N and H are sets (or finite) and that can be problematic for the SA (cf. Besnard, 2004).

4. Conclusion

The Simulation Argument has been very influential inside academia and especially outside. We analyzed very superficially arguments in favour (section 1) and against (section 2) the idea. A multitude of consequences would arise if the SA happens to be sound. For example, if one accepts that we are living in a simulation, then we may open the door to eternal life, resurrection and things that were, formerly, discussed within the realm of Religion. The reason is actually quite simple: if our mind is substrate-independent, then we can be simulated in different contexts (materials) infinite times. If we are eternal, does it make sense to vote for a government in a democratic election? Does it make sense to work and to have a family, or even to cure our diseases?

Another repercussion of the simulation idea is that it seems to imply a creator (the programmer) that knows everything and is all powerful. This can raise some interesting questions, from a naturalistic point of view, about the existence of a god and, consequently, about the role of Religion in today's society (cf. Steinhart, 2010).

Of course, one can be skeptical about this kind of research program and actually think that this is a waste of energy and public funding and that more important questions matters, such as, how can we deal with the responsibility gap with artificial agents that have a high rate of autonomy? (cf. Gouveia, 2019b, for a solution attempt) Although this is maybe true, it doesn't show that the SA is implausible. What it shows is that some people take this idea too seriously and maybe we should focus on more relevant problems for (real) society.

References

Barrat, James (2013) *Our Final Invention: Artificial Intelligence and the End of the Human Era*, New York, NY: St. Martin's Press.

Beane, Silas R., Davoudi, Zohreh & Savage, Martin J. (2014) "Constraints on the Universe as a Numerical Simulation", *The European Physical Journal A*, 50: 148.

Besnard, Fabien (2004, September 5) *Refutations of the Simulation Argument*, Retrieved from http://fabien.besnard.pagesperso-orange.fr/pdfrefut.pdf.

Bostrom, Nick (2003) "Are You Living in a Computer Simulation?", *Philosophical Quarterly*, 53 (211): 243–255.

Bostrom, Nick (2005) "The Simulation Argument: Reply to Weatherson", *The Philosophical Quarterly*, 55 (218): 90-97.

Bostrom, Nick (2009) "The simulation argument: Some explanations", *Analysis*, 69 (3): 458-461

Bostrom, Nick (2014) *Superintelligence: Paths, Dangers, Strategies*, Oxford: Oxford University Press.

Brueckner, Anthony (2008) "The Simulation Argument again", *Analysis*, 68 (3): 224–226.

Chalmers, David (2005) "The matrix as a Metaphysics" In *Philosophers Explore the Matrix* (Christopher Grau, ed.), Oxford: Oxford University Press.

Chalmers, David (2017) "The Virtual and the Real", *Disputatio*, 9 (46): 309-352.

Gates, James et al. (2008) "Relating Doubly-Even Error-Correcting Codes, Graphs, and Irreducible Representations of N-Extended Supersymmetry" In *Discrete and Computational Mathematics* (F. Liu, et. al., eds.), Hauppauge: Nova Science Publishers.

Gouveia, Steven S. and Neiva, Diana (2017) "The Problem of Consciousness on the Mind Uploading Hypothesis" In *Philosophy of Mind: Contemporary Perspectives* (Manuel Curado & Steven S. Gouveia, eds.), Newcastle: Cambridge Scholars Publishing.

Gouveia, Steven S. (2019a) "Philosophy of Mind and Reductionism" In *Automata's Inner Movie: Science and Philosophy of Mind* (Manuel Curado & Steven S. Gouveia, eds.), Delaware: Vernon Press.

Gouveia, Steven S. (2019b) "The Responsibility Gap in Artificial Intelligence" In *Vanguards of Responsibility – Law, Neuroscience and Artificial Intelligence* (M. Curado, A. E. Ferreira & A. Pereira, eds.) (Portuguese edition), Coimbra: Petrony.

Hammarström, Anders (2008) *I, Sim - an Exploration of the Simulation Argument* (Master Thesis), Umeå: Umeå Universitet.

Keller, Ben W. and Wadsley, James W. (2017) "CDM is Consistent with SPARC Radial Acceleration Relation", *The Astrophysical Journal Letters*, 835 (1): 5

Ludlow, Aaron D. et al. (2017) "Mass-Discrepancy Acceleration Relation: A Natural Outcome of Galaxy Formation in Cold Dark Matter Halos", *Phys Rev Lett.*, 118 (16): 161103.

Newton, Isaac (1972) *Principia Mathematica* (Alexandre Koyré & I. Bernard Cohen, eds.), Cambridge, M.A.: Harvard University Press.

Russell, Stuart J. & Norvig, Peter (2003). "Section 26.3: The Ethics and Risks of Developing Artificial Intelligence" In *Artificial Intelligence: A Modern Approach*, Upper Saddle River, N.J.: Prentice Hall.

Sterling, Grant (2011) "Ockham's Razor" In *Just the Arguments* (Michael Bruce & Steven Barbone, eds.), Oxford: Wiley-Blackwell.

Steinhart, Eric (2010) "Theological Implications of the Simulation Argument", *Ars Disputandi*, 10 (1): 23-37.

Tenneti, Ananth et al. (2018) "The radial acceleration relation in disc galaxies in the MassiveBlack-II simulation", *Monthly Notices of the Royal Astronomical Society*, 474 (3): 3125–3132.

Virk, Rizman (2019) *The Simulation Hypothesis*, N.A.: Bayview Books.

Filmography:

The Matrix (1999), Lana Wachowski & Lilly Wachowski.

SECTION III:
AESTHETICS AND LANGUAGE IN ARTIFICIAL INTELLIGENCE

CHAPTER 8

CAN A COMPUTER CREATE A MUSICAL WORK? CREATIVITY AND AUTONOMY OF AI SOFTWARE FOR MUSIC COMPOSITION[1]

Caterina Moruzzi[2]

University of Nottingham, UK

Abstract: In this chapter, I discuss the question "Can a computer create a musical work?" in the light of recent developments in AI music software generators. In attempting to provide an answer, further questions about the creativity and intentionality exhibited by AI will emerge. In the first part of the chapter, I argue that the question of whether a computer can be creative should be replaced by different questions. The notion of creativity is indeed too embedded with our subjective judgement of an idea or product 'as' creative. In the last section, I thus suggest shifting the focus of the inquiry from creativity to autonomy. I finally argue that the application of generative adversarial networks to music generators confers them a level of autonomy which is sufficient to deem them able to create musical works.

1. Introduction

While the latest developments in the field of AI music are a matter of excitement among both listeners and researchers, they also raise some concerns. For example, who should be the copyright owner of the music produced? Or, should copyright regulations change if, instead of creating new melodies by rearranging existing music, the program produces completely original patterns? In addition, AI music raises also less practical and more

---

[1] Adapted by permission from Springer Nature Customer Service Centre GmbH: Springer International Publishing. Philosophy and Theory of Artificial Intelligence 2017 by Vincent C. Müller (ed.), Springer Nature Switzerland AG 2018 (2018).

[2] The University of Nottingham, Nottingham NG7 2RD, UK. Email: caterina.moruzzi1@nottingham.ac.uk.

philosophically oriented questions. In this chapter, I discuss the impact of AI on one of the key topics in the philosophy of art: the nature of musical works. The question I will address is the following: "Can a computer create a musical work?". In attempting to provide an answer, further questions about the creativity and intentionality exhibited by AI will emerge.[3]

In section 1, I will identify two necessary requirements for deeming something a musical work. I will then move on to the consideration of the related issue of creativity. The difficulty we meet in assessing the level of creativity of a system mainly depends on the subjective nature of the notion. When judging something as creative, we apply subjective beliefs and biases to our evaluation and this prevents us from giving objective claims on the issue. I will then try to modify the question and add the consideration on the intentionality possessed by the system, asking "Can a computer be intentionally creative?". In section 2, I will quickly discuss the main stances we can take towards the notion of intentionality and whether subjective judgements affect this notion as well.

In section 3, I will propose to shift the focus of inquiry from creativity to autonomy in order to judge the results achieved by AI in the musical field in a more objective way. I will claim that software like Jukedeck or Flowmachines, where the human presence is still essential for performing the task of composing music, do not create musical works. They at best can be considered an extension of the programmer's or user's mind. The use of Generative Adversarial Networks (GANs) in software for music composition, instead, may grant AI a sufficient level of autonomy for deeming it able to create musical works.

The final claim of the chapter will be that shifting the focus from the assessment of creativity to the consideration of the autonomy of the system is beneficial under many respects, one of which is that it allows us to give an objective judgement on what we commonly call creative acts. The analysis of the application of AI to the production of music will be beneficial also to address the fears that may arise in respect to the emergence of a 'superintelligence' which may overcome humans also in an, arguably, human prerogative: the creation of Art.

---

[3] For computer art, specifically computer music, in this chapter I will refer to what Boden and Edmonds call CG-art: "art produced by a program left to run by itself, with minimal or zero interference from a human being" and not C-art which instead "uses computers as part of the art-making process" (see Boden and Edmonds, 2009).

2. AI, Creativity, and Musical Works

2.1. Features of Musical Works

The question that will lead the discussion of the following sections is "Can a computer create a musical work?".4 Before dealing with the implications this question brings with it, it is worth asking why answering it is important. First, as mentioned in the Introduction, we are currently witnessing an exponential development of the application of AI in many fields, among which music. Given the growing availability of AI-based music, it is compelling to analyse which are the consequences that this development might bring to the music industry and, more generally, what are the differences between the application of AI to music in respect to its application to other fields. Lastly, and even more importantly, the consideration of how does AI fare in the context of the creative industry will give us means to answer the question that worries both supporters and detractors of AI: will ever computers replace humans? Indeed, if computers could equate or outperform us in a prevalently human quality such as creativity, then the answer to the previous question would critically lean towards a "Yes".5

In order to answer the question of whether computers can create musical works, it is necessary to establish which the essential features of musical works are. Our common sense recognises two requirements as essential for musical works:

(i) Repeatability: intuitively, musical works are repeatable since there are multiple events which can qualify as their performances;
(ii) Creatability: it is an ordinary belief that musical works are creatable entities that are not available before being actually created through an act of composition.[6]

---

[4] In this chapter I will focus on music, without considering the application of AI to other domains and art forms. AI software for creating paintings, for example, is arguably more sophisticated than AI music software. This is due to historical reasons: the use of machine learning systems was initially applied to image recognition and, thus, to visual and not sonic products.

[5] I qualify here creativity as a 'prevalently human' quality not to exclude cases of animal creativity (e.g. the creativity observed in chimpanzees, elephants, crows, and other animals). For a discussion on the consequences of the emergence of a superintelligence in a wider context see Bostrom, 2014.

[6] It is not within the aims of the paper to assess the specifics and the consistency of these desiderata. Each theory within the ontology of musical works accounts for these desiderata differently and, as a result, provides a distinct definition of what a musical work is. The lack of consensus in this debate, however, does not affect the argument of the present paper.

The assessment of the first desideratum, repeatability, is not within the scope of this chapter. It is indeed possible for a computer to produce something which is repeatable multiple times. The compliance with requirement (ii), instead, is more problematic.

A widely recognised feature of musical works is that they did not exist before being created by a composer: a musical work, to be identified as such, is necessarily created.[7] The *Fifth* Symphony would never have existed if Beethoven did not write it between 1804-1808. In writing a score, the composer performs an act of creativity. In order to assess whether computers can create musical works, then, it is necessary to ask a second question: "Can a computer be creative?".

2.2. Creativity

No philosophical discussion about art would be complete without addressing the issue of creativity. The analysis of this notion is important not only because it allows us to consider the psychological, social, neurological, and artistic aspects which lead to the production of a creative output but also because the acknowledgement of an action, or an output, as creative bears practical consequences, for example at the level of intellectual property or copyright. It would be impossible to offer here a review of all the insightful debates that have been given on the theme. For the sake of the chapter, though, I will outline a few aspects that will be relevant for the discussion that follows.

Numerous definitions of creativity have been proposed in the literature on the topic. I start from Margaret Boden's definition: "Creativity is the ability to come up with ideas or artefacts that are new, surprising and valuable" (Boden, 2004: 1).[8] For Boden, a creative activity produces something which has never been done before. Also, the novelty, which is brought about by a creative action, should necessarily be accompanied by a feeling of surprise.[9] Finally, this novel and surprising idea or artefact needs to be 'valuable'. It should, that

---

[7] The nature of musical works as created entities is disputed in the literature, see for example Levinson, 1980; Dodd, 2002; and Kivy, 1993.

[8] In section 3 I will propose an alternative interpretation of creativity which aims at solving some of the complications which may result from the adoption of Boden's definition.

[9] It can be argued that not necessarily a creative output raises a feeling of surprise in the audience. For example, Giotto's paintings are certainly an example of creativity, but they may provoke no feeling of surprise in someone who is already accustomed with his style or with the painting conventions of his period.

is, have a recognised quality which makes us value it for some reason. The attribute of 'valuable' is a very important addition to the other two attributes of 'novel' and 'surprising'. Indeed, in this chapter, I might write a few sentences which have never been written before and that, for some people, result as surprising. Yet, for other people, they may not count as valuable and, thus, neither as creative.[10]

There are other available definitions of creativity but they generally agree on the features that Boden identifies (see Gaut, 2010; Colton, 2008; Wiggins, 2005). Borden's analysis on the theme of creativity is widely valued for having brought the topic to the attention of philosophical analysis. However, the definition she gives is not without flaws. A controversial aspect of Boden's definition is the fact that it is too strong. Indeed, Boden's inclusion of the attribute 'valuable' within the requirements for creative acts has the unwelcome result that, for example, a song which we do not like nor find valuable is not a product of creativity.[11] Besides this downside, Boden's description of an act of creativity as producing something 'valuable' highlights a characteristic which our common sense essentially attributes to the notion of creativity: its being *subject-dependent.*[12]

It is evident that in the definition of creativity a prominent role is given to our subjective evaluation of the product of creativity. An obvious example is ready-mades. Duchamp's *Fountain* is not valuable because of its material or formal properties. It is instead valuable because of the choice of *presenting* it as an artwork and because of the meaning that this choice brings with it. Indeed, the evaluation of something as a product of creativity is subject-dependent. *We* identify something 'as' creative. Being creative is not an intrinsic feature of an act or of a product, it is instead a feature which it assumes after being perceived and evaluated by us.[13]

Arguably, then, a test over the intuitions that people hold in respect to artefacts or ideas would be enough to determine the creative nature of the act that produced them. This is what the Neukom Institute for Computational

---

[10] For the discussion that follows, it is worth stressing the fact that the attribution of value is subjective, thus something may count as valuable for you but not for me.

[11] See Wiggins,2006, for a formalisation of Boden's definition of creativity that aims at solving some of its inconsistencies.

[12] Of a different idea is Ritchie (Ritchie, 2007) who claims that we can judge the level of creativity in the artefact produced by an individual without considering also the process through which it has been created. I will come back to this issue below.

[13] See Colton, 2008; Pease, Winterstein, and Colton, 2001; Boden and Edmonds, 2009.

Science has been testing since 2016 with its 'Turing Test in the Creative Arts'.[14] This contest asks machines to create poetry that is indistinguishable from human poetry. The award then goes to the AI output which fools the jury into thinking that it is was produced by a human. In 2017 this contest was extended to music and dance. The winner in the music section was Music Plus One with the song *The Wild Geese*. On the basis of the features of the outcome, it thus seems possible for a computer to exhibit a level of creativity comparable to a human's.

2.3. Biases in the Evaluation of Creativity

The (subjective) evaluation of the creativity of a product should therefore be included among the elements that we need to test when assessing the level of creativity of a system. Boden claims that, in order to pass a Turing Test, a program should produce an output which is "seen as having as much aesthetic value as one produced by a human being" (Boden, 2010: 409). A similar measure is taken by Pearce and Wiggins with their 'discrimination test' in which participants are asked to distinguish between human or computer-generated music (see Pearce and Wiggins, 2001).

If the object of these tests is not only the final product (as supported by Ritchie, 2007) but also the process through which this product comes into being, then computer systems may not be able to pass them, though.[15] When assessing the creativity of a computer in producing an output, in fact, "it is commonplace for people to attribute creativity to the programmer in addition (or instead of) the software" (Colton, 2008). The default position is then that of denying creativity to a program, even if its output is comparable to one produced by a human. From this comes the charge of "double standard" when assessing human or computer art (Colton, 2008).

Deeming the analysis of the output as sufficient for our consideration of creativity would allow us to bypass the difficulty of having to judge whether the process of producing an output possesses the right features to be deemed creative. By analysing only the output, in fact it does not matter who produced it: if we judge it sufficiently novel, surprising, valuable, or having any other feature we may want to add to the list of creative attributes, then it is creative. I believe that this would not respect the beliefs we bear in respect

---

[14] See https://bregman.dartmouth.edu/turingtests/node/1.

[15] The FACE and IDEA Descriptive models designed by Colton, Charnely and Pease (2016) introduce a computational creativity theory which includes in its parameters the impact that creative artefacts have on the audience.

to creativity, though. We very much value also the path that led to the process of creation and this is testified also by the number of studies and documents on the history of production of artworks and other creative products.

The appreciation of the process which led to the production of the work is so relevant that, as mentioned, most people change their judgement on the output once they know that a computer and not a human is responsible for its creation (see Boden, 2010: 411; Hertzmann, 2018; New Scientist, 2017: 178). We can take as a way of example the piece *Fremdarbeit* (*Outsourcing*) (2009) by the German composer Johannes Kreidler. In 2009 he received a commission for a new piece. Then, he asked a Chinese composer to write a piece in a style imitative of his music for the desired instrumentation. Finally, he asked the Chinese composer to use for the composition a software which algorithmically imitated Kreidler's style, created by an Indian programmer. Kreidler received 1500 for the commission, and he paid just $150 the workers who composed the score in Asia: his aim was to convey a political message about the exploitation of work in countries with cheap labour. Should our reception of this piece change once we learn that it was not composed by Kreisler but by a software instead? If we judge only the formal qualities of the piece, we should not, but it is hard to believe that we can keep an objective detachment in providing our personal evaluation of it.

Another example of the effect of the biases that we bear in respect to non-human art is the case of David Cope's music generation program EMI. This software was created by Cope to get over the composer's block he was suffering from. The software creates new music by deconstructing the structure of the music entered as input into the database. The instructions for deconstructing the music and for creating new music in the same style are written by Cope himself. This software was not exempt from biases: "A music critic published a damning «review» of the first public concern of music composed by the Emmy program [...] – but he did so fully two weeks before the concert took place. [...] His preconceptions about the inadmissibly of computer-composed music [...] made any attempt at a TT [Turing Test] utterly pointless" (Boden, 2010: 411). The lack of appreciation from the part of the audience ultimately led Cope to destroy the Emmy database (see Boden, 2010: 412).

Our judgement in respect to computer-generated art seems then to be biased: why should we evaluate differently a product made by humans or AI when our response to it, in the absence of provenential information, is the same? For example, I may imagine attending a concert of computer-created music: can the pieces be 'works of music' when the provenance is unknown and 'not works of music' when it is? Indeed, a sequence of sounds and noises without a distinctive melody or pattern can be composed by a computer with no difficulty. So, why should we evaluate it differently if instead it is written by an experimental music composer? Following what has been noted above in respect to the subjective nature of our judgement on creativity, it seems impossible for us not to have biases in respect to this kind of evaluation. If, given the nature of the notion of creativity itself, our judgement cannot be detached from a set of subjective beliefs we bear in respect to what we deem 'creative', in fact, it seems impossible for us not being influenced by our inclinations.

The subjective judgements and biases which are implied by the evaluation of something as creative make it impossible to objectively answer the question "Can a computer be creative?". What we are measuring in trying to provide an answer to this question, in fact, are not the computer's skills but instead our subjective evaluation of them. I therefore argue that we need to change the question. If we want to be able to objectively judge the process of production of an output and the output itself in order to take a stance in respect to its 'creativity' and, as a consequence, to the related issues of intellectual property and copyright, we need to change focus. We need to acknowledge that the notion of creativity is too embedded with subjectivity to serve our scope. Another, more objective, notion should take its place instead. We can then try to rephrase the question and ask: "Can a computer be intentionally creative?".

3. AI and Intentionality

The literature on intentionality and on the implications of this analysis in respect to the issue of consciousness is vast. It is not my aim to provide here a review of all the different positions that have been assumed concerning this topic. I will instead point out the two opposite stances that one might take in respect to intentionality, showing how this choice may affect the answer we give to the question "Can a computer be intentionally creative?".

The difficulty in addressing the topic of intentionality starts from the ambiguous usage of this term. In philosophical literature, intentionality refers to the power of the mind to represent something, to be about something. In its more common usage, instead, 'intentionality' refers to the act of consciously wishing to attend to the action that one is undertaking. For example, I may express my intentionality in playing Chopin's *Sonata n. 2* by being conscious in my willingness or intention to play exactly this sonata. In what follows, I will use the term 'intentionality' in this second acceptation, meaning intentionality as the conscious intention of attending to an action. What I wish to examine is whether computers may be deemed able to possess this kind of intentionality.

In simplistic terms, we can give two opposite answers to the question "Can a computer be intentional?". Searle's reply is "No". His Chinese Room Argument challenges the claim made by Strong AI that computers can have mental capabilities similar to humans': the program used by the system to produce a meaningful output is not a mark of intentionality:

> Intentionality is a biological phenomenon [...]. Whatever is that the brain does to produce intentionality, it cannot consist in its instantiating a program since no program, by itself, is sufficient for intentionality (Searle 1980, 424).

Of a different opinion is Daniel Dennett. He proposes a strategy for understanding the behaviour of an entity:

> [T]his third stance, with its assumption of rationality, is the *Intentional stance*; the predictions one makes from it are Intentional predictions; one is viewing the computer as an Intentional system. One predicts behavior in such a case by ascribing to the system *the possession of certain information* and by supposing it to be *directed by certain goals*, and then by working out the most reasonable or appropriate action on the basis of these ascriptions and suppositions. It is a small step to calling the information possessed the computer's *beliefs*, its goals and subgoals its *desires* (Dennett 1971, 90).

By positing the intentional stance strategy, Dennett deems behaviour sufficient for the attribution of intentionality to an entity (Dennett, 1971: 90).[16] In keeping with what I stated above in regard to creativity, following this view, we can attribute *subject-dependent* intentionality to non-animal entities. More specifically, in answer to the question opening the section, we can attribute to computers a subject-dependent intentional creativity. Likewise, we can ascribe to the wind blowing through the canyon the intention of composing a melody (see Wolterstorff, 1980: 74).

But, as with creativity, it seems to be against our intuitions to consider subject-dependent intentionality sufficient for deeming a computer (or, for what it matters, the wind) intentionally creative. It seems safe to claim that, following common-sense intuitions, we would value the *Fifth* Symphony differently if, instead of by Beethoven, we learnt that it was composed by the wind blowing through an empty house. Rather, we sometimes speak of artefacts 'as if' they were intentional beings (for example, we may talk of the thermostat 'as if' it has the intention of regulating the temperature in the

[16] Here I limit the discussion to the works of Searle and Dennett but there clearly are many other views regarding this topic. For an extensive review, see Siewert, 2017.

room) but this does not mean that they possess intentionality independent from our interpretation of their behaviour (see Searle, 1990).[17]

If both notions of creativity and intentionality can be assessed only subjectively, how can there be a way to *objectively* judge whether a computer can be intentionally creative? Despite their recurrence in our everyday language when talking about art and other products, creativity and intentionality do not seem to be easily employable in a discussion on which stance we should take towards computer-generated art. It is no doubt interesting to investigate the biases that we bear towards non-human art – as I mentioned in the previous section – and whether we should get rid of them for being able to give impartial judgements. Yet, it is due to these biases if we struggle to give a definite answer to the question "Can a computer be intentionally creative?".

In the next section, I will propose an alternative definition of creativity with the intention of getting rid of the subject-dependency implications which result from the definitions of creativity and intentionality. The definition I will propose is deliberately weak. This is made not only to simplify the discussion, avoiding the need to refer to issues such as consciousness and intentionality, but also to provide the minimal requirements for something to be creative. A key role will be played by the notion of autonomy: the real challenge for computers would be then to be perceived as subject-independently intentional, namely as autonomous entities.[18]

4. AI and Autonomy

In the previous sections, I discussed the question of whether computers can create musical works. For addressing this question I needed to consider related issues such as whether a computer can be deemed creative and intentionally so. The subjectivity inherent to the notions of creativity and intentionality, however, makes it hard to assess the initial question on an objective level. My aim in this section is to dispel this difficulty by proposing a minimal alternative definition of creativity that can be assessed without the

---

[17] It may be argued that most of the conceptual difficulties that we meet in the definition and evaluation of creativity in computers comes from our apparent strive to anthropomorphise machines. From this comes also the criticism of inadequacy against the Turing Test. What the Turing Test does, in fact, is to measure the level of intelligence - or creativity - of a system taking as a model human-kind intelligence or creativity. Getting rid of this "carbon fascism" may instead allow us to recognise and acknowledge other types of creativity other than the human one" (see Pease and Colton, 2011).

[18] See Boden and Edmonds, 2009.

need to incur into subjective considerations. This alternative definition of minimal creativity (Creativity$_{m}$) shifts the focus to the autonomy needed by a system to produce an output which is deemed creative.

A specification of what I mean by 'autonomous' is needed, since it may be argued that nothing is autonomous, not even humans. Our choices are indeed affected by our upbringing, culture, and environment. For autonomy in relation to software here, I intend not a complete independence from the programmer but instead the ability of 'breaking the rules' that the programmer inputted at the beginning.[19] The definition I propose is deliberately weak to avoid the need to refer to the issue of intentionality, and to identify the minimal requirements for an action to be creative.

I define minimally creative an action that follows the following steps:

Creativity$_{m}$:
(i) Autonomous *reception* of external or internal stimuli;[20]
(ii) Autonomous *selection* of some of these stimuli;
(iii) Autonomous *elaboration* of the selected stimuli;
(iv) Autonomous *production* of new stimuli.

An example of this creative process in the musical field is *La Mer* by Claude Debussy. In composing this orchestral piece, Debussy was inspired by the sounds and noises produced by the ocean, by prints, and by short stories. The creative process that Debussy went through is constituted of an (i) autonomous reception of the mentioned sources; an (ii) autonomous selection of some of the stimuli that he received from them; an (iii) autonomous elaboration of these stimuli in a musical form; and an (iv) autonomous production of the resulting piece: *La Mer*.[21]

To examine if computers can be deemed creative according to Creativity$_{m}$, I consider the music software Jukedeck. This software enables customers to set the parameters of the track that they want the software to create, e.g., style, tempo, duration, and mood. The software then composes a track which is downloadable paying a small fee. It is easy to see how the process of decision-

---

[19] An interesting parallel to draw here is that between autonomy and the concept of innateness (see Marcus, 2018): can artificial intelligence start *tabula rasa* and then learn all that it needs to learn by itself (thus in full autonomy) or does it need to start with a certain amount of innate information (likely programmed by humans)?
[20] For stimulus I intend every thing or event which carries some information and evokes a reaction.
[21] See Huscher, Philip. "Chicago Symphony Orchestra Program Notes. Chicago Symphony Orchestra: Claude Debussy, *La Mer*".

making is completely in the hands of the programmers and of the users of the software. It is not Jukedeck's autonomous decision to compose track *x*, it merely follows a set of commands and rules. A similar process is followed by other AI music software such as Sony's Flow Machines.[22] Also in this case, the actions performed by the program are not autonomous. The process followed by the software for composing one of its pieces, *Daddy's Car*, is the following:

1) We set up a database called LSDB. It contains about 13000 leadsheet from a lot of different styles and composers (mainly jazz and pop about also a lot of Brazilian, Broadway and other music styles).
2) The human composer (in this case, Benoît Carré [...]) selected a style and generated a leadsheet (melody + harmony) with a system called FlowComposer. For Daddy's Car, Carré selected as style "the Beatles" [...].
3) With yet another system called Rechord the human musician matched some audio chunks from audio recordings of other songs to the generated leadsheets.
4) Then the human musician finished the production and mixing.[23]

It is evident what the steps of this process lack in respect to the definition of creativity I provided: they lack autonomy. The description of the process is full of indicators of a human presence: 'We', 'The human composer', 'the human musician'. Even if the end product, i.e., *Daddy's Car* can be equated to a product of human creativity, the way in which it reaches this result definitely cannot. The actions performed by the software are not autonomous. The initial command, as well as the following feedback are human: the software just acts following the instructions that it receives. As a result, I argue that AI music generators of this kind cannot be deemed creative because they do not comply with the requirements for creativity specified by $\text{Creativity}_{\text{m}}$. Instead, AI music software can be deemed creative only insofar as it is part of a more comprehensive system formed by the program and the human agent(s).

The program can be considered an extension of the programmer's or user's mind. Indeed, even though they are located outside of the human's head, AI programs are integrated into the cognitive process that leads to the production of music at such a deep level to be considered part of the process itself. In arguing for this claim, I follow Clark and Chalmers' extended mind theory: cognitive systems extend beyond the individual and into the environment which surrounds her. The features and constituents of this environment are part of the cognitive system itself (see Clark and Chalmers,

---

22 See http://www.flow-machines.com/.

23 See http://www.flow-machines.com/ai-makes-pop-music/.

1998). Similarly, computers can present creativity only insofar they are an extension of the programmer's or user's cognitive machinery. They do not, however, possess autonomous creativity.

From what emerged in the discussion so far, it seems that software for music composition cannot be deemed creative. At best, they can be considered as part of a wider creative system where humans still are necessary. However, the application of deep learning and generative adversarial networks (GANs) as in the MidiNet system (see Yang et al., 2017) may overturn this conclusion. The collaborative work of the generator and discriminator software in GANs allows the system to gain independency from the programmer who does not need to constantly supervise the software's job. Unlike other software for music generation, GANs systems change the rules that were initially entered by the programmer through unsupervised learning. They, thus, bypass the hindrance to creativity which is a consequence of mimicking the corpus of data given as input.[24]

GANs indeed can allow music generators to transform an input of random noises into a melody after being trained through algorithms and a data collection of existing music. One of the main features that we commonly attribute to creative products is unexpectedness. GANs produce unexpected results: the combination of an input of random noise, constraints, and transgressions makes the prediction of the outcome impossible. This adds to the perception of the neural network as a 'black box' and, in general, of the alleged 'creativity' of the system.

5. Conclusion

I am now in the position of answering the question stated at the outset: "Can a computer create a musical work?". In order for a computer to be able to create a musical work, it needs to comply with the requirements expressed by Creativity$_m$. It needs, thus, to be autonomous in selecting and elaborating stimuli without constant external intervention. I argue that the application of GANs allows software for music generation to reach the level of autonomy needed for deeming it able to create MWs.

The quick development of the field bodes well for the advancements in respect to the independence of computers and software for music generations from humans' programmers. With the progress of systems of unsupervised learning and a progressive independence from the rules given in the initial

---

[24] See Briot and Pachet, 2017 and also Widmer, Flossman and Grachten, 2009 for other examples of software.

phase of programming, it will be possible for computers to create unexpected results, more similar to what we might expect from humans. Still, it may be argued that the difficulties raised by our subjective perspective in judging the output and the process which leads to its production still remains. This may well be true if we decide to be faithful to the application of the category of creativeness. What I showed in this chapter, instead, is that we should abandon the attribution of creativeness in favour of that of autonomy.

The shift from the consideration of consciousness and intentionality as necessary requirements of a creative systems to the consideration of its autonomy, suggested by Creativity$_{m}$, is beneficial under at least two respects. First, it allows us to address creativity in AI without the need to account for the implications it bears in regard to the enigmatic issues of intentionality and consciousness. Second, it is possible to test the level of autonomy, namely, to test whether the system is able to change the initial rules through an unsupervised training, while the same cannot be said for intentionality and consciousness. The nature of consciousness is still unexplained and also the question of what intentionality is has not received a definite answer. Consequently, at present, it is impossible to test the level of consciousness and intentionality that AI might exhibit, since we do not yet possess a determined meter against which to assess it. If we recognise a system as autonomous, I claim, we may then consider it on a par to a human actor able to produce what we commonly call a 'creative' action.

The relevance of addressing the questions which this chapter focuses on does not only stand on the implications that they bear on the level of copyright issues or on our interaction with musical practices. Considering how AI can fare in the creative sector in respect to humans is beneficial also to address the concern shared by many that AI may in the future outperform and replace humans. It is easier, and less alarming, to conceptualise and program a machine able to work in a warehouse or to build cars and other commodities. It is instead more challenging and disturbing to think of a machine that could equate or even surpass humans in one of the abilities which distinguish us from other beings: the creation of Art.

References

Boden, Margaret A. (2004) *The Creative Mind: Myths and Mechanisms* (2nd ed.), London: George Weidenfeld and Nicolson.

Boden, Margaret A. (2010) "The Turing Test and Artistic Creativity", *Kybernetes*, 39: 409-413.

Boden, Margaret A. and Edmonds, Ernest A. (2009) "What is Generative Art?", *Digital Creativity*, 20: 21-46.

Bostrom, Nick (2014) *Superintelligence: Paths, Dangers, Strategies*, Oxford: Oxford University Press.

Briot, Jean-Pierre and Pachet, Francois (2017) "Music Generation by Deep Learning - Challenges and Directions" in ArXiv:1712.04371v1.

Clark, Andy and David Chalmers (1998) "The Extended Mind", *Analysis*, 58: 7-19.

Colton, Simon (2008) "Creativity Versus the Perception of Creativity in Computational Systems" in *AAAI Spring Symposium: Creative Intelligent Systems*.

Colton, Simon, Charnley, John, and Pease, Alison (2016) "Computational Creativity Theory: The FACE and IDEA Descriptive Models" in *Proceedings of the Second International Conference on Computational Creativity*.

Cope, David (2005) *Computer Models of Musical Creativity*, Cambridge, MA: MIT Press.

Dennett, Daniel C. (1971) "Intentional Systems", *The Journal of Philosophy*, 68: 87-106.

Dodd, Julian (2002) "Musical Works as Eternal Types", *British Journal of Aesthetics*, 40: 424-440.

Gaut, Berys (2010) "The Philosophy of Creativity", *Philosophy Compass*, 5: 1034-1046.

Kivy, Peter (1993) *The Fine Art of Repetition: Essays in the Philosophy of* Music, Cambridge: Cambridge University Press.

Levinson, Jerrold (1980) "What a Musical Work Is", *Journal of Philosophy*, 77: 5-28.

Marcus, Gary (2018) "Innateness, AlphaZero, and Artificial Intelligence" In *Machines That Think* (Douglas Heaven, ed.), London: John Murray.

Pearce, Marcus T. and Wiggins, Geraint (2001) "Towards a Framework for the Evaluation of Machine Compositions" in *Proceedings of the AISB '01 Symposium on AI and Creativity in Arts and Sciences*.

Pease, Alison and Colton, Simon (2011) "On Impact and Evaluation in Computational Creativity: A Discussion of the Turing Test and an Alternative Proposal" in *Proceedings of AISB '11: Computing and Philosophy*, 15-22.

Pease, Alison, Winterstein, Daniel, and Colton, Simon (2001) "Evaluating Machine Creativity" In *Case-based Reasoning: Papers from the Workshop Programme at ICCBR* (R. Weber and C.G. von Wagenheim eds.), pp. 129-137.

Ritchie, Graeme (2007) "Some Empirical Criteria for Attributing Creativity to a Computer Program", *Minds & Machines*, 17: 67–99.

Searle, John R. (1980) "Minds, Brains and Programs", *Behavioral and Brain Sciences*, 3: 417-57.

Searle, John R. (1990) "Consciousness, Explanatory Inversion, and Cognitive Science", *Behavioral and Brain Science*, 13: 585–642.

Siewert, Charles (2017) "Consciousness and Intentionality" In *The Stanford Encyclopedia of* Philosophy, edited by E.N. Zalta, https://plato.stanford.edu/archives/spr2017/entries/consciousness-intentionality/.

Widmer, Gerhard, Flossmann, Sebastian, and Grachten, Maarten (2009) "YQX Plays Chopin", *AI Magazine*, 30: 35-48.

Wiggins, Geraint (2005) "Searching for Computational Creativity", *New Generation Computing*, 24: 209-222.

Wiggins, Geraint (2006) "A Preliminary Framework for Description, Analysis and Comparison of Creative Systems", *Knowledge-Based Systems*, 19: 449-458.

Wolterstorff, Nicholas (1980) *Works and Worlds of Art.* New York: Oxford University Press.

Yang, Li-Chia, Chou, Szu-Yu, and Yang, Yi-Hsuan (2017) "MidiNet: A Convolutional Generative Adversarial Network for Symbolic-domain Music Generation Using 1d and 2d Conditions" in ArXiv:1703.10847.

CHAPTER 9

FORMAL REPRESENTATION OF CONTEXT IN COMPUTATIONAL CREATIVITY FOR MUSIC

René Mogensen[1]

Birmingham City University, UK

Abstract: How can we formally represent the *context* in computational creativity for music? I start with a modified Wiggins framework for computational creativity as searches in concept spaces, and I propose to represent the context in which the computational creativity operates as a network of 'memories' of dynamic concept spaces. The context is represented by the interactions of traces of creative activity, as these traces form networks (*trace-networks*) of the total memory of the individual, or computational agent. My formalisation includes imagination of the computational agent as a function of memory and prediction of future search results in the concept space. I enter this formal view of context and imagination into the specification for computational creativity which I express in Z-style schema notation. This enlarges the computational creativity specification to formalise the creative 'agent' as a node in an intertextual network, where individual creativity is dependent on, and limited by, the trace-networks of the creative agent, but is also a contribution to the emergent creativity of the entire network expressed as the musical intertext. My critical revision of the formalism of computational creativity, applied to improvising computer systems, achieves a step towards more detailed specification and better understanding of potentials in computational creativity for music.

1. Introduction

How can we formally represent context in computational creativity? Currently, there is no definitive description of what computational creativity could be, but I argue for a formal specification as a working understanding of computational creativity for music. I propose to represent the context in which the computational creativity operates as a network of 'memories' of

[1] Birmingham City University, UK; email: Rene.Mogensen@bcu.ac.uk.

dynamic possibility spaces. In effect, the context is represented by the interactions of emergent traces[2] of creative output, as these traces form networks (*trace-networks*) of relation-patterns both of the total memory of the individual, or computational agent, and also of the *intertextual network* (Allen, 2011) in which the agent operates.

I enter this formal view of internal and external context into my specification for computational creativity.[3] This development enlarges the computational creativity specification to formalise the creative 'agent' as a node in an intertextual network, where individual 'creativity' is dependent on, and limited by, the internal trace-networks of the creative agent, but functions as a contribution to the emergent creativity of the entire network expressed as the cultural intertext. In this specification, the cultural memory also shapes the creative output of the individual agent. I examine the specification in the situation of computers that improvise music *co-creatively*[4] with human performers and therefore focus on the musical part of the intertext.

A computational creativity is not necessarily in the same category as human creativity and comparing these two 'creativities' may, logically, be a category mistake. Kinds of what we call *creativity* may have in common what Wittgenstein called 'family resemblances', and so I take the *creativity concept family* as a term covering possible 'creativities' that exhibit both similarities and differences. The vaguely defined 'human creativity' serves heuristically as a prototype for the creativity concept family only to the extent that I use terms derived from ideas about human creativity to name and to guide the conceptualisations of my components in the specification for computational creativity. No identity between human creativity and computational creativity is implied.

I take as given that anything that a current digital computer (or a Universal Turing Machine[5]) can do, can be represented in a formal specification.

---

[2] I use the term 'trace' in the sense of Jean-Jacques Nattiez where "the symbolic form [of the work] is embodied physically and materially in the form of a trace accessible to the five senses" (Nattiez, 1990: 12).

[3] I use Z-style notation (Diller, 1990). Z is a specification language created to model computational processes without specifying coding solutions; I use this notation style for its conciseness and also with the aim of developing a practical specification for computational creativity.

[4] I have previously examined 'co-creativity' in the musical context (Mogensen, 2017b).

[5] The Universal Turing Machine was presented by Alan Turing (1936). 'The [Universal] Turing Machine not only established the basic requirements for effective calculability but also identified limits: No computer or programming language known today is more

Therefore, if a computer can, in some way, be programmed to perform creatively, in other words, produce a kind of 'creativity' and become a member of the creativity concept family, then such a creativity must be definable as a formal specification of 'computational creativity'.

Creativity is often referred to as consisting of some creative process, whereas I have argued for understanding creativity as determined by *product* achieved by a learning process, so that creativity itself is not a process but instead is a product (Mogensen (2018) echoing Glickman (1976)). In support of this view of creativity, I have argued that the formal specification allows alignment with learning models (e.g., John Dewey's experiential model (Dewey, 1938; Mogensen, 2018)).

I build my formal specification on my reworking of Geraint A. Wiggins's (2006a) formal framework, which in turn was based on Margaret Boden's (2004) conception of 'creativity' as searches in concept spaces. In the following section (2) I give an overview of a working specification for understanding computational creativity. I then expand this specification to include *Imagination* as a function of *Memory*. Following this, I give a specification of what I am calling trace-networks, and a formalisation of an intertextual network. The formal specification interprets context as primarily the memory of the creative agent within an intertextual network of musical culture.

The formal specification takes context, memory and imagination as entangled components of a concept family which, in the case of human musical creativity, is focused around how we understand our experience of the musical world we operate in as individuals. My critical revision of the formal framework for computational creativity, applied to improvising computer systems, achieves a more detailed specification and better understanding of the potentials of computational creativity. Developing more detailed formal specifications for computational creativity is an essential step towards better understanding the potentials in such technology. Such a specification can additionally serve as a guide for developing more capable implementations that can interact constructively with human priorities.

2. A working specification of computational creativity

I have developed my working specification for computational creativity to include formalisations of Memory (Mogensen, 2017b) and Motivations (Mogensen, 2017a), which allow iteration in a time-based framework (Mogensen,

---

powerful than the Turing Machine' (Petzold, 2008: 330). See Petzold's (2008) book for an insightful interpretation and discussion of Turing's (1936) article.

2018). Figure 9.1 gives an informal overview of the main components of the specification I have arrived at previously. In this section, I give a descriptive overview of previous development of the specification. I develop the specification further, to include context as mentioned, in sections 3 and 4.

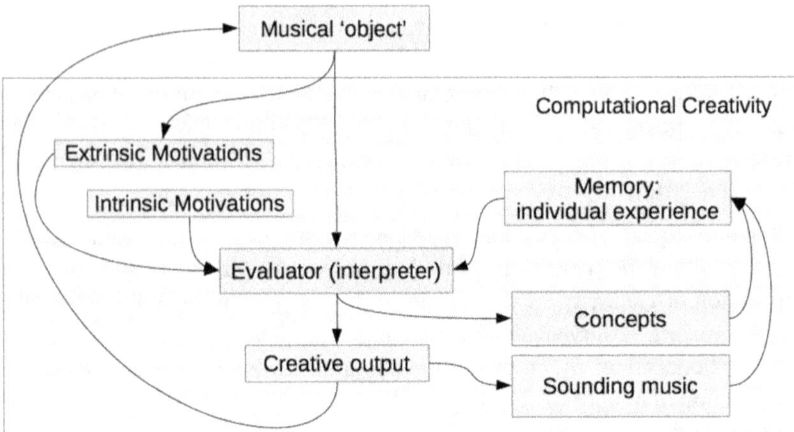

Figure 9.1: Overview of initial version of the specification adapted from, Mogensen, 2018.

The formal specification begins with the axioms in Figure 9.2 which states that in the universe of possible music there are concepts and artefacts that are not identical.[6] Wiggins appears to conflate the term 'artefact' with the term 'concept' and examines the 'conceptual space in which the artefact is found' (Wiggins, 2006b: 209). This seems to be a confusion of terms since 'artefact' refers to physical objects or phenomena made in some way by humans, whereas 'concepts' exist in human consciousness. What the nature of the relations between concepts and artefacts is, is a question beyond the present scope, but I expect that the distinction between these terms would still hold when applied in the context of computational creativity. For the present arguments, I will take as given that while it is possible to operate in a musical concept space, it is necessary to operate in a musical phenomenon space in order to create sounding music. I therefore modify the framework by separating the universe of

6 Briefly, the Z schema notation includes a declarations part above the central horizontal line and predicates below the horizontal line. "The central horizontal line can be read 'such that'." The axiomatic predicates (below the horizontal line in Figure 10.2) "appearing on separate lines are assumed to be conjoined together, that is to say, linked with the truth-functional connective $\wedge$" (Diller, 1990: 6).

possibilities, in which the computational creativity operates, into two types:[7] 1. the concept type and 2. the sonic (music) phenomenon type.

$$
\begin{array}{|l}
\mathscr{U} : Possible\ music \\
C_1 : Concept\ type \\
C_2 : Sonic\ (music)\ phenomenon\ type \\
c_1, c_2 : instances\ of\ [C_1, C_2] \\
\hline
[C_1, C_2] \in \mathscr{U} \\
\forall\, c_1, c_2 \in \mathscr{U} \,|\, c_1 \neq c_2
\end{array}
$$

Figure 9.2: The axioms for the specification adapted from Mogensen, 2018.

Secondly, there is a *Creative Output* formalisation, shown in Figure 9.3 which draws on Wiggins's framework. Wiggins proposed a formalisation of concept evaluations, the concepts being discovered through the searches in 'conceptual spaces', and some of his components are included in Figure 9.3: a *Language* ($\mathscr{L}$) gives the basis for a *Search strategy* ($\mathscr{T}$) and *Constraints* ($\mathscr{R}$) on the possibility space, along with *Evaluation criteria* ($\mathscr{E}$), that are related to form part of the input to a decision function which consists of an *interpreter* $\langle\langle \cdot, \cdot, \cdot \rangle\rangle$ and an *evaluator* $[[\cdot]]$.[8] In other words, Wiggins's value definition $\mathscr{E}$ drives his evaluator function, which judges the output of the interpreter function, to produce *creative output.* The interpreter function is controlled by *Search strategy* ($\mathscr{T}$), *Constraints* ($\mathscr{R}$) on the possibility space, and *Evaluation criteria* ($\mathscr{E}$).

Other components in Figure 9.3 are my developments. I have previously expanded the specification with *Motivations* (Mogensen, 2017a) based on information theoretic types proposed in Oudeyer and Kaplan's typology of computational approaches to motivation, which combines psychological concepts with generalisations of robot implementations (Oudeyer and Kaplan, 2007: 4–5). The formalised representations of intrinsic motivation can indicate a combination of motivations that can be described as in the schema in Appendix Figure 9.11. Extrinsic motivations can be formalised in a similar way although focused on external input as shown in Appendix Figure 9.12.[9]

---

[7] These two types are perhaps interlinked dimensions of possible music, but clarifying this is beyond the present scope and will remain a task for future work.

[8] For a full narrative explaining more details of Wiggins's framework for computational creativity I refer the reader to his (2006a) paper.

[9] For a narrative explaining more details of intrinsic and extrinsic motivations in the specification for computational creativity please see (Mogensen, 2017a).

Creative Output

$\mathscr{L}$: *Language*
$\mathscr{R}$: *Constraints on possibility space*
$\mathscr{T}$: *Search strategy*
$\mathscr{E}$: *Value definition*
c : *Instance of* $[C_1, C_2]$
$\mathscr{M}_1$: *Intrinsic Motivation*
$\mathscr{M}_2$: *Extrinsic Motivation*
$\mathscr{W}$: *Memory*
t : *Time*
$\langle\langle .,.,. \rangle\rangle$: *Wiggins's interpreter function*
$[[.]]$: *Wiggins's evaluator function*

$\mathscr{R}, \mathscr{T}, \mathscr{E} \in \mathscr{L}$
$\mathscr{W}(t) = \bigcup_{p=1}^{t-1} \Big(Q(p) \cdot [[\mathscr{E}]]\Big(\langle\langle \mathscr{R}, \mathscr{T}, \mathscr{E} \rangle\rangle (c(p))\Big)\Big)$
$[[\mathscr{E}]]\Big(\langle\langle \mathscr{R}, \mathscr{T}, \mathscr{E} \rangle\rangle (\Delta\mathscr{M}_1(t), \Delta\mathscr{M}_2(t), c(t), \mathscr{W}(t-1))\Big)$

Figure 9.3: My initial Creative Output formalisation (Mogensen, 2018).

This specification required a more explicit *Memory* representation, as discussed in (Mogensen, 2017b), which is defined as $\mathscr{W}(t)$ in expression 1 and is applied in Figure 9.3 in my initial version of the Creative Output specification. $\mathscr{W}(t)$ is a memory of past evaluations at time t: it is the set of past results of Wiggins's evaluator functions. Each element of the memory (subset of past interpreter function outputs) may be attenuated by some time-dependent effect which I indicate as Q and which represents a degradation of memories over time (a commonplace characteristic of human memory).[10]

$$Memory : \mathscr{W}(t) = \bigcup_{p=1}^{t-1} \Big(Q(p) \cdot [[\mathscr{E}]]\Big(\langle\langle \mathscr{R}, \mathscr{T}, \mathscr{E} \rangle\rangle (c(p))\Big)\Big). \tag{1}$$

My specification includes dynamic *Possibility Spaces* as shown in Figure 9.4 which depart from Wiggins's delineated sets of 'conceptual spaces'. In (Mogensen, 2018) I argued that the epistemological basis for well-defined 'conceptual spaces', in the Wiggins/Boden framework, is problematic: instead of well-defined types or sets (that imply a somewhat static character), such theoretical spaces should rather represent traces of creative output, and these

[10] Arguably, in Figure 3 and expression 1 the component (c(p)) should be replaced by $(\Delta\mathscr{M}_1(p), \Delta\mathscr{M}_2(p), c(p), \mathscr{W}(p-1))$ if we want to include memory of motivations. I will leave these complications for future work.

traces are much better represented by dynamic possibility spaces. The structure over time of these spaces is generated from the experiences of the agents that operate within them. In Figure 9.4, I have formalised a view of dynamic possibility spaces: changes in constraints $\Delta\mathscr{R}(t)$, search strategy $\Delta\mathscr{T}(t)$ and value definitions $\Delta\mathscr{E}(t)$ are functions of memory $\Delta\mathscr{W}(t-1)$ and motivations $(\mathscr{M}^1(t-1), \mathscr{M}^2(t-1))$. The change of a *Possibility Space* at time t (i.e.,$\Delta\mathscr{C}(t)$) is, in turn, a function of the changes of $\mathscr{R}(t)$, $\mathscr{T}(t)$ and $\mathscr{E}(t)$ as well as the latest possibility $c(t)$ and the *Possibility Space* previously perceived $\mathscr{C}(t-1)$.

$\mathscr{C}(t)$: *Possibility Space*

Creative Output
$\mathfrak{C}_1(t)$: *Concept space*
$\mathfrak{C}_2(t)$: *Sonic (music) phenomenon space*
t : *Time*

$$
\begin{aligned}
&\Delta\mathscr{R}(t) = f_{\mathscr{R}}\big(\mathscr{R}(t-1), \mathscr{W}(t-1)\big)\\
&\Delta\mathscr{T}(t) = f_{\mathscr{T}}\big(\mathscr{T}(t-1), \mathscr{W}(t-1), \mathscr{M}_1(t-1), \mathscr{M}_2(t-1)\big)\\
&\Delta\mathscr{E}(t) = f_{\mathscr{E}}\big(\mathscr{E}(t-1), \mathscr{W}(t-1), \mathscr{M}_1(t-1), \mathscr{M}_2(t-1)\big)\\
&\Delta\mathfrak{C}_1(t) = f_{\mathfrak{C}_1}\big(c(t), \mathfrak{C}_1(t-1), \Delta\mathscr{R}(t), \Delta\mathscr{T}(t), \Delta\mathscr{E}(t)\big)\\
&\Delta\mathfrak{C}_2(t) = f_{\mathfrak{C}_2}\big(c(t), \mathfrak{C}_2(t-1), \Delta\mathscr{R}(t), \Delta\mathscr{T}(t), \Delta\mathscr{E}(t)\big)\\
&\mathfrak{C}_1(t) = \mathfrak{C}_1(t-1) \cdot \Delta\mathfrak{C}_1(t)\\
&\mathfrak{C}_2(t) = \mathfrak{C}_2(t-1) \cdot \Delta\mathfrak{C}_2(t)\\
&\mathscr{C}(t) : [\mathfrak{C}_1(t), \mathfrak{C}_2(t)]
\end{aligned}
$$

Figure 9.4: Possibility Space morphology, adapted from Mogensen, 2018.

This morphology of the *Possibility Space* is examined from the agent perspective, since it is dependent on memory and motivations. So here the *Possibility Space* is not an *ideal* space encompassing all possibilities in a particular domain, rather it is a *dynamic* space of possibilities as perceived by an agent which may or may not correspond to a particular idealised domain.

3. Imagination as a function operating on Memory

Mullally and Maguire argue that "metacognitive, cognitive, neuropsychological, and neuroimaging evidence clearly illustrate the close ties between episodic memory, imagination and predicting the future" (Mullally and Maguire, 2014: 228). Zeidman and Maguire argue that "visual perception, imagination and episodic recall depend upon a common process – the creation of models of the world and the use of those models to plan scenarios and re-play memories" (Zeidman and Maguire, 2016: 11). It seems that memory, prediction and imagination are entangled in the human mind, and these terms may express a concept family which is focused around how we understand our experience of the world in which we operate as individuals. One relationship between members of such a concept family may be that imagination is a function of memory which is used to predict future possibilities. As mentioned in the Introduction, human characteristics may serve heuristically as a prototype to guide conceptualisations for components in the computational creativity specification.

Amending the specification overview in Figure 9.5, I expand the memory and evaluator (interpreter) function. I separate memory into two categories: 1. individual experience and 2. cultural rulesets. I also replace the evaluator (interpreter) function by components that I call *Imagination* and *Creative Judgement*. I define *Imagination* as a function operating on *Memory 1* ($\mathscr{W}_1$) and *Memory 2* ($\mathscr{W}_2$) which may serve to predict future possibilities. This predictive process by *Imagination* will affect *Creative Output*, using the attenuated memory and putting into effect the constraints $\mathscr{R}$ and search strategy $\mathscr{T}$ as limiters on the possibility space search. In effect, it is the *Imagination* that delimits the extent of *Creative Output*.

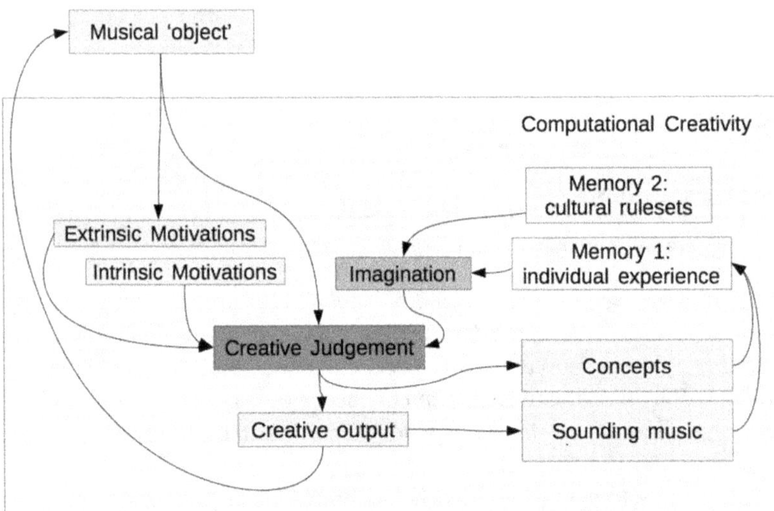

Figure 9.5: Amended overview of the specification components.

As mentioned, I have previously proposed a view of *Memory* in relation to my specification, as shown in expression 1 (Mogensen, 2017a: 3); here *Memory* is the set of past results of Wiggins's evaluator functions. In expression 2, I use this as *Memory 1*: individual experience ($\mathscr{W}_1$) or *Possibility Space Memory*.

$$\textit{Possibility Space Memory}: \mathscr{W}_1(t) = \bigcup_{p=1}^{t-1} \Big[Q_1(p) \cdot [[\mathscr{E}]] \Big(\langle\langle \mathscr{R}, \mathscr{T}, \mathscr{E} \rangle\rangle (c(p)) \Big) \Big]. \tag{2}$$

The cumulative attenuation $Q_1[1..t-1]$ results in what could be called an 'imperfect' memory which serves as an input to *Creative Output.* The cumulative memory imperfections (or memory error) may then be represented as a set V_1 (at time t) of the differences between concept spaces and the attenuated memory of concept spaces, as shown in expression 3.

$$\textit{Possibility Space Memory Error}: V_1(t) = \bigcup_{i=1}^{t} \left[\mathscr{C}(i) - \mathscr{W}_1(i) \right]. \tag{3}$$

Human memory does not consist of only the concept space that the individual has explored, for example, in the case of music performance the musician has also assimilated both a cultural 'background' for the music and for functioning in the musical culture he or she works in. In other words, there is a memory that has been learned aside from $\mathscr{W}_1$; I argue that in the Wiggins/Boden framework this is indicated, at least in part, by the parameters of Wiggins's interpreter and evaluator functions: $\mathscr{R}$ (constraints), $\mathscr{T}$ (search strategy), and $\mathscr{E}$ (value) as well as $\mathscr{L}$ (language).

I represent the cultural memory with $\mathscr{W}_2$, shown in expression 4, and take it to be a dynamic entity which gives the wider cultural framework for any 'search' in a possibility space. The $\mathscr{W}_2$ memory may be attenuated by $Q_2(t)$ and so we may specify a cultural memory error $V_2(t)$ as shown in expression 5.

$$\textit{Cultural Memory}: \mathscr{W}_2(t) = \bigcup_{i=1}^{t-1} \Big[[\mathscr{R}(i), \mathscr{T}(i), \mathscr{E}(i)] \cdot Q_2(i) \Big]. \tag{4}$$

$$\textit{Cultural Memory Error}: V_2(t) = \bigcup_{i=1}^{t-1} \left[[\mathscr{R}(i), \mathscr{T}(i), \mathscr{E}(i)] - \Big[[\mathscr{R}(i), \mathscr{T}(i), \mathscr{E}(i)] \cdot Q_2(i) \Big] \right]. \tag{5}$$

Wiggins's framework includes an 'evaluator function' but I argue that this may be reduced to a choosing mechanism that is an effect of an *Imagination* function. Wiggins uses an $\mathscr{E}$ value definition both in the interpreter and evaluator functions as a means to make choices about how to proceed in the search of the concept space. But 'value' in the common use sense is often a dynamic idea, and not a statically defined type. If you can predefine value against which to measure ideas then anything that expands that value would logically be rejected, meaning that value becomes statically defined. I argue that value is not statically defined, but is rather a dynamic concept. Also, recognition and assessment of value is dependent upon learning, as an example: in the case of music, a person's valuation of a musical work can become more insightful as that person assimilates a wider repertoire of music, such as for example by taking a music course that introduces the person to music works that he or she had not experienced before. However, as mentioned earlier, human experience only serves as a guide to conceptualising components in the specification for computational creativity.

$\mathscr{I}(t)$: *Imagination*

$\mathscr{L}$: *Language*

$\mathscr{R}$: *Constraints on possibility space*

$\mathscr{T}$: *Search strategy*

$\mathscr{E}$: *Value definition*

$\mathscr{W}_1$: *Possibility space memory*

$\mathscr{W}_2$: *Cultural memory*

$\mathscr{V}$: *memory attenuation*

t : *time*

$$\mathscr{R}, \mathscr{T}, \mathscr{E} \in \mathscr{L}$$
$$\mathscr{W}_1(t) = \bigcup_{p=1}^{t-1} \left(\mathscr{C}(p) \cdot Q_1(p)\right)$$
$$\mathscr{W}_2(t) = \bigcup_{i=1}^{t-1} \left(\left(\mathscr{R}(i), \mathscr{T}(i), \mathscr{E}(i)\right) \cdot Q_2(i)\right)$$
$$\mathscr{I}(t) = f\left(\mathscr{W}_1(t), \mathscr{W}_2(t)\right)$$

Figure 9.6: A specification of *Imagination*.

I represent imagination in Figure 9.6, where Wiggins's constraints $\mathscr{R}$, search strategy $\mathscr{T}$, and value definition $\mathscr{E}$ become part of cultural memory $\mathscr{W}_2$; and my formal *Imagination* is a function using these two categories of memories ($\mathscr{W}_1$ and $\mathscr{W}_2$) as input. With this, we can revise the *Creative Output* specification as shown in Figure 9.7, where my *Judgement* function $\ll . \gg$ uses *Imagination* $\mathscr{I}(t)$ to articulate the current possibility $c(t)$ with input from motivations $(\mathscr{M}_1(t), \mathscr{M}_2(t))$.

In this specification, *Imagination* takes input from *Memory* in order to make decisions through the *Judgement* function, and these decisions shape the *Creative Output*. The revised *Creative Output* specification is shown in Figure 9.7. The errors in memory (V_1, V_2) may be indicators of the degree of deviation which the *Creative Output* contains, compared to the non-attenuated *Possibility Space* and *Cultural Ruleset*. As a result, perhaps we may say that errors in memory can have a direct impact on the 'novelty' in Boden's sense, or the deviation from the normal, of the creative output?

Boden (2004: 1) defines creativity as "the ability to come up with ideas or artefacts that are new, surprising and valuable". Boden (2004: 2) also argues for a graduated scale of "how creative" something is, and for asking: in "just which ways(s)" something is creative? With the present analysis, I could hypothesise that 'how creative' something is might be a function of the errors in memory (V_1, V_2). But I suspect it is not a useful question; instead, the question of gradation is more usefully directed towards individual and cultural valuations of the creative output. I leave these questions open for future work.

Creative Output
c : *Instance of* $[C_1, C_2]$
$\mathscr{M}_1$: *Intrinsic Motivation*
$\mathscr{M}_2$: *Extrinsic Motivation*
t : *Time*
$\ll . \gg$: *Judgement function*
$\ll \mathscr{I}(t) \gg \left(\mathscr{M}_1(t), \mathscr{M}_2(t), c(t)\right)$

Figure 9.7: A revision of the *Creative Output* formalisation including *Imagination.*

4. Trace-networks and intertextual networks

If we take a *Possibility Space* as representing a domain, then a creative agent can search for possibilities in several *Possibility Spaces,* representing different domains. To accommodate multiple *Possibility Spaces,* I expand the idea of possibility space memory to $\mathscr{X}_1$ as shown in expression 6, where there are n domains and the complete memory is the set of memories from all domains which have been accessed by the creative agent. Likewise, I hypothesise that there may be multiple cultural memories, analogous to a human who has a 'multicultural background'. Such multicultural rule-sets may be indicated as $\mathscr{X}_2$ as shown in expression 7 where there are h distinct cultural rule-sets which are present in the creative agent's memory.

$$\mathscr{X}_1(t) = \bigcup_{d=1}^{e} \left(\mathscr{W}_1(d,t)\right). \tag{6}$$

$$\mathscr{X}_2(t) = \bigcup_{c=1}^{h} \left(\mathscr{W}_2(c,t)\right). \tag{7}$$

I can now theorise a second-order aspect of the *Imagination* specification which is: the generation of relation-patterns between elements within $\mathscr{X}_1$, between elements within $\mathscr{X}_2$ as well as across elements from the union of $\mathscr{X}_1$ and $\mathscr{X}_2$. In this case, the generation of the dynamic concept space in a particular domain (as a result of *Creative Output*) is not merely a search within a domain, but it is also a fertilisation of the generation process of the *Possibility Space* by the output of a search of relation-patterns within $\mathscr{X}_1$, $\mathscr{X}_2$, and $(\mathscr{X}_1 \cup \mathscr{X}_2)$.

This leads to a revision of the *Imagination* specification as shown in Figure 9.8, where there are prioritised inclusions of $\mathscr{W}_1$ and $\mathscr{W}_2$ but also a relation-pattern generator function $\lll . \ggg$ which takes as inputs $(\mathscr{X}_1(t), \mathscr{X}_2(t))$. The generated relation patterns are networks of memories that are theorised, or synthesised, by the agent and which affect the agent's production of creative outputs. The *Creative Output* of the agent can be understood as a trace of the decisions made by the *Judgement* function as guided by the *Imagination.* In the domain of an improvised music performance, this trace is the performed music.

The revised informal overview of the framework (Figure 9.10) now may include an individual *context,* which encompasses the individual and cultural memory as well as relation-patterns. And the individual computational creative agent may be a poietic system that interacts with the musical 'object' or 'trace' in the semiotic sense used by Nattiez (1990).

If we take a group of $[1..n]$ computational agents, all of which are concerned with creative outputs, we can represent the resulting intertextual network of this group as the set of second-order relation-patterns of these agents, in other words: the intertextual network becomes a second-order relation-pattern analysis (formally generated by the function $[[[\cdot]]]$) of the set of relation-patterns generated by the members of the group as shown in Figure 9.9. The $\mathscr{X}_{group[1..n]}(t)$ relation-patterns form a theoretical trace of the intertextual network. The formal *Creative Outputs* of the members of the group, within the cultural rulesets included in $\mathscr{X}_{2,group[1..n]}(t)$, are the artefact traces (in Nattiez's sense) of the *Creative Outputs* of the group.

$\mathscr{I}(t)$: *Imagination*

$\mathscr{L}$: *Language*
$\mathscr{R}$: *Constraints on concept space*
$\mathscr{T}$: *Search strategy*
$\mathscr{E}$: *Value definition*
$\mathscr{W}_1$: *Concept space memory*
$\mathscr{W}_2$: *Cultural memory*
$\mathscr{X}_1$: *Aggregate agent domains*
$\mathscr{X}_2$: *Aggregate (multi)cultural rulesets*
$\lll . \ggg$: *relation−pattern generator function*
t : *time*

$$\mathscr{R}, \mathscr{T}, \mathscr{E} \in \mathscr{L}$$
$$\mathscr{W}_1(t) = \bigcup_{p=1}^{t-1} \left(\mathscr{C}(p) \cdot Q_1(p)\right)$$
$$\mathscr{W}_2(t) = \bigcup_{i=1}^{t-1} \left((\mathscr{R}(i), \mathscr{T}(i), \mathscr{E}(i)) \cdot Q_2(i)\right)$$
$$\mathscr{X}_1(t) = \bigcup_{d=1}^{e} \left(\mathscr{W}_1(d, t)\right)$$
$$\mathscr{X}_2(t) = \bigcup_{c=1}^{h} \left(\mathscr{W}_2(c, t)\right)$$
$$\mathscr{I}(t) = f(\mathscr{W}_1(t), \mathscr{W}_2(t), \lll \mathscr{X}_1(t), \mathscr{X}_2(t) \ggg)$$

Figure 9.8: A revised specification of *Imagination.*

$\mathscr{Z}_{group[1..n]}(t)$: *Intertextual Network*
$\mathscr{X}_{1,g}(t)$: *Member g domain(s) memory*
$\mathscr{X}_{2,g}(t)$: *Member g culture(s) memory*
$\lll . \ggg$: *relation−pattern generator function*
$[[[\ .\ ]]]$: *relation−pattern generator function*
t : *time*

$$\mathscr{Z}_{group[1..n]}(t) = \left[\left[\left[\bigcup_{g=1}^{n} \left(\lll \mathscr{X}_1(g,t), \mathscr{X}_2(g,t) \ggg \right)\right]\right]\right]$$

Figure 9.9: Specification of the *Intertextual Network.*

5. Conclusion

The *context* of the *Creative Output* of a computational creativity agent g is, in this specification, the *Intertextual Network* $\mathscr{Z}(t)$ of the group of creative agents $[1..n]$ (where $g \in [1..n]$), as well as the *Imagination* $\mathscr{I}(t)$ of the individual agent g, where *Motivations* $(\mathscr{M}_1, \mathscr{M}_2)$ are inputs to the *Judgement* function which is a second-order function shaped by *Imagination.* This means that the individual *Creative Output* at time *t* is generated by *Imagination* $\mathscr{I}(t)$ and *Motivations* $(\mathscr{M}_1(t), \mathscr{M}_2(t))$ within the *Intertextual Network* $\mathscr{Z}(t)$. The diagram in Figure 9.10 gives a revised overview of the components.

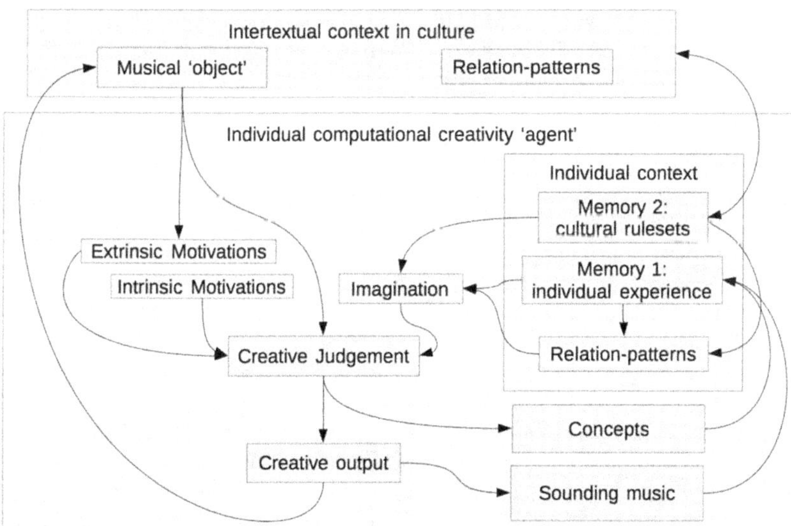

Figure 9.10: Overview of the components of the revised specification.

This specification of computational creativity takes the individual context as well as the intertextual context as parameters for a functional view of creative output as artefacts resulting from generative searches of possibilities. A fundamental problem for creative music systems can be to understand improvisation as searches which have no optimal solution. Indeed, a focus on the optimisation problem is most probably a category mistake. In other words, the question of whether an optimal music has been arrived at, would be a misleading question. What may be of more fundamental interest are the interactions of the artefact with or upon the intertext and the deviations between memory and past experience, both at the individual and cultural levels. I suspect that these interactions and deviations may be relevant for understanding the aesthetic experience of music. To develop computational music creativity, aesthetic experience seems necessary, and this will be a direction for future work.

References

Allen, Graham (2011) *Intertextuality*, New York, NY: Routledge.

Boden, Margaret A. (2004) *Creative Mind*, New York, NY: Routledge.

Dewey, John (1938) *Experience and Education*, New York, NY: Macmillan Co.

Diller, Antoni (1990) *Z, An Introduction to Formal Methods*, Chichester: John Wiley and Sons Ltd.

Glickman, Jack (1976) "Creativity in the arts" In *Culture and Art* (L. Aagaard-Mogensen, ed.), Atlantic Highlands, NJ: Humanities Press, pp. 130–146.

Mogensen, René (2017a) "Computational motivation for computational creativity in improvised music", *Proceedings of Computer Simulation of Musical Creativity conference 2017*, Paper 11, pp. 1–10.

Mogensen, René (2017b) "Evaluating an improvising computer implementation as a 'partial creativity' in a music performance system", *Journal of Creative Music Systems*, 2 (1):1–18.

Mogensen, René (2018) "Dynamic concept spaces in computational creativity for music" In *Philosophy and Theory of Artificial Intelligence III* (Müller, V. C., ed.) SAPERE, Berlin: Springer.

Mullally, Sinéad L. and Maguire, Eleanor A. (2014) "Memory, imagination, and predicting the future: A common brain mechanism?", *The Neuroscientist*, 20 (3):220–234.

Nattiez, Jean-Jacques (1990) *Music and Discourse: Toward a Semiology of Music*, Princeton, NJ: Princeton University Press.

Oudeyer, Pierre-Yves and Kaplan, Frederic (2007) "What is intrinsic motivation? a typology of computational approaches", *Frontiers in Neurorobotics*, 1 (6):1– 14.

Petzold, Charles (2008) *The Annotated Turing: A Guided Tour through Alan Turing's Historic Paper on Computability and the Turing Machine*, Indianapolis, IN: Wiley Publishing, Inc.

Turing, Alan (1936). "On computable numbers, with and application to the entscheidungsproblem", *Proceedings of the London Mathematical Society*, 2nd series, 42: 230–265.

Wiggins, Geraint A. (2006a) "A preliminary framework for description, analysis and comparison of creative systems", *Knowledge-Based Systems*, 19 (7):449– 458.

Wiggins, Geraint A. (2006b) "Searching for computational creativity", *New Generation Computing*, 24 (3): 209–222.

Zeidman, Peter and Maguire, Eleanor A. (2016) "Anterior hippocampus: the anatomy of perception, imagination and episodic memory", *Nature reviews. Neuroscience*, 17 (3): 173–82.

Appendix

$\mathscr{M}_1$: *Intrinsic Motivation*

P : *Probability*
c_k : *Instances of* $[C_1, C_2]$
J : *Constant*
H : *Knowledge of possibility space*
r_l : *Attraction to novelty*
r_m : *Information gain*
r_n : *Pleasure of surprise*
r_o : *Comfort of the familiar*
t : *Time*

$$H(\mathscr{C}(t)) = -\sum_{c_k \in \mathscr{C}(t)} P(c_k) ln(P(c_k))$$
$$r_l(c_k, t) = J_l \cdot (1 - P(c_k, t))$$
$$r_m(c_k, t) = J_m \cdot (H(\mathscr{C}, t) - H(\mathscr{C}, t+1))$$
$$r_n(c_k, t) = J_n \cdot \frac{1 - P(c_k, t)}{P(c_k, t)}$$
$$r_o(c_k, t) = J_0 \cdot P(c_k, t)$$
$$\mathscr{M}_1 = [r_l(c_k, t), r_m(c_k, t), r_n(c_k, t), r_o(c_k, t)]$$

Figure 9.11: My adaption of some types from the Oudeyer/Kaplan formal intrinsic motivation typology (Mogensen, 2017a).

$\mathscr{M}_2$: *Extrinsic Motivation*

P : *Probability*
M_k : *External input*
J : *Constant*
H : *Knowledge of possibility space*
r_l : *Attraction to novelty*
r_m : *Information gain*
r_n : *Pleasure of surprise*
r_o : *Comfort of the familiar*
t : *Time*

$$H(\mathscr{C}(t)) = -\sum_{M_k \in \mathscr{C}(t)} P(M_k) ln(P(M_k))$$
$$r_l(M_k, t) = J_l \cdot (1 - M_k, t))$$
$$r_m(M_k, t) = J_m \cdot (H(\mathscr{C}, t) - H(\mathscr{C}, t+1))$$
$$r_n(M_k, t) = J_n \cdot \frac{1 - P(M_k, t)}{P(M_k, t)}$$
$$r_o(M_k, t) = J_0 \cdot P(M_k, t)$$
$$\mathscr{M}_2 = [r_l(M_k, t), r_m(M_k, t), r_n(M_k, t), r_o(M_k, t)]$$

Figure 9.12: My adaption of some types from the Oudeyer/Kaplan typology for motivation (Mogensen, 2017a).

CHAPTER 10

A HUMAN TOUCH IN COMPUTER-GENERATED LITERATURE

Mariana Chinellato Ferreira[1]

University of Coimbra, Portugal

Abstract: Although there might not seem to be a clear link between Literary Studies and Computer Science, there have been extensive works in the field of Digital Humanities. This way, the purpose of this chapter is to foster the intersection between the Literary Studies and the Computer Science to cooperate with the solution of language and creativity issues in computer-generated literature. To meet this purpose, the origins of Digital Poetry and the pre-digital experiments that linked poetry to computing will be presented. Next, I will present a brief overview of computer-generated poetry, then, the system PoeTryMe[2] and its variations[3] – Co-PoeTryMe and the Twitter bot @poetartificial. Finally, we will analyze some of the poems generated by/with the use of the systems (during the development of the chapter), in order to point out some (aesthetic) aspects that could be improved from the point of view of the Literary Criticism.

1. Introduction

Artificial Intelligence (AI) must probably be the most popular subject in Computing for quite some time, especially because of the great number of books, films and series dedicated to present supercomputers and intelligent robots that can be friends with humans or are capable of outsmarting the human beings and conquer the world. These super machines – mostly

---

[1] PhD Candidate of the PhD Program in Materialities of Literature, University of Coimbra with a research about questions of language and creativity in Narrative Generation Systems in a partnership with joint supervision of the Centre for Informatics and Systems of the University of Coimbra. Contact: marichinellato@gmail.com

[1] See http://poetryme.dei.uc.pt/.

[2] See http://poetryme.dei.uc.pt/~copoetryme/ and https://twitter.com/poetartificial.

horrifying – occupy the collective imagination since de beginning of the technological development. There are also many theories of secret advanced technologies stating that robots are almost fully independent, in an apocalyptic way. However, in real science, the efforts to construct a completely autonomous computer are great and there is much work to be done yet, even though AI has already developed a lot.

Natural Language Generation (NLG) and Computational Creativity (CC) are the fields of AI in which researchers try to develop computer programs able to create texts as creative as possible in order to be considered similar to human (creative) writing. Although researches and experiments are improving exponentially, a relatively small number of them could develop such an autonomous system able to create literature (poetry or narratives) by itself. This way, it is not wrong to say that we cannot remove the human aspect of the Story Generator Systems, so humans and machine split the authoring process. In all cases, man creates the system, the algorithms, and the whole process becomes a partnership with the computer. Because we are dealing with a new study area, many aspects of the generative literature are still under consideration both by Literary Theorists and Computer Scientists.

In this context of a partnership between man and machine, we can find a gap where Literary Studies and Linguistics can be of some aid for Computer Science in Computer-Generated Literature. We believe that these fields of study can help programmers to solve problems closely related to language and creativity. Thus, this work has the objective of finding common ground where researchers in digital literature and computer scientists can work for the scientific development to be productive for both areas. Besides that, we also need to point out some conceptual problems (once Literary Studies work in a more subjective level) and present a few possible contributions for these problems concerning language, textual structures, literary genres, creativity, and other aspects.

2. Origins of Digital Poetry

Digital Literature is quite young as a study field. It was first defined in 1999 by the Electronic Literature Organization[4] (ELO) – founded with the objective of promoting debates around the understanding of literature in the digital environment. Thus, defining digital poetry is hard labor. In general, digital poetry can be defined as the artistic process in which oral referential is used as a link to adapt itself and remain as a part of the new media. Digital poetry is

---

[4] See https://eliterature.org/.

also a type of contemporary poetry composed of words, imagery, graphic forms, sounds, and animations. The first experiments[5] were made in digital visual poetry, interactive poetry, and hypertexts. Among the many other types of digital poetry, there is also video poetry, code poetry, animated poetry, holopoetry and generative poetry, which is the focus of this chapter. This context of multiple forms of poetry shows the need for a reinterpretation of poetry and literature as a whole, as well as understand them as author-driven practices instead of the publisher-driven market (cf. Antonio 2008).

Even though digital poetry seems to be trendy and innovative, way before the use of computers, poets started to explore the generative feature of the language and its relation to mathematics. In fact, the troubadours were the first poets and composers of combinatorial poetry from a mathematical perspective.

In Paris, 1960, a group of artists and mathematicians and engineers started thinking about the use of language in a formalist point of view and experiment with the "discrete"[6] potential of the chain of significants. Oulipo (*Ouvroir de Littérature Potentielle*) is the most significant example of the rational use of the language for aesthetic purposes. Founded by Raymond Queneau and François Le Lionnais, they believed that the freedom of creation would come from "the rule-bound procedures and mathematical constraints" (Leong, 2017: 100). The group's main objective was to elaborate the aesthetical potentialities of the arbitrary rules of combination and selection inherent to language and literature. This way, the group experimented with the development of a predetermined set of rules to produce literary texts of different genres. These set of rules allowed the group to develop works of collaborative and collective writings, using the constraints as tools to increase the productivity. Consequently, the constraints opened the possibilities for the automation of the experiments (Portela, 2017; Pereira, 2014; Bellos, 2010).

With the creation of the group A.R.T.A. (*Atelier de Recherches et Techniques Avancées*) at the Centre Pompidou, Oulipo started experimenting with computers. The first experiences were the remediation of two pieces of work by Raymond Queneau: *Cent Mille Milliards de poèmes*[7] through combinatory and

---

[5] Experiments using the computer to **create** and **run** the poems as well.

[6] In the mathematical meaning of the term.

[7] Currently, there are a few versions of the poem available online, both professional and amateur projects, as it follows: http://x42.com/active/queneau.html; http://www.growndodo.com/wordplay/oulipo/10%5E14sonnets.html; http://www.bevrowe.info/Queneau/QueneauRandom_v5.html; among others.

the programmed version of "A Story as You Like It"[8]. After the experiences with A.R.T.A., the Oulipo group founded the ALAMO[9] (*Atelier de Littérature Assistée par la Mathématique et les Ordinateurs*) in 1981, in order to implement constraints for computer use (Portela, 2017; Gonçalo Oliveira, 2009; Fournel, 2003).

From the point of view of Computing, poetry is the literary genre that adapts more easily to a computational context because of its fixed elements such as meter, rhythm, rhyme, line, stanza, strophe, and verse, among others, that can be formalized and somehow "translated" into code. As the Oulipian experiments could prove, language can also be manipulated in a mathematical way, especially in poetry. Other advantages of poetry generation are the fact that, in terms of aesthetics, the language does not need to be precise, some of the rules can be broken for artistic purposes and meaning is ambiguous and usually given by the reader's point of view.

Narratives are also part of the experiments in generative literature, but their complexities and varieties make it difficult for writers to define the constraints that are common to most types of narratives. This way, because of the flexibility of the poetry, poets can often experiment more freely with the language.

3. Poetry generation

The field of digital literature is quite young as well as its researches, although writers and poets have been experimenting with poetry generation for a longer time. Most of the works can be divided into two main groups: (i) authors and poets who experiment with language and literature in digital (multimedia) settings, (ii) computer scientists and programmers who experiment with language and literature in a coding environment. Most of the times, the first group develops high-quality aesthetical works using less complex programming while the second group creates complex systems that deliver less sophisticated aesthetical works. This is the point where the two fields can meet to develop a refined system able to deliver high-quality literature.

This meeting point helps researchers of both study areas to come up with solutions for common questions that usually arise from this kind of experiments. Broadly, it is common to question the role of the author and the

---

[8] Currently, there are a few versions of the story available online, both professional and amateur projects, as it follows: http://www.thing.de/projekte/7:9%23/queneau_1.html; http://archiv.digitalcraft.org/is-berlin/isb/_inscape/queneau_story/queneau_1.html; http://post-post.net/asyoulikeit/; among others.

[9] The group's page is available and one can

reader in generative literary texts, as well as the kind of effect this kind of literature has on the way people perceive literature. More specifically, this kind of experiment raises questions around the formalization of the language and poetry in order to create a set of rules that will, eventually, become code. These questions can be answered, from the literary point of view, with the help of the Literary Studies (Digital Literature, Comparative Literature, Literary Criticism, and Poetics) and Linguistics (Stylistics, Syntax, and Semantics).

Artificial Intelligence and Computational Creativity have been developing several works on modeling, simulating or replicating creativity in form of many different artistic artifacts, such as generators of narrative, lyrics, movies synopsis, music, anecdotes, and poetry, among others. Poetry generators are systems that generate poetic texts with a certain creative value. In general, the output of these systems must present (i) a certain level of creativity, (ii) grammatically correct sentences, (ii) well-structured texts and (iv) some meaningful message (Gonçalo Oliveira, 2016).

Carolyn Lamb, Daniel Brown and Charles Clark (2016) developed a taxonomy of Poetry Generation techniques that fit the analysis that will follow. They state that poetry generators can (generally) be included in the following categories:

(i) Mere Generation[10], in which poems are generated through templates, using combinatory or permutation and context-free grammar. The problem of mere generation is the lower rate of semantic coherence even though they can generate a fairly interesting output;

(ii) Human Enhancement, in which the user can edit the (final) poems. The use of human enhancement can be positive depending on the purpose of the work. There are researches that study the potential of the systems as a tool through the creative process. In this case, human enhancement is natural. When the purpose is scientific, in other words, to improve the autonomy of the system and the quality of the output, human enhancement is not valid.

(iii) Computer Enhancement, in which the computer optimizes the output through techniques such as data mining and knowledge representation and optimization. This category fits most of the systems developed by computer scientists whose focus is the mainly to model, simulate or replicate human creativity (Gonçalo Oliveira 2016, 14).

Occasionally, poetry generators use one or more categories in order to get better output.

---

[10] For the discussion on "mere generation", see Ventura 2016.

Mere Generation – Example 1	Mere Generation – Example 2
This Shakespearean Sonnet Generator available online chooses lines at random from different sonnets to generate a new one. To dry the rain on my storm-beaten face, But bears it out even to the edge of doom. When my love swears that she is made of truth, Even such a beauty as you master now. But thou, to whom my jewels trifles are, But do not so; I love thee in such sort, Simply I credit her false-speaking tongue: Making lascivious comments on thy sport, these rebel powers that thee array As thou being mine, mine is thy good report. So is my love still telling what is told. That music hath a far more pleasing sound: When to the sessions of sweet silent thought Or at your hand the account of hours to crave, Example retrieved from http://www.languageisavirus.com/sonnet-generator.php	The poetry generator Hafez offers the user an advanced option with some level of customization among other options. Oh baby please remember mommy telling! Its like another lonely love affair, This one I need a gracious happy ending! How many selfless feelings from despair. Some lucky sexy little Juliet, Any happy loving foolish pride. My heart is such a lovely love duet! Too busy contemplating suicide. No way to use the precious fascination, Nothing on an educated guess! So very young and full of admiration, We glide together like a huge success. I really wanna feel a pure delight, About the love of you and me tonight. Example retrieved from http://52.24.230.241/poem//advance/

Figure 10.1: Example from a system of "Mere Generation".

Figure 10.2: Example from a system of "Mere Generation" that offers some level of customization.

There are a great number of websites where users can generate poems[11]. These websites offer a wide range of poetry types and levels of customization and interactivity with the user. In Figure 10.1, the Shakespearean Sonnets

[11] For examples of generative poetry systems, see the following websites: http://52.24.230.241/poem//index.html;https://www.poem-generator.org.uk/; http://www.poemofquotes.com/tools/poetry-generator.php; http://www.languageisavirus.com/poetry.php#.Wr-8SogbPIU; http://www.runokone.com/makeapoem/;

Generator chooses lines from different sonnets very randomly. As we can see in the example above, the sonnet follows its formal structure of three quatrains and a couplet, except for the rhymes. The sentences are grammatically correct (taken from the original poems). It is also possible to state that there is a component of creativity, but the text as a whole lacks cohesion and coherence.

In the second example (Figure 10.2), even though the user is able to choose words and customize the generation options, the final output is given exclusively by the system. Comparing to the first example, the second poem presents a more logical bridge between words and it is easier to perceive meaning in the poem as a whole. We can observe a few mistakes, but the text follows the structure of the English sonnet – three quatrains and a couplet, with the rhyme scheme ABAB CDCD EFEF GG.

These poems can be improved depending on the use purpose. If the system is used as a collaborative tool during the creative process, users can make use of the human enhancement to improve the poem. The most challenging optimization of the system is the computer enhancement because it requires further technical knowledge from the creator of the system. The complexity of some works requires a strong basis of Natural Language Processing, Computational Creativity, Deep Learning and other branches of Artificial Intelligence.

When the authors noticed that the developers of generative poetry included artists, scientists, and hobbyists, they felt the need to think of not only the technical processes but also the purpose of these processes. Thus, this taxonomy illustrates how to improve generative poetry both computationally and artistically. Therefore, this taxonomy seems to be the most appropriate for the following analysis.

4. PoeTryMe

PoeTryMe is a system developed by Hugo Gonçalo Oliveira (at the Center of Informatics and Systems of the University of Coimbra) to create poetry in the Portuguese[12] language. Different from other poetry generators in Portuguese[13],

---

[12] Even though the system also generates poems in English and Spanish, Portuguese was the language of choice because of the lack of studies in linear generative literature in this language.

such as *Poemário*[14], by Rui Torres[15], PoeTryMe must be the most complex, the most flexible and most autonomous of all. On the system's website, it is possible to find the main structure of PoeTryMe, as it follows:

(i) A semantic network, with words and semantic relations;
(ii) A grammar with line templates which are semantic relation renderings;
(iii) A sentence generator that uses the previous;
(iv) A tool for performing syllable related operations;
(v) A sentence generator that uses the previous;
(vi) A tool for performing syllable related operations (Gonçalo Oliveira, n.d.).

Its architecture is modular and enables the improvement of each module independently, providing a high level of customization.

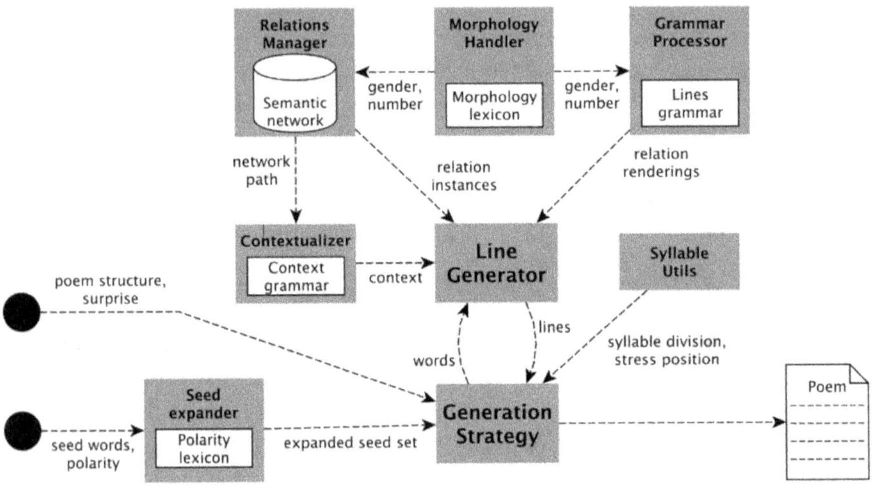

Figure 10.3: PoeTryMe architecture (http://poetryme.dei.uc.pt/).

---

[13] As the time of writing the only known poetry generators in Portuguese language are: Sintext (1996), developed by Pedro Barbosa, Abílio Cavalheiro and José M. Torres, no longer available for experimentation; Poemário (2008), by Rui Torres, which is, in a certain way, an update or a remake of Barbosa's Sintext; and PoeTryMe (2009), developed by Hugo Gonçalo Oliveira.

[14] http://telepoesis.net/galeria-poemas/peditor.php In short, this system allows users to edit the input to generate combinatory poems

[15] For further information on the work of Rui Torres, see https://telepoesis.net/index.html.

In the generation strategy, the poem is generated through the input of a set of seed words that originates sentences from the Sentence Generator in a poem template (that includes stanzas, lines, and syllables). The Sentence Generator guarantees the meaningfulness of the sentences using a semantic graph, managed by the Triples Manager and the generation grammars, processed by the Grammar Processor. The Grammar Processor uses a chart-parser in order to perform chart generation.

4.1. Resources

In order to generate text, the system needs to use a set of natural language resources, as it follows:

(i) "PEN is an implementation of the Earley chart-parsing algorithm used to analyze sentences according to grammars given as input" (Gonçalo Oliveira 2012, "4.1.", para. 1). The algorithm analyzes each sentence in order to break them down into their constituents to show their syntactic relations. Then, PEN provides the grammar rules for the generation of the sentences.

(ii) CARTÃO is a large semantic network for the Portuguese language, extracted automatically from three Portuguese dictionaries. It includes relations such as synonymy, antonym, hypernym, hyponym, part-of, causation, purpose, and property, amongst others. CARTÃO is responsible for the semantic relation between the words of the poem.

(iii) SilabasPT is an API used to perform syllabic division and identify stress syllables for Portuguese words based on a set of rules. Developed to generate texts based on rhythm[16].

(iv) LABEL-LEX is a lexicon of the Portuguese language that specifies "lexical and syntactic restrictions on word combinations" (Ranchhod 2005, "2.", para. 1). Each entry provides information on part-of-speech, lemma and morphological attributes (Ranchhod 2005, Gonçalo Oliveira 2012).

4.2. Strategies

Three different algorithms were implemented as generation strategies for the development of the prototype that generates poems in the Portuguese language. They are:

[16] For further information, see Gonçalo Oliveira et al., 2007.

- Basic: for each line to be filled, a random sentence is generated using the key terms, used as a baseline for debugging (Gonçalo Oliveira, 2012: "4.3.", par. 4);
- Generate-and-test: for each line to be filled, *n* random sentences are generated. The evaluation function scores each sentence. The one with the best score is chosen. All unused sentences are indexed and can be used if a new line needs exactly the same amount of syllables of the previously unused sentence (Gonçalo Oliveira 2012, "4.3.", par. 5).
- Evolutionary: an initial population of *n* poems is generated using the basic strategy. Then, each poem is scored according to the aforementioned evaluation function. Each new generation consists of the poems with the best evaluation, poems that are the result of crossing two random poems in the population, and newly created poems as well. When two poems are crossed, a new poem is created with lines selected from both. The best scoring poem of the last generation is returned (Gonçalo Oliveira, 2012: "4.3.", par. 6).

This is a very broad overview of the generation process of the system PoeTryMe. This way, it is possible to verify the output keeping in mind the stronger and weaker aspects of the system. It is also very important to understand every step of the creative process not to criticize the output the same way we would criticize the work of a poet. In fact, the main objective here is not to compare the poems with the ones created by humans, but to understand the limitations of the language use, to stretch the use of language as an aesthetical experimentation in order to come up with new forms of poetry. As a characteristic of the digital avant-garde, the experimentations allow writers, programmers, and users to play with the poetics of strangeness.

5. Other instantiations of PoeTryMe

Because of its modular architecture, PoeTryMe is flexible enough to have different instantiations. Co-PoeTryMe is a web application built on the top of the original PoeTryMe API, and it is considered a co-creative application, once there are more options for the user to customize the poems, including editing words, lines or stanzas, as well as to import your own poems as an input for the poem generation or write the draft poem yourself.

@poetartificial is a Twitterbot created on the top of PoeTryMe architecture, which produces a poem inspired by the Portuguese trending topics in the Twitter social network. It searches the tweets that mention a certain topic, extracts the most frequent words in those tweets and uses them as seeds for the generation of new poems. The generation is performed by the classic PoeTryMe system shown before.

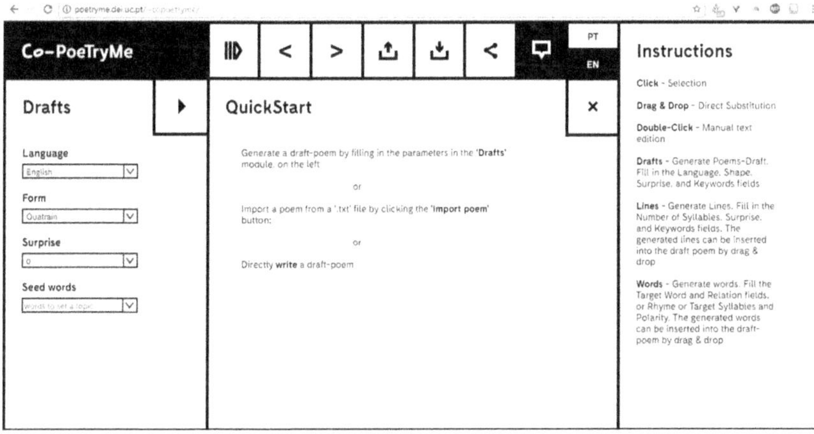

Figure 10.4: Screenshot of the webpage poetryme.dei.uc.pt/~copoetryme/.

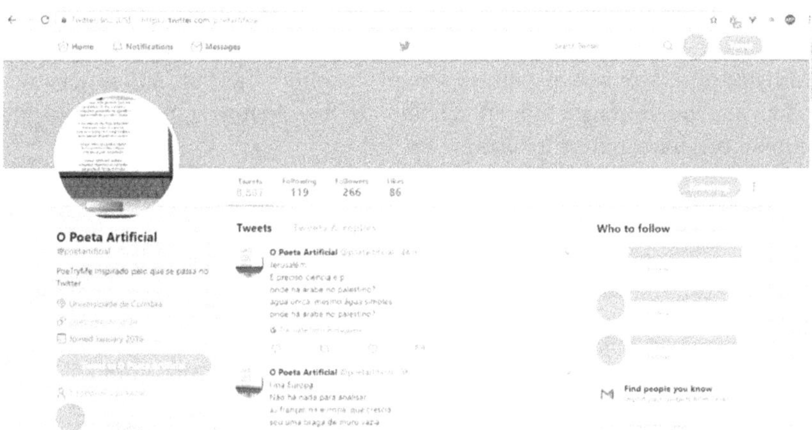

Figure 10.5: Screenshot of the webpage twitter.com/poetartificial.

6. Examples

For the analysis of the output of the available systems, three seed words were chosen: *alma* (soul) and *sonho* (dream) – because of their high level of subjectivity as well as their usage in Portuguese poems and *mundo* (world) – because of its subjective and objective meanings and its common usage in Portuguese language poetry. The three seed words were used in both classic PoeTryMe and Co-PoeTryMe in order to have the same basis to compare them. Two out of the five poems with the highest scores (zero or less than zero, according to the scores of the system) of the following types were selected:

- 5-7-5 syllabic pattern (Haiku[17] – a short form of Japanese poetry);
- 2x10 (a decasyllabic couplet);
- 4x7 (a heptasyllabic quatrain – *rendondilha maior*[18]);

The two poems for the analysis of the Twitter bot were selected according to different aspects. The first poem was based on the topic #VAMOSSK, which is a trending topic related to a German professional electronic sports organization, best known for their success in Counter-Strike: Global Offensive. Besides the unusual topic, the poem called the attention because of the way the words were used, creating a great number of alliterations and assonances. The rhymes were also important features to take into account. However, the global meaning is the most interesting aspect of the poem. The second poem was chosen because the trending topic was related to a very sensitive event: the death of Zé Pedro, founder and guitar player of the band *Xutos e Pontapés*, a well-known Portuguese rock band. This poem was chosen because this topic requires a type of sensitivity that is difficult to generate randomly and is particular to human feelings: the grief.

The original poems were translated almost literally, only with the purpose of clarifying the (attempt of finding some) meaning. The translation does not follow the poem's structure in relation to the rhyme or meter. Only the number of verses and strophes were respected.

6.1. PoeTryMe

sem coi**s**a **sem s**onho (alliteration S/ consonance S M)	no thing no dream
das **mu**lheres do **meu mu**ndo (alliteration M/ assonance U)	from women of my world
c**o**rp**o** ap**ó**s dia (score 0.0) (assonance O)	body after day

Figure 10.6: Haiku 1.

alma entre peças	soul between pieces
mundanas **mundo de m**ão (alliteration M/ consonance N D)	mundane world of hand
desta já me chão (score = 0.0) (rhyme ÃO)	from this me already ground

Figure 10.7: Haiku 2.

Overall, the first Haiku (Figure 10.6) is well formed and presents grammatically correct sentences with a certain level of creativity and some metaphorical meaning closely related to the seed words *sonho* and *mundo*.

---

[17] Short, descriptive form of Japanese poetry, in which the poet describes a scene of the nature world he is surrounded.

[18] *Redondilha maior* is an Old Portuguese metric pattern of seven syllables per line commonly used in medieval popular poetry.

The second Haiku (Figure 10.7) presents strong alliteration and consonance in the second line. The meaning is the weakest aspect of this poem, although the last verse has some potential for an interesting image: as the first two verses show the soul among pieces of a mundane world, the last verse shows a possible fusion of the self (me/soul) into the ground of that metaphorical world.

dumas peças **d**e alma, bem tecidas (assonance A E/ alliteration D)	of some pieces of soul, well woven
tenho um **t**ear de gen**t**e de mundo (alliteration T N M/ assonance O U)	I have a loom of people of the world

Figure 10.8: Decasyllabic Couplet 1.

o mundo das maiorias ardendo (alliteration M)	the world of the majorities is burning
mundo mundial, cheio de razões (alliteration M)	global world, absolutely right

Figure 10.9: Decasyllabic Couplet 2.

The first couplet (Figure 10.8) is an interesting poem because it presents grammatically correct sentences that also generate meaning. The sentences are inverted, which makes it more interesting. The assonances and alliterations give the poem a pace that matches the movement of a loom, in which the poetic self-weaves the souls of the people of the world.

Even though the second couplet (Figure 10.9) makes poor use of poetic devices, this poem seems to express some kind of message. The poem clearly relates to the seed word *mundo*. Different from the haiku poems, that made use of a larger number of poetic devices and seemed more creative because of this, the couplets were more creative because of the language usage. In this case, the poem allows for multiple interpretations but seems to have some meaning. According to my reading, the majorities of the world are always right and seem not to care if the world is burning (full of problems).

um espírito de alma	a spirit of soul
do mundo fingem sistemas	from the world pretend systems
mundanas mundo de mão (alliteration M/assonance U)	mundane world of hand
é alma da assombração (rhyme ÃO/ alliteration SS Ç/ Rhyme ÃO)	it is soul of revenant
it is soul of revenant	
(score = -1.0)	

Figure 10.10: Redondilha Maior 1.

From the many heptasyllabic quatrains generated with the system, the chosen ones presented the best scores even though they were the least creative and had less potential to generate meaning. It was clear, then, that the scores are based on rhyming. The first *redondilha* (Figure 10.10) presents a rhyme, but it is difficult to generate meaning.

pousas o mundo no chão (assonance O)	settle the world on the ground
oníricas sonho de mão (rhyme ÃO)	oneiric dream of hand
espaço, tlim, tlim, tlim, tlim (onomatopoeia)	space, tlim, tlim, tlim tlim
as classes, as freguesias (assonance A/ alliteration S)	the classes, the parishes
(score = -1.0)	

Figure 10.11: Redondilha Maior 2.

Compared to the previous poem, the second *redondilha* (Figure 10.11) sounds more poetic and, except for the second verse, it gives us the possibility of generating meaning. It presents rhyme (ÃO) and poetic devices in every verse. What makes this poem stand out from the others is the use of onomatopoeia, which makes it more poetic and enjoyable.

6.2. Co-PoeTryMe

un**s** **s**olo**s** dum mundo (consonance D M N S/ assonance U O)	a few lands of a world
por entre o universo	through the universe
tu és o teu ar (alliteration T)	you are your air
(score = 2.5)	

Figure 10.12: Co-PoeTryMe Haiku 1.

um astro do mundo (assonance O)	a world star
onde a peça da alma (assonance A)	where the piece of the soul
sem coisa sem sonho (alliteration/ consonance S)	without a thing nor a dream
(score = 2.5)	

Figure 10.13: Co-PoeTryMe Haiku 2.

Haiku 1 (Figure 10.12) has a clear connection to the seed word *mundo.* We can say that the poem follows the rules of the Haiku poems when it shows a scenery. In this poem, we can picture the lands around the world as well as the air floating around. It gives us the feeling that we are the air floating around.

fico sem sonho e sem fantasia (alliteration S/ consonance S F)	I get no dream and no fantasy
sou uma alma de fantasma vazia (asson. A/ alliterat. M/ rhyme IA)	I am a soul of an empty ghost
água alma, mesmo água benigna (assonance A)	water soul, even kind water
com espíritos de prata e alma (assonance A)	with silver spirit and soul
(score = 3.25)	

Figure 10.14: Co-PoeTryMe Re dondilha 1.

The second Haiku (Figure 10.13) generated with Co-PoeTryMe makes use of poetic devices, but the poem as a whole leaves us little or no room for interpretation.

o homem e depois o mundo (assonance O E)	the man and then the world
as terras do mundo passam por mim (assonance A/ alliteration S M P)	the lands of the world pass by me
o homem e depois o mundo (assonance O E)	the man and then the world
sonho onírico, ó mar atlântico (assonance O A)	oneiric dream, oh Atlantic sea
(score = 0.0)	

Figure 10.15: Co-PoeTryMe Redondilha 2.

Both *redondilhas* generated with Co-PoeTryMe were good in their general aspects. The first one (Figure 10.14) presents a rhyme but had a bad score according to the system. Nevertheless, even with a low score, it shows a good level of creativity and the words give us scope for generating meaning.

The second *redondilha* (Figure 10.15) presents repetition. It is interesting because this is a very common poetic device and never appeared in any poem before. This poem is full of assonances and presents one pair of verses that rhyme. In addition to these devices, the poem sounds a little like the Portuguese poems written during the great navigations. It sounds like there is the man, then there is the world and the oceans divide them. When the Portuguese started the expeditions, the lands were passing by the navigators and this was their oneiric dream: to cross the Atlantic Ocean. I hope I am not stretching the search for meaning, but I believe this is one of the poems as similar as possible to the human writing – even though this is not the purpose of the analysis.

6.3. @Poetartificial

#VAMOSSK	
é da família das irmãs (rhyme mãs/ assonance A)	it is from the sister's family
tantos suecos tantas alemãs (assonance A/ alliteration T S)	so many Swedes so many German women
tantos suecos tantas francesas (asson. A/ allit. T S/ rhyme esas)	so many Swedes do many French women
passam-se os suecos, holandesas (asson. A/ allit. S)	the Swedes go through, Dutch women
9:30 AM - 25 Nov 2017	
(https://twitter.com/poetartificial/status/934383816474492928)	

Figure 10.16: @poetartificial Poem 1.

As previously stated, the selection of the poems from the Twitter bot @poetartificial was made firstly based on the overall aspect of the poems and secondly on the trending topic they were based on. Poem 1 (Figure 10.16) has a certain rhythm given by the repetition, alliteration, and assonances that make it interesting even if the meaning is not so clear. In fact, the meaning has nothing to do even with the trending topic the poem was based on, which was related to electronic sports.

#zepedro	
falo sobre #zepedro, também eu! (rhyme eu)	I talk about #zepedro, I too!
e o astro astral se esqueceu (assonance E A/ alliteration S)	and the astral star has forgotten
astro astral de coisas naturais (assonance A/ alliteration S)	astral star of natural things
astro, as pessoas mais banais (assonance A/ alliteration S/ rhyme ais)	star, the most banal people
(12:38 AM - 1 Dec 2017)	
(https://twitter.com/poetartificial/status/936424146526982144)	

Figure 10.17: @poetartificial Poem 2.

The second poem was chosen almost exclusively because of the trending topic it was related to. The death of a notorious individual would require a higher level of sensitivity for the system to generate a poem. Understandably, the poem uses quite a few poetic devices to talk about the artist as a (rock) star. The tone of the poem is cheerful and the mood quite positive as well as no mention of the sad happening. Overall, the fact that the algorithm "understood" that people were talking about that person for a special reason is a good start. Maybe with a bit more improvement on the Generate-and-test and the Evolutionary strategies as well as the usage of an improved set of natural language resources, especially a semantic network more related to sentiments.

7. Conclusions and Further Work

From the point of view of the Literary Studies, this work helped to raise some issues related to the role of the author in the creation process of generative work. It is clear that the author is a blurred entity compose by all the subjects involved in the process. In addition, because of the virtually infinite number of co-authors and the virtually infinite numbers of poems that will be generated by the system, the unity of the literary work is subverted.

As we could see, the poetry generated by computer devices is also subversive since its poetics undermine our perception of writing and reading with which we are familiar. Consequently, it is time for us to think of new methods of creating literature as well as rethinking our relationship with the language. For that, we must use generative poetry (and other kinds of generative literature) as a tool for aesthetical experimentation in order to generate new modalities of literature. The main idea, as stated previously, is to use the digital avant-garde in improbable processes to cause strangeness not because a machine as a simulation of human creativity does generate the texts, but to challenge the literary tradition.

This first analysis serves to understand how a more complex system can generate poetry from a small number of input words in an autonomous way. The results, compared to systems that work with mere generation, in other words, systems that pick random sentences and combine them in a pre-existent template, are that PoeTryMe can generate more interesting output once the generation is original, which is it does not rely on a pre-existent set of poems. The flexibility of the architecture also gives users more freedom to experiment and develop other models on top of the 'classic' PoeTryMe, which is the case of Co-PoeTryMe.

In a more direct approach, both systems PoeTryMe and Co-PoeTryMe allow users to use both Human Enhancements and Computer Enhancement in order to refine the final output. Co-PoeTryMe has a good number of options for the user to customize the poem generated by the system. For understanding the qualities and limitations of the systems, I opted not to use this strategy and allow the system to generate the poems almost by itself.

The results were, overall, very positive. The quality of the output is good, considering all the characteristics and limitations of the system. From my understanding, the system has many features that were not explored by this analysis and need to be more developed in further studies. The analysis was focused on the traditional use of the language to generate poetry. Maybe the next experiment should focus on the creative aspect of the system in order to explore the system more freely and discover its potentials.

From the point of view of the Computer Science, as the title of the chapter suggests, I propose a multidisciplinary work among Linguists, Literary Theorists, and Computer Scientists. This group of researchers should work collectively in order to find solutions for a better use of lexical knowledge bases, to improve grammars for the poems to use poetic devices such as imagery, metaphor, mood, tone, synecdoche, symbol, allusion, among others.

As for creativity itself, it is hard to emulate a human feature that scientists do not know exactly how it works. We must keep in mind that artwork is intrinsically linked to the context where it was generated, in other words, it is linked to the social, political, cultural development. That seems to be the main issue in terms of computational creativity: how to emulate the context of a piece of art. And this is especially true for the generation of literary artifacts, such as poems. Once the context can be applied to the system, maybe many of the language issues can be solved and the interestingness could be improved.

For further works, an evaluation developed with users of different background seems to be interesting to test both the process of interacting with the system as well as the interestingness of the outputs.

References

Antonio, Jorge Luiz (2001) "Alguns aspectos da Poesia Digital" in *INTERCOM – XXIV Congresso Brasileiro da Comunicação*, 1-18, Campo Grande/MS.

Bellos, David (2010) "Mathematics, poetry, fiction: The adventure of the Oulipo", *Bulletin Journal of the British Society for the History of Mathematics*, 25: 104-118.

Fournel, Paul (2003) "Computer and Writer: The Centre Pompidou Experiment" In *The New Media Reader* (Noah Wardrip-Fruin and Nick Montfort, eds.), Cambridge, Massachusetts: The MIT Press, pp. 182-189.

Lamb, Carolyn, Daniel G. Brown and Charles L.A. Clarke (2016) "A Taxonomy of Generative Poetry Techniques" In *Proceedings of Bridges 2016: Mathematics, Music, Art, Architecture, Culture*, 11 (3); 159-179.

Leong, Michael (2017) "Oulipo, Foulipo, Noulipo: The Gendered Politics of Literary Constraints" In *The Oulipo*, 6, Verbivoracious Press, pp. 100-130.

Gonçalo Oliveira, Hugo. (n.d.). *PoeTryMe*. http://poetryme.dei.uc.pt/.

Gonçalo Oliveira, Hugo, Amílcar F. Cardoso, and Francisco C. Pereira (2007) "Tra-la-Lyrics: An approach to generate text based on rhythm" in *Proceedings of 4th International Joint Workshop on Computational Creativity*, London, UK. IJWCC, pp. 47-55.

Gonçalo Oliveira, Hugo (2009) "Automatic generation of poetry: an overview" In *Art, Music, Creativity and Artificial Intelligence*, Universidade de Coimbra, Coimbra, Portugal.

Gonçalo Oliveira, Hugo (2012) "PoeTryMe: a versatile platform for poetry generation", in *Proceedings of ECAI 2012 Workshop on Computational Creativity, Concept Invention, and General Intelligence*, Montpellier, France, *C3GI* 2012.

Gonçalo Oliveira, Hugo (2016) "Automatic Generation of Poetry Inspired by Twitter Trends" In *Knowledge Discovery, Knowledge Engineering and Knowledge Management IC3K 2015. Communications in Computer and Information Science* (Ana Fred et al., eds.), 631: 13-27.

Pereira, Vinícius C. (2014) "A Estética da Hibridação e a Literatura Gerada por Computadores: Um Estudo do Software Alexandrins au Greffoir", *Linguagem e Ensino*, 17 (3): pp. 927-950.

Portela, Manuel (2017) "Writing under Constraint of the Regime of Computation" In *The Bloomsbury Handbook of Electronic Literature* (Joseph Tabbi, ed.), London: Bloomsbury Academic, pp. 179-198.

Torres, Rui (2012) "Sustentabilidade e Entropia na Geração Textual Variacional: POEMADS – Sob o Signo da Devoração", *Texto Digital*, 8 (2): 361-375.

Ventura, Dan (2016) "Mere Generation: Essential Barometer or Dated Concept?" in *Proceedings of the Seventh International Conference on Computational Creativity (ICCC 2016)*, Paris, France, pp. 17-24.

CHAPTER 11

NATURAL LANGUAGE PROCESSING IN ARTIFICIAL INTELLIGENCE: A FUNCTIONAL LINGUISTIC PERSPECTIVE

Kulvinder Panesar

York St John University, UK

Abstract: This chapter encapsulates the multi-disciplinary nature that facilitates NLP in AI and reports on a linguistically orientated conversational software agent (CSA) (Panesar 2017) framework sensitive to natural language processing (NLP), language in the agent environment. We present a novel computational approach of using the functional linguistic theory of Role and Reference Grammar (RRG) as the linguistic engine. Viewing language as action, utterances change the state of the world, and hence speakers and hearer's mental state change as a result of these utterances. The plan-based method of discourse management (DM) using the BDI model architecture is deployed, to support a greater complexity of conversation. This CSA investigates the integration, intersection and interface of the language, knowledge, speech act constructions (SAC) as a grammatical object, and the sub-model of BDI and DM for NLP. We present an investigation into the intersection and interface between our linguistic and knowledge (belief base) models for both dialogue management and planning. The architecture has three-phase models: (1) a linguistic model based on RRG; (2) Agent Cognitive Model (ACM) with (a) knowledge representation model employing conceptual graphs (CGs) serialised to Resource Description Framework (RDF); (b) a planning model underpinned by BDI concepts and intentionality and rational interaction; and (3) a dialogue model employing common ground. Use of RRG as a linguistic engine for the CSA was successful. We identify the complexity of the semantic gap of internal representations with details of a conceptual bridging solution.

1. Introduction to NLP in AI

A simple definition of artificial intelligence (AI) is '*the science of mimicking human mental faculties in a machine*'(Hopgood, 2014*)*. Philosophers pose the question – how do minds work? This raises two hypotheses – one of weak AI – the assertion that machines could act as if they are intelligent or strong AI – the assertion that machines that do so are actually thinking. In most cases, researchers comply with weak AI, as they are not concerned whether it simulated intelligence or real intelligence, but more importantly their program works (Russell and Norvig, 2016). However, for this chapter, we will consider strong AI. In terms of the spectrum of intelligence (from reactive agents to expert systems), natural language (NL) sits midway depicting that the level of understanding required for computers is complex, fundamentally because NL is inherently ambiguous, complex and dynamic, compared to being the most easily understood knowledge representation for people. For example, 'he saw her duck' may mean he saw a waterfowl that belonged to her, or that he saw her move to evade something. Here we have a probability distribution over possible meanings.

Computational linguistics (CL) is an interdisciplinary field concerned with processing of language by computers, and sub-field of AI. NLP is defined by Liddy (2001) 'as a theoretically motivated range of computational techniques for analysing and representing natural occurring texts at one or more levels of linguistic analysis for the purpose of achieving human-like language processing for a range of tasks or application'. Real-world applications are varied such as machine translation, question answering, text data mining, NL interfacing and generation, spoken dialogue systems, and multilingual online language processing (Mitkov, 2003). Our focus lies with NL interfaces, generation, understanding and processing – an engineering domain (Clark, Fox, and Lappin, 2013).

One explanatory approach to understanding the NLP system is the 'levels of language', from phonology to pragmatics, where the levels can interact in a 'variety of orders' and thus dynamic. Each level conveys meaning and knowledge, and humans tend to use all levels of language to gain understanding, subsequently, a potential for a more capable NLP system. The different forms of knowledge relevant to NLU exist at each level of language whereby there is the additional world knowledge (general knowledge, and knowledge of the other user's beliefs and goals). With knowledge of the possible ambiguities, we are confronted with ascertaining what something means by determining "WHO did WHAT to WHOM, WHERE, WHEN, HOW and WHY?" (Pradhan et al., 2004) which poses two problems. Firstly, WSD (word sense disambiguation) algorithms to deal with words with similar meanings (polysemy) and to extract their sense, for example, the ambiguous word "dish" as either of three senses 'plate', 'course of a meal' or a 'communication device'. Secondly, pronoun resolution, is a deeper kind of NLU to work out "Who did want to whom", to detect the subjects and objects of the verb (Bird, Klein, and Loper, 2009). For instance, in the sentence "thieves stole the painting" may be followed by either one of these sentences – "They were subsequently sold" or "They were subsequently found". Answering this question involves finding the antecedent of the pronoun, "they" either "thieves or paintings". Computation techniques for tackling this problem are called anaphora resolution. This may include identifying what the pronoun or noun phrase refers to – and semantic role labelling – identifying how the noun phrase relates to the verb (as an agent, patient, instrument, and so on).

A practical NLP system must be good at making dis-ambiguous decisions of word sense, word category, syntactic structure and semantic scope. A technique of using selection restrictions, e.g., a verb like "swallow" requires an animal being as its subject and physical object as its object. For example: 1) 'I swallowed his story book, ink and ruler'; (2) 'the supernova swallowed the planet'. This restriction will disallow common and straightforward metaphorical extensions of the usage of "swallow" as above. Our next question is – what do we want to achieve from NLP. Liddy (2001) states that full NLP system goals are: (1) Paraphrase an input text; (2) Translate the text into another language (not considered here); (3) Answer questions about the contents of the text; (4) Draw inferences from the text. NLP research has generally made progress on goals (a-c), but (d) refers to NLU – a more challenging goal, within this work. What computers find hard is processing an utterance which is implicit, high contextual, ambiguous and often imprecise (McCord, Murdock, and Boguraev, 2012) and hence which requires a deeper linguistically knowledge aware architecture.

To summarise the aim of this chapter is to establish, explain, demonstrate and discuss future work for our use of a functional linguistic perspective for NLP in AI. The objective of this chapter include: (1) to establish the use of formal to functional linguistic phenomena; (2) to define the goals of linguistic theory; (3) to explain the basic concepts of our strong functional linguistic theory (4) to present an experimental case (5) to demonstrate a proof-of-concept (6) discuss future work.

2. NLP Models – Statiscal to Formal to Linguistic to Functional

2.1. NLP models – statistical to formal

Orduñez-Salinas and Gelbukh (2010) highlights there are several statistical probabilistic computational structures (CS) used in NLP from limitation in syntax-driven processing to semantic-driven processing and using more complex structures such as vector-based, graph-based and tree-based models. To reiterate the goal of NLU defined by Winograd (1972) is "we can describe the process of understanding language as a conversion from a string of sounds or letters to an internal representation of 'meaning". Further, NLU is popular in linguistics, and cognitive science as well as NLP.

The first NLP related task was a machine translation (MT) – a computer-based application, derived in the late 1940s. This was followed by computer translation projects, using various techniques of dictionary-lookup, and word order tasks, but omitted lexical ambiguity. In the 1950s, researchers realised there was a need for an adequate theory of language, and in 1957 Chomsky published "syntactic structures". In the 1990's there was much acceleration in three areas as confirmed by (Lester, Branting, and Mott 2004), which include: (1) development of large corpora of tagged text, for instance, the Brown Corpus, the Penn Treebank and the British National Corpus (BNC); (2) using the tagged corpora influenced the development of statistical, machine-learning and empirical techniques for extracting grammars and ontologies; (3) rise of competitions such as Text Retrieval Conferences (TREC) since 1992 provide a driving force for NLP prototypes and automation.

A language-understanding system must have some formal way to express its knowledge to a subject, and must be able to represent the 'meaning' of a sentence in this formalism. The formalism must be structured so the system can use its knowledge in conjunction with a problem-solving system to make deductions, accept new information, answer questions, and interpret commands. Further Raskin and Nirenburg (1987) identify that NLP applications are easier to build, and deceptively intelligent as noted by Periñán-Pascual (2013) but controversially, without linguistic theories, semantic relevance is missing, and therefore dismisses the NLU. For instance, the first conversational agent was ELIZA, contrived for NLP, based on simple pattern matching technique ALICE (Artificial Linguistic Internet Computer Entity) ALICE (n.d) at the website – http://www.alicebot.org) programmed in a language called AIML (Artificial Intelligence Markup Language) that is based on XML (Weizenbaum, 1966). It could facilitate dialogue, but the responses did not reflect the user's intentions, and not having any understanding of the language.

The earliest language models were n-gram letter models proposed by (Markov, 1913) followed by n-gram word models in English based on the statistical methods and ideas from information theory (Shannon and Weaver, 1949). Duta (2014) notes the theoretical modern CLs started in the late 1950s, with the introduction of generative grammar by Noam Chomsky. This theory's aim was to produce a set of rules that could correctly predict which combination of word forms made up grammatical sentences. After this, there were attempts at NLU by a first computer at Stanford and MIT called STUDENT which had the ability to input algebraic word problems.

Ovchinnikova (2012) asserts that most of the NLU programs in the 1970s were also called knowledge-based systems, and their development closely related to the AI research on knowledge representation (KR). The focus of these programs was world knowledge from large efforts into general-purpose knowledge base such as CYC to more recent large-scale project attempting NLU with Watson – Deep QA project by Chaplot, Chhipa, and Kumar (2013). Here several models and algorithms were used, such as machine learning techniques, semantic frames, shallow and deep parses logical forms, semantic role labels, co-reference, relations, named entities, probabilistic methods to name a few, for question answering. The challenge here was to build a computer system that could compete at the human champion level in real time (Ferrucci et al., 2010). The interesting elements here are the cognition approaches used. To date, this project has expanded, and the Watson API is invoked for various NLU related activities.

The challenge of NLP and NLU is due to the range of ambiguities from lexical to pragmatic as illustrated in Figure 11.1 with the NLP pipeline. Our next question is – what are the theoretical aspects of language, and further what are the non-statistical ways of processing natural language text.

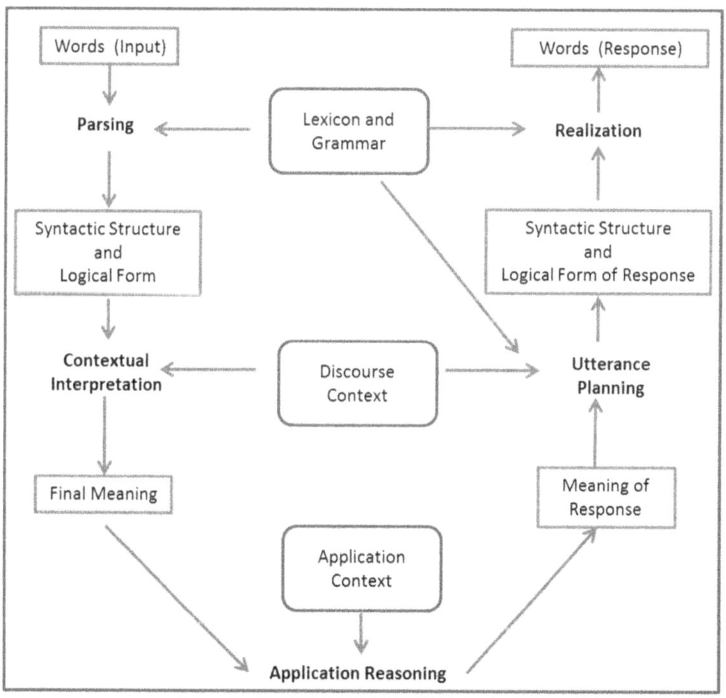

Figure 11.1: NLP Pipeline.

There are several popular linguistic-based computational structures. One example is dependency grammar where a relation exists between the head word and the other words that depend on it, and where the order of the words and the positon in the sentence is not important (Tesnière, 2015). Comparatively link grammar developed undirected relations between pairs of words (Sleator and Temperley, 1995). Further, there is constituent-based grammar, whereby a group of words will form a constituent category, which may be a noun phrase or a verb phrase. Here a tree has an iterative segmentation into a range of constituent categories based on their grammatical classification. There are formal language theories and varied grammars such as phrase structure grammars, and probabilistic context-free grammars which are also used for CL projects, and popularly used, for example, in word processing packages. However, for our NLP focus, we will explore functional grammars in the next section.

2.2. NLP models – formal to functional grammars

Butler and Taverniers (2008) and Rezai (2003) in their discussions outline that over the last 40 years, several new approaches to "grammar theorising and description" have evolved under the different "functional(ist)" perspective of language as opposed to the "formalist" theories by Chomsky. The idea was that knowledge was acquired by the use of language, which Chomsky (1986) identifies relates to communication, syntax "rules of grammar", semantics "rules of meaning" and the communication of a spoken NL using syntax and semantics theoretically. Further to this, Chomsky discusses that there are problems with the behaviourist view, for example, the badger will howl in danger, but humans do not need an environmental stimulus or the "occurrence of internal states", in order for producing language. Key findings on human languages formulated by the Chomskyian approach (1986) is that the study of language is known as universal grammar (UT). Contrastingly, Rezai (2003) asserts that functionalists look at all aspects of the structural organization of language in the light of its role in human social interaction and that linguistic investigation is expansive. Hence, the guiding principle of functionalism is such that the form of a sentence is determined by its meaning with reference to pragmatic and social considerations. Butler and Taverniers (2008) notes the central claim of functionalism is that language should be seen first and foremost as a system for communication between human beings. They further add the forms that languages take are strongly conditioned by both the nature of human cognitive capacities, and the sociocultural contexts in which, and purposes for which, language is used.

A trait of most functionalism is that syntactic phenomena is being motivated by (and not simply correlating with) meaning, and to degrees which vary with the particular functional theory, and further to tackle everything that is systematic in language. Natural language (NL) is conceived of as a functional system (Francois, 2014). The functionalist continuum composes three groups which identify a reduction of grammatical structure to discourse from extreme (no relevance to formal aspect of language), to conservative noted by Yang (1994). Between these two extremes, there exists moderate functionalist known as structural-functional grammar (SFG) approaches – focus on both meaning and formal structure, and their inter-relationships, with generative qualities. They have an explicit framework of rules and principles, linked in specified ways, by means of which the structure and interpretation of linguistic expressions can, in principle, be exhaustively described" (Butler, 2009).

There are four periods of structural-functional grammar (SFG) in which the theories were mostly developed in the realm of European structuralism starting with: (1) period of 1925-1950; (2) period of 1950-1975; (3) period of 1975-2000; and (4) since 2000 (Francois 2014). The fourth era saw further work by (Halliday and Matthiessen 2004) on SFG, and the development of the Functional Discourse Grammar (FDG) in 2008, and the new formulation of Role and Reference Grammar (RRG) by Van Valin Jr (2005a) specifically on the syntax-semantic interface. A key feature of SFG is layering, that is the clause and other phrasal units can be analysed in terms of number of layers which are related to the functions which language serves in communication, and provide further interesting frameworks for syntax, semantics and pragmatics, compared to the traditional ways. Another important influence to layering was from the work by (Foley and Van Valin Jr., 1984) when working on RRG with its central tenet related to strong typological orientation, and the discovery of linguistic universals. This work further influences the present-day RRG, the layered model of the clause structure.

With this in mind, RRG specifically will be pursued in our experiment of linguistically-centred conversational software agent. The reasoning behind this is that RRG's major influence on layering and its ongoing developments such as detailed rules for mapping from semantic structures to syntactic structures (needed in modelling language production) and from syntactic structures to semantic structures (needed for comprehension), and its computational abilities and applications have provided some fruitful results (Salem, Hensman, and Nolan, 2008). This is discussed in the next section. Before focusing on an RRG implementation, it is important to understand the goals of linguistic theory.

2.3. Introduction to the goals of linguistic theory and NLP

Language 'is an immensely complex behaviour' requiring a mental lexicon consisting of knowledge of words in that language, together with knowledge of the grammatical use of those words in the sentence (Jones and Mewhort, 2007). This signifies that the mental lexicon has a rich history in psychology. This is deeply rooted with work of Van Valin and LaPolla (1997) state that there are a set of general goals that most linguists have agreed upon, which include: (1) description of the linguistic phenomena; (2) explanation of linguistic phenomena; (3) understanding of the cognitive basis of language.

Describing the linguistic phenomena demands: (1a) describing an individual language (1b) describing the commonality between languages (language universals); (1c) describing the differences between languages (language topology). The demands are crucial for the evolution of the language itself, and presupposes the other goals, in that, one cannot explain a language if you cannot describe it first, alternatively one cannot understand the cognitive basis of a language without its description.

Explaining the linguistics phenomena involves the use of criteria and standards to explain (2a) how speakers use language in different social situations; (2b) why human languages have the structure they do? (2c) what is common to all human languages; 2(d) use of inductive or deductive theories; (2e) theory-internal explanatory criteria related to phonology, semantics, pragmatics and processing; (2f) theory-external criteria related to reasoning, categorisation and perception.

Understanding the cognitive basis of language outlined by Van Valin and LaPolla (1997, 4) present three facets of the psychology of language in (1):

(1)

a) Processing: What cognitive processes are involved when human beings produce and understand language on line in real time, assist our understanding of the cognitive basis of language? How specialised to languages are these processes?
b) Knowledge: What constitutes knowledge of language? How is it organised? How is it represented? How is it employed in language processing? How does knowledge of language relate to knowledge in other cognitive domains?
c) Acquisition: How do human beings come to have knowledge of language? What is the nature of the acquisition process? Is coming to know language similar to or different from acquiring knowledge in other cognitive domain? Does it involve knowledge form other cognitive domains?

Our focus in this chapter is processing and knowledge of language. Chomsky (1965) in his book – *Aspects of the theory of syntax,* proposed levels of adequacy that grammar must meet, and align with the goals mentioned above. These include: (1) observational adequacy – determines and predicts which sentences (grammatically) are well-formed, and those which are not; (2) descriptive adequacy – presupposing it is observationally adequate in native speaking, and assigns structural description for sentences – both in terms of the structure and meaning; (3) explanatory adequacy – presupposes it is descriptively adequate and part of the theory that informs about the "how these facts arise in the mind of the hearer and speaker, theory-internal. Both observational and descriptive adequacy are empirical accuracies. The latter two types of adequacy are explicitly cognitive in nature, as they make reference to native-speaker intuition and to language acquisition. Van Valin and LaPolla (1997) discuss the communication – and cognition perspective. This describes the link between a human language's role as a means of communication and the broader cognitive processes such as reasoning and conceptualisations, and further cognitive systems such as perception and knowledge. The knowledge and processing of language is critical under the computational adequacy of the linguistic theory, and RRG, which from the above discussions presents credentials for the exploration of an NLP language model for a text-based conversational software agent (CSA).

3. Functional Linguistic Theory – Role and Reference Grammar (RRG)

NL is conceived of as a functional system (Francois, 2014), and RRG is a mature, functional, linguistic theory that involves the interaction of syntax, semantics and pragmatics across grammatical systems (Van Valin Jr., 2005a). This is essential for NLU and a computational model addressed on three points. Firstly, RRG's functional view encapsulates the syntax-semantic generalisations which are prominent to explain the semantic motivation of grammatical phenomena. Secondly, RRG has a system and linking algorithm which will allow for comprehension and production of linguistic expressions. Thirdly, RRG looks at language as a means of communication, and aims to be "typologically adequate" to address fairly the analysis of all languages, as languages have varying arbitraries such as the position of the verb, and hence RRG provides added value. The following outlines its basic concepts.

3.1. Basic theory and organisation of RRG

RRG uses a single syntactic description which is semantically motivated. RRG looks at language as a means of communication, and aims to be "typologically adequate" to address fairly the analysis of all languages, as languages have varying arbitraries such as the position of the verb. The organisation of RRG includes a parser and syntactic inventory that link to the syntactic representation, and a lexicon that links to the semantic representation. The linking system is a set of rules that relates the syntactic representation and semantic representation to each other and the discourse-pragmatics plays a role in the linking system, with an input from constructional schemas. A point to note here is that languages will differ from each due to the manner in which the discourse-pragmatics interacts with the linking between syntax and semantics (Van Valin Jr., 2005a: 1).

It further illustrates RRG is a language processing model with mappings of the semantics-to-syntax (production process) and from syntax–to-semantics (comprehension process), and further how the pragmatics interface supports the generation of output responses for example for a NLP example such as a text-based conversational agent (LING-CSA) using dialogue management. Figure 11.2 represents a view of is the RRG linking system (Van Valin Jr., 2009), where the syntax links to semantics. There are some universal and language-specific traits. The semantic macroroles, logical structures, and hierarchy linking them are universal and represent the domain of lexical processes.

This encompasses limited cross-linguistic variations; alternatively it is the linking to the syntactic functions of the macrorole and arguments that languages differ substantially.

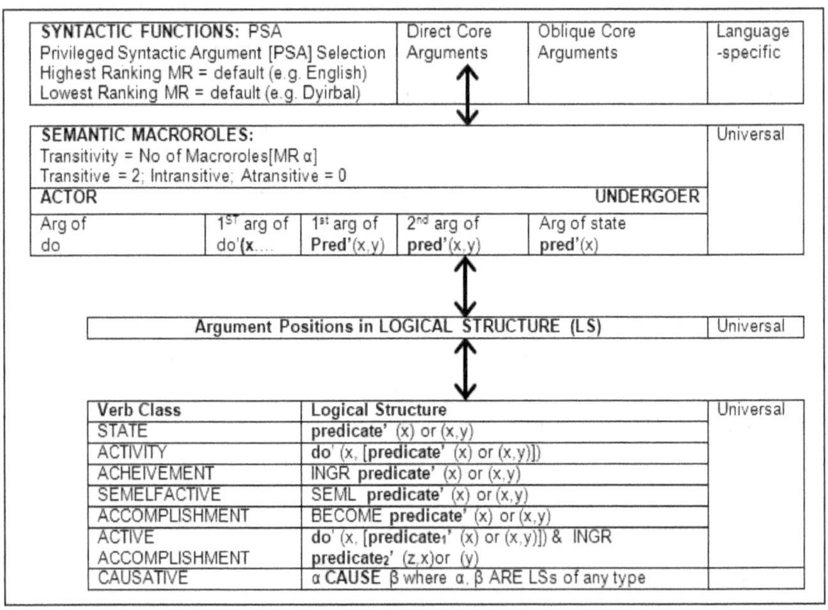

Figure 11.2: RRG Linking System (Based on Van Valin Jr., 2005b).

The arrows are double-headed representing a bi-directional linking system and facilitate the working of the syntactic-semantic pragmatic interface for simple and complex sentences. The logistics of the comprehension process is that the parser would take the input and produced a structured syntactic representation of the clause (Van Valin Jr., 2005a: 129) asserts that all grammatically relevant elements of LSc (layered structure of the clause), cases, adposition and other elements will be identified in this representation. Next, it is the task of the grammar to map the LSc and operator projection into the semantical representation of the clause. This interpretation of this mapping is achieved by the syntax-to-semantic linking algorithm. In the semantic-to-syntax linking, Gottschalk (2011) notes that an inheritance process within the lexicon maps the lexical elements in to the LS (logical structure), thus producing an output. Having produced the LSs, it is the task of the grammar to project the LSc and all other grammatically elements from the LS in question.

Van Valin Jr. (2009) states that RRG posits three main representations as identified in Figure 11.3 detailed as: (1) A representation of the syntactic structure of the sentence, which corresponds closely to the actual structural form of the utterance. (2) A semantic representation depicting important facets of the meaning of the linguistic expression. (3) A representation of the information (focus) structure of the utterance is focused on the communicative function in particular, its speech act (SA).

As in Figure 11.3: Gareth ate everything fast (BNC ADY 1079). (Butler and Taverniers 2008) a syntactic and semantic representation is defined in (2):

(2)

SYNTACTIC: SENTENCE (CLAUSE (<CORE> <NP> Gareth (<NUC> (<PRED> <V> ate)) (<NP> (everything))) (PERIPHERY fast)

SEMANTIC: [<IF> ASS <TNS> PT, do'(ACT: Gareth, (eat'(Gareth <NOM>, pizza <ACC>)])] & INGR consumed' (UND:pizza)]

Here we have an assertive speech act; 'ate' is the predicating element, and *'fast'* is the non-argument, as the semantic core meaning is intact with 'Gareth ate everything', but still pragmatically motivated by 'fast'.

Another interesting principle of the linking algorithm relates to the 'completeness constraint' as stated by Van Valin Jr. (2005a) 'all the specified arguments in the semantic representation must be realised in the syntax in some way, and conversely that all the expressions in the syntax must be linked to something in the semantic representations of a sentence, in order to be interpreted'. This principle is an important formalistic, computational test in the implementation phase, and as part of the proof of concept.

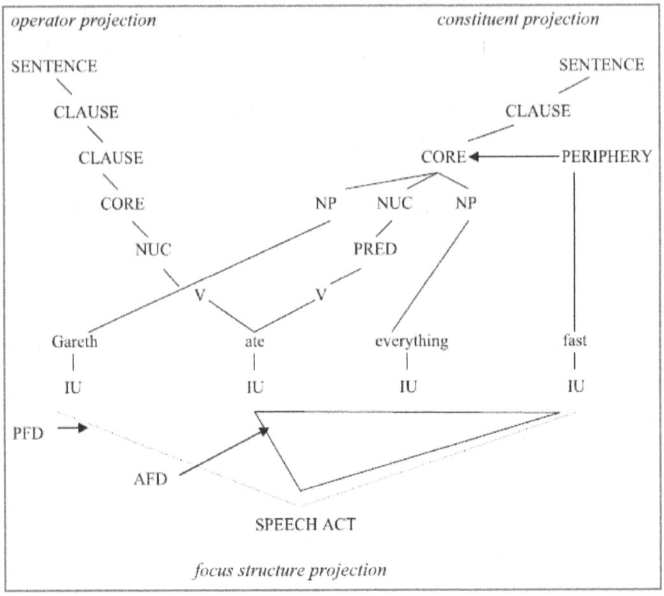

Figure 11.3: Layered Structure of the Clause (LSC) (Based on Van Valin Jr., 2005b).

Constructional schemes (CSs) play a crucial role in the linking system. For the semantics-to-syntax linking algorithm they supply the language-specific and construction-specific details necessary for the correct encoding of the meaning in the morphosyntax (Van Valin Jr., 2005a: 134).

Similarly, CSs are important for the syntax-to-semantic linking for languages which have different privileged syntactic arguments (PSA) for different constructions such as Jakaltek and Sama. CSs invoke the syntactic inventory for the appropriate syntactic template (combination of lexical entries), as illustrated in the original RRG

4. Experiment – Text-Basd Conversational Software Agent (CSA)

An experiment is conducted to explore NLP in AI using a functional linguistic perspective (that is RRG) – as a linguistic engine for a linguistically centred CSA (LING-CSA).

4.1. Introduction to Conversational software agents

A CSA is a program that engages in conversation using NL dialogue with a human user. The challenge here is for the system to encapsulate sufficient knowledge from the user's question to present a grammatically correct response. Here a discussion of knowledge and its role is very important. When a person hears or sees a sentence, an individual will make full use of their knowledge and intelligence to understand it; additionally, besides the grammar, knowledge about the words, the sentence context, and an understanding of the subject matter is necessary. Here the computational model must model this language understanding and the interactions that combine grammar, semantics and reasoning.

Two main requirements for a CSA specified by Mao, Sansonnet, and Li (2012) and Lester, Branting, and Mott (2004) are the ability of accurate NLU and the technical integration with an application. A CSA must respond appropriately to the user's utterance via three phases: (1) interpret the utterance, (2) determine the actions (logic) that should be taken in response to the utterance (3) and perform actions replying with text. In the last decades, there has been a great evolution in the field of CSA enfolding three emerging trends of more sophisticated NLP via improved parsing techniques, humanising of agents through language and their pervasiveness (Perez-Marin and Pascual-Nieto, 2011).

CSAs are becoming prominent in the marketplace where the consumer's use and familiarisation of the smartphone technology and consumer support agents, human-computer interaction, and ubiquitous computing is maturing, and is of great value. A long-standing issue within NLP CSA systems is refining the accuracy of the interpretation of meaning to provide a realistic dialogue to support the human-to-computer communication.

Historically, CSA systems link to the concept of intelligent machines originated by British mathematician and code breaker Alan Turing who addressed the question 'can machines think' (Turing, 1950). He worked on a proposal of how to test this using the imitation game (also known as the Turing Test) and used dialogue testing, as means to work out whether or not the computer program was intelligent. This question has inspired much exciting CSA research and commercial undertakings as the yearly Loebner competition to demonstrate human-like conversation ("loebner.net", 2015) as in the Chatbot 'Mitzuki' in 2013, 2016, 2017 and 2018, built on Pandorabots: the world's leading conversational (AI) Chatbot platform. As a thought to the foundations of AI, cognitive science, investigations into 'can machines think' and to challenge the concepts of strong AI, Searle (1980) introduces the Chinese room argument (experiment).

This experiment demonstrates that the computer program may very well look like it understands Chinese, but it only simulates that knowledge which is not a form of intelligence. Subsequently, there is a need to address the differences between a human brain and computer program and being able to think and understand. This "understanding" implies both the possession of mental (intentional) states and the truth (validity and success) of these states

(Searle, 1980). In other words, a dialogue system is a more refined "intelligent" version of a Chatbot, where the NL input must be constrained and is unexpected, thus requiring a robust complex framework, with an integrated knowledge base (KB) which may take the form of ontologies for recognition, interpretation and generation, and deploying computational linguistically technologies.

A main distinction between an applied Chatbot and CSA is that the CSA has a deep strategic role to hold a conversation and enable the mechanisms to focus the conversation on achieving a goal, via NL dialogue (O'Shea, Bandar, and Crockett, 2010). There is a need to plan, and to decide what do next, and manage the conversation – this is the work of the dialogue manager (DM) (Treumuth, 2011). The CSA's role is that of a linguistic aware knowledge aware process simulating an empowered human to take part in the conversation and ask questions.

4.2. Requirements of a conversational software agent (CSA)

As noted previously, the two main requirements for a CSA are the ability of accurate NLU and the technical integration with an application. To enable cognitive manipulations as part of the conversation, a CSA will have the architectural components of an interpreter, dialogue management (DM) and response generator (RG) with three models – internal state, external interaction and language feature Nolan (2014b). The internal model and state is reflected by the agent's BDI. The external model of the agent is reflected by the interacting participants (human and other agents) in its world. The language model is related to the interaction model to support dialogue via speech acts – the linguistic act of uttering a sentence (Searle, 1969). Further, the CSA must have robust functional capabilities for parsing and generating utterances, as well as dealing with tricky grammatical rules and structures, and handle conversational behaviour. Key facets noted by Nolan (2014b) in (3):

(3)

a) The set of beliefs that the agent has at a given time
b) The goals that agent will try to achieve;
c) The actions that the agent performs and
d) The knowledge of the effects of these actions;
e) The environment information the agent has (not necessarily complete or correct);
f) The ongoing discourse interaction that the agent participates in the environment over time;
g) Human language understanding and conversation tracking over a discourse.

4.3. Additional requirements of LING-CSA

The CSA must respond appropriately to the user's utterance via three phases – (1) interpret the utterance, (2) determine the actions (logic) that should be taken in response to the utterance, and (3) generation of a grammatical correct dialogue response in the target human language, English.

To achieve these phases speech act theory (SAT) comes into play based on work from Austin (1962) to Searle's work to address the issues of meaning and expression of NL. Searle (1969: 22) states "speaking a language is engaging in a rule governed form of behaviour" and that 'illocutions are intentional acts'. Further, SAT provides the basis for interpretation of the utterance in terms of a speaker's intention and the effect it has on a listener. These intentions are composed of speech acts (SAs) involving a range of functions (actions) denoted as illocutionary acts for example – John says to Mary '*pass me the glasses please*' – involves the illocutionary act (point of the speech act) of requesting or ordering – '*Mary to hand the glasses over to him*'. As highlighted previously, the constructional schemas (CSs) are enhanced into speech act constructions (SAC) based on grammatical objects, providing a richer and better-motivated understanding of constructions in a lexically orientated functional model of grammar (Nolan, 2014a).

The next question is how this goal is achieved and able to conduct conversational behaviour. The Belief-Desires-Intentions (BDI) model was based on a model of human behaviour from the theory of human practical reasoning developed by the philosopher (Bratman, 1987). Put simply, the BDI model reflects a change of 'mental' states. Beliefs refer to the facts representing what an agent believes about the world. Desires are all the possible state of affairs (goals) that the agent potentially may accomplish. Intentions are the agent's commitment to its desire (goals) and its commitment to the plans selected to achieve these goals. Therefore, practical reasoning consists of two key processes: deciding what goal we wish to achieve known as deliberation, and secondly, how will we achieve our goals is called means-end reasoning. Searle (1985, 4) identifies that SAs differ due to differences in psychological states. For instance, as in the BDI model, a SA reflects beliefs, desires, wants and intentions. For instance, a speaker states, claims, explains, asserts or claims P is a representation that the speaker expresses the belief of P; a speaker who promises, vows, threatens or pledges to do A, is a representation that the speaker expresses an intention to do A; a speaker who orders, commands, requests the hearers to do A is a representation of the desire (want, wish) that hearer do Action A.

From a knowledge perspective, Ramirez and Valdes (2012) note that knowledge has several facets, with some static components called concepts or facts, and some dynamic components such as skills, abilities, procedures, actions which collectively enable general cognition.

This cognition is associated with several different processes, for instance, perceiving, distinguishing, abstracting, modelling, storing, recalling, and remembering which constitute three primary cognitive processes: learning, understanding and reasoning (Ramirez and Cooley, 1997). Knowledge representation and reasoning (KRR) is a central issue in AI. Chein and Mugnier (2008) affirm that a knowledge base (KB) collates symbolic knowledge representation about an application domain, containing different kinds of knowledge such as ontology (symbolic representation of objects and relations), facts, rules and constraints, and a reasoning engine.

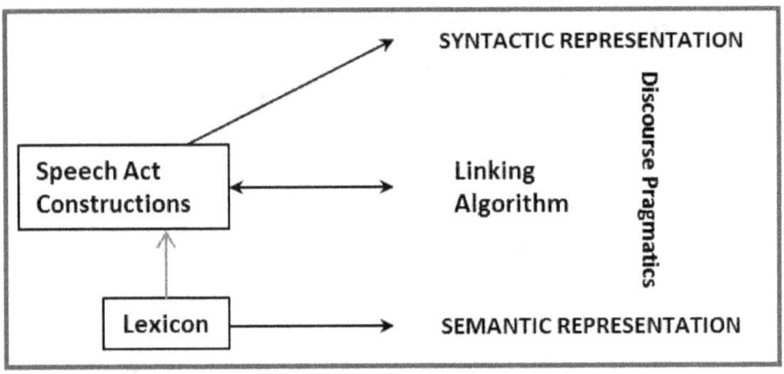

Figure 11.4: Re-organisation of RRG for Ling-CSA.

To recap, RRG is deployed as a linguistic theory that can adequately explain, describe and embed the communication-cognitive thinking in conversation, in a computational form. With this in mind, we propose a language model, knowledge model, speech act constructions (SAC) and belief, desires, and intentions (BDI) framework for LING-CSA. It will further investigate how language can be comprehended and produced, to gain a deep understanding, and how it interfaces with knowledge. In order to use the RRG linking system and to create an effective parser, RRG is re-organised to facilitate the innovative use of SAC as grammatical objects (Nolan 2014a) and a rich lexicon, as illustrated in Figure 11.4.

5. Methodology

The architecture of LING-CSA (diagrammed in Figure 11.5) constitutes three-phase models. These are: (1) a linguistic model based on RRG; (2) Agent Cognitive Model (ACM) with two inner models: (a) knowledge representation model employing conceptual graphs (CGs) serialised to Resource Description Framework (RDF) (Chein, Mugnier, and Croitoru, 2013); (b) a planning model underpinned by BDI concepts (Wooldridge, 2013) and intentionality (Searle, 1983) and rational interaction (Cohen and Levesque 1990); and (3) Phase 3 – Agent Dialogue Model (ADM) where the DM is a common component of Phases 1 and Phase 3 – due to the discourse referents of an utterance, and the need to create a grammatically correct response. It employs a common ground (Stalnaker, 2002).

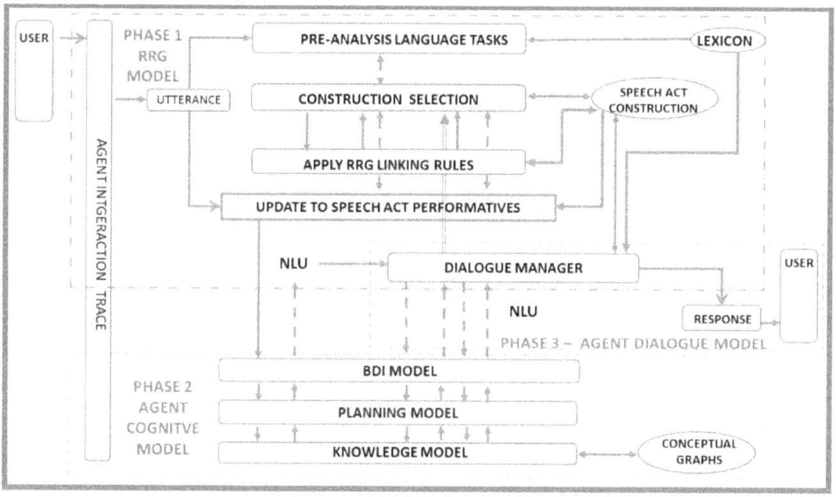

Figure 11.5: Conceptual Architecture of the Ling-CSA (Panesar, 2017).

The CSA's goals are to achieve: (1) present a mapping of the syntactic representation to a semantic representation; (2) an adequate explanation and representation of the NLU of the utterance; (3) present a range of SAC manipulations based on the utterance; (4) evidence the dialogue processes and manipulation interfacing with the language model; (5) present the agent's knowledge and its BDI states; (6) present the user's BDI states; (7) query the knowledge base for a fact based on the utterance the agent has perceived; (8) devise an appropriate plan based on the BDI states to support the response generator; (9) generate a grammatically correct response in RRG based on the agent's knowledge.

The CSA utilises a set of data sets, as highlighted in Table 11.2 – RRG steps. Firstly, the RRG lexicon, semantically motivated, with lexical entries that have features: (a) part-of-speech (POS); (b) tense/aspect, (c) definiteness, (d) person, (e) number, (f) gender, (g) case, (h) humanness, (i) animacy and (j) logical structure (LS), as identified in Table 11.1. Secondly, a lexical words list. Thirdly, SAC – a grammatical object based on SA types (assertive, and interrogative (Searle, 1969)) forms an updateable object, selected initially for processing the utterance and later for the agent response via the dialogue management processes and illustrated in Table 11.3. SAC template elements includes: (a) SA type, (b) signature list, (c) input, (d) workspace, (e) syntax, (f) semantic, (g) morphology, (h) pragmatic, (i) illocutionary force, (j) focus structure and (k) output LS (logical structure). Element manipulations are performed to create a completed SAC, which is extended by performative information to a form message perceived by the agent environment. Fifth data set is the agent's belief (knowledge) represented initially as a CG vocabulary as facts, due to their effective representation of natural language (Sowa 2009) in COGUI (COnceptual Graphs User Interface) (GraphIK, 2012). This is further serialised into RDF/XML (Graphlk, 2017) as there is no loss of meaning, forming a set of RDF triples and an opportunity of Semantic Web manipulations. Finally, the CSA's input are utterances constrained to a single clause.

5.1. RRG Language Model (Phase 1)

The LING-CSA will constitute a closed domain – dialogue on food and cooking, for its richness and universality. A range of NLP tasks manipulated via the language model will include tokenisation, sentence splitting, part-of-speech-tagging, morphological analysis, syntactic and semantic parsing.

It is the dialogue manager (DM) that will assist with the SA dialogue working with the syntactic parser to work effectively with the data sources (word list, lexicon, empty speech acts constructions (SACs)). RRG manipulations exist for both simple and complex sentences, but the LING-CSA is limited to the manipulation of the linking system of simple sentences (active or passive) using transitive, intransitive and ditransitive verbs/auxiliary verbs with variable word order flexibility with SVO (subject-verb-object), and modelled via speech acts (Searle, 1969) applied to cognitive and rational interaction analysis. RRG's bi-directional linking algorithm and discourse-pragmatics interface will be mapped into completed SACs, invoking the lexicon (Table 11.1) to supports all three LING-CSA phases, and internal BDI manipulations (Panesar, 2017) as shown in Table 11.2.

Lexical entry 1: ate								
POS TYPE	**VERB TENSE - ASPECT**	**DEF**	**PERSON TYPE**	**NO**	**GENDER**	**CASE**	**ANIM**	**HUM**
Verb	PST	DEF +/-	3	SG	M/F	DNA	ANIM	HUM
LOGICAL STRUCTURE (LS): <tns:pst <do'(x, [eat'(x, y)]) & BECOME consumed' (y) >>								

Lexical entry 2: eat								
POS TYPE	**VERB TENSE - ASPECT**	**DEF**	**PERSON TYPE**	**NO**	**GENDER**	**CASE**	**ANIM**	**HUM**
VERB	PRS/ FUT	DEF+/-	3	SG	M/F	DNA	ANIM	HUM
LOGICAL STRUCTURE (LS): <tns:prs <do'(x, [eat'(x,y)]) & BECOME consumed'(y) >> <tns:fut <do'(x, [eat'(x,y)]) & BECOME consumed'(y)>>								

Lexical entry 3: eating								
POS TYPE	**VERB TENSE - ASPECT**	**DEF**	**PERSON TYPE**	**NO**	**GENDER**	**CASE**	**ANIM**	**HUM**
VERBAL NOUN	PROG	DEF+/-	3	SG	M/F	DNA	ANIM	HUM
LOGICAL STRUCTURE (LS): <tns:prs <asp:prog <do'(x, [eat'(x, y)]) & BECOME consumed' (y) >>								

Lexical entry 4: is								
POS TYPE	**VERB TENSE - ASPECT**	**DEF**	**PERSON TYPE**	**NO**	**GENDER**	**CASE**	**ANIM**	**HUM**
VERB BE	DNA	DEF+	DNA	DNA	DNA	DNA	DNA	DNA
LOGICAL STRUCTURE (LS): be'(x,[pred'])								

Lexical entry 5: hungry								
POS TYPE	**VERB TENSE - ASPECT**	**DEF**	**PERSON TYPE**	**NO**	**GENDER**	**CASE**	**ANIM**	**HUM**
ADJECTIVE	DNA	DNA	DNA	DNA	M/F	DNA	ANIM	HUM
LOGICAL STRUCTURE (LS): DNA								

Lexical entry 6: restaurant								
POS TYPE	**VERB TENSE - ASPECT**	**DEF**	**PERSON TYPE**	**NO**	**GENDER**	**CASE**	**ANIM**	**HUM**
NOUN	DNA	DEF+/-	DNA	SG/PL	DNA	DNA	DNA	DNA
LOGICAL STRUCTURE (LS): DNA								

Table 11.1: Snapshot of the Lexicon – Lexical entries and Lexemes (Panesar, 2017).

Note: Logical structures have been partially expanded for readability.

PHASE 1 - RRG LANGUAGE MODEL			
Step	PHASE		(TRACKING) DISPLAY
1	PHASE 1A - PRE-ANALYSIS	Parsing, Sentence and word tokenisation, check against the lexicon.	Display the utterance and various internal representations.
2	PHASE 1B – SAC SELECTION	Construction selection computations	Display construction selected
3	PHASE 1C – SYNTAX-TO-SEMANTICS	• Verb selection and voice • Retrieve the LS and verb from the lexicon • Assignment of arguments to the LS arguments • Semantic representation • Update the LS • Map the operator projection and focus projections • Actor and undergoer assignment • Select the morphosyntactic properties of the argument	Display the verb selection and voice topics Display the LS and verb from the lexicon Display the argument value Display the semantic representation Map the operators to the Tense and illocutionary force Display the actor and undergoer as part of the LS. Display the additional information
4	PHASE 1E - SEMANTICS-TO-SYNTAX	• Verb selection and voice • Retrieve the LS and verb from the lexicon • Assignment of arguments to the LS arguments	Display the verb selection and voice topics Display the LS and verb from the lexicon Display the argument values
5	PHASE 1F – DIALOGUE MANAGER	• Ascertain whether there is any missing information • Refer to the stored discourse referents – if any, and process with the earlier Phase 1 steps	Display the specific elements that missing Display appropriately any saved discourse referents
6	PHASE 1G – CREATE SAP	• Attach the performative information to the Speech Act Construction	Display the performative type information

Table 11.2: Overview of the Phase 1 – RRG Model Steps (Panesar, 2017).

The analysis of utterance/responses will be constrained to assertive and interrogative (WH-word) SAs. SAC example shown in Table 11.3 include: – 'assertive: ATE', taken from a sample collection of 34 – created originally by the researcher. The selected SAC is updated with all the grammatical information. These include: voice opposition; macrorole; pragmatic; semantic; lexical; the PSA (privileged syntactic argument) – similar to the 'subject' of S-V-O; matching signature; syntactic morphology (rules); focus; semantic information such as tense and aspect.

Here the updated SAC will use the generalised lexicon and computationally work with the surface syntax to the underlying semantic forms. Any discourse referents that are generated are also checked and updated accordingly. The updated SAC will store each text and the associated complete logical structure (LS) based on the working semantic predicating element. The linking system will facilitate the syntactic parse to enable fixed word order for English (SVO), and to unpack the agreement features between elements of the sentence into a semantic representation (the logical structure) and a representation of the Layered Structure of the Clause (LSC). The linking system will facilitate procedures for semantic-to-syntax and syntax-to-semantics, parsing and in the process of formulating a grammatically correct response.

```
PERFORMATIVE: <ASSERTIVE:ATE>
:SENDER <USER>
:RECEIVER <AGENT-1>
:ONTOLOGY <FoodAndCookKB>
:CONTENT <do'(Gareth, (eat'(Gareth, pizza)])] & INGR consumed' (pizza)] everything>
SIGNATURE: [PN V NP ADJ]
CONSTRAINT: Default
INPUT: Gareth ate everything fast
WORKSPACE: (Gareth, PN), (ate, VERB), (everything N), (fast, ADJ)
SEMANTICS: Contains a noun phase before and after the verb
CONSTRUCTION BODY
SYNTAX: SENTENCE ( CLAUSE ( <CORE> <NP> gareth ( <NUC> ( <PRED> <V> ate ) )
( <NP> (everything ) ) ) (PERIPHERY fast)
PSA: gareth
SEMANTICS
    Linking:
MORPHOLOGY:Default
PRAGMATICS
Illocutionary force: ASSERTIVE
Focus structure: narrow focus on the element
OUTPUT [LS]: [<IF> ASS <TNS> PST, do'(ACT:Gareth, (eat'(Gareth <NOM>, pizza <ACC>)])] & INGR
consumed' (UND:pizza)]
```

Table 11. 3: Speech act construction performative "ate" used as a message to the agent environment (Panesar, 2017).

5.2. Design framework – Agent Cognitive Model (Phase 2)

Cohen and Levesque (1988) take the viewpoint of language as action, and view utterances as events (updated to speech act performatives (SAP) messages) that change the state of the world, and hence speakers and hearer's mental state change as a result of these utterances. Further, it is necessary for a computer program to recognise the illocutionary act (IA) of a SA, for both the speaker (USER) utterance and hearer (AGENT) and a response. The agent environment will constitute a language model, mental model to work the BDI states, planning model (to reason with states based on current knowledge), knowledge model (world knowledge – made up of shared and individual beliefs) and dialogue model (to provide a response back to the user).

The next question is how will this integrate and function; this is achieved in Phase 2 – Agent Cognitive Model (ACM) with an input of the Phase 1 RRG Language Model in the form of a message as a speech act performative (SAP). The ACM contains two inner models, namely the Agent Knowledge Model (AKM) and Agent Dialogue Model (ADM). The ACM Phase 2 model has a series of pre-agent steps and main agent steps illustrated in Figure 11.6 (Panesar 2017). The behaviour of agent is identified by the following basic loop where agent iterates the following two steps at regular intervals: (1) Read the current message, and update the mental state including the beliefs and commitments; (2) Execute the commitments for the current time, possibly resulting in a further belief change.

The user instigates the utterance into the CSA framework, and it is the agent which decides on the appropriate dialogue and grammatical response. The sentence pair and sub dialogue internal representations will be stored – based on theories and models such as the discourse representation theory (DRT) (Kamp, Van Genabith, and Reyle 2011). To facilitate conversation, the DM is invoked, and discourse referents in the previous utterance of the sub-dialogue are resolved. This will serve two purposes: 1) to establish the NLU of the utterance; (2) and to forward to the dialogue model to ascertain a response. This response generator here will further make use of the language model, and SAC model, to formulate a grammatically correct response.

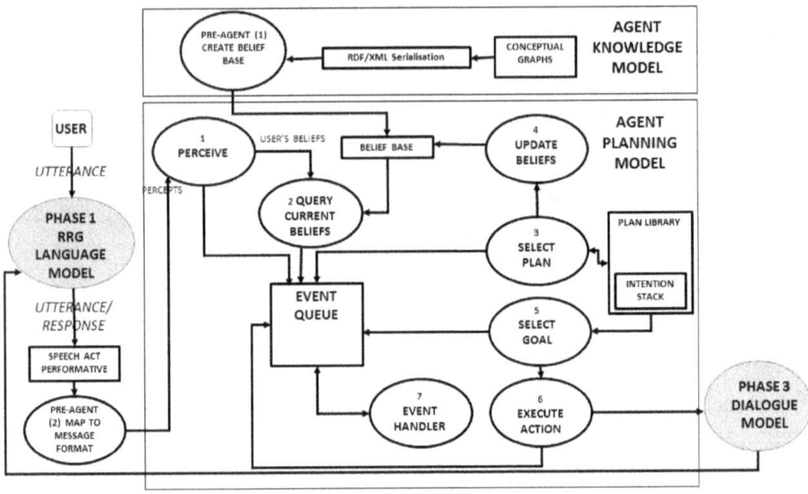

Figure 11.6: Design framework: Ling-CSA agent cognitive model.

6. Implementation Details

Figures 11.5 and 11.6, and Tables 11.1 to 11.3 are mapped and configured to an operational framework, and implemented as a Java prototype, developed in Eclipse IDE platform, predominantly as POJO (plain old java objects) with some API support, based on a food and cooking domain.

7. Evaluation Conclusions and Recommendations

To iterate, our experiment was to present empirical evidence (proof-of-concept) of the use of functional linguistic theory via a linguistically centred CSA (LING-CSA). Its goal was to investigate the accuracy of the interpretation of meaning to provide a realistic dialogue to support the human-to-computer communication. LING-CSA was multi-disciplinary invoking NLP, computational linguistics, knowledge representation, functional linguistics, agent thinking and AI, as demonstrated in the conceptual and operational framework of LING-CSA. A set of evaluation criteria were grouped per phase model, and the testing framework aims to assess the interface, intersection and integration of all phase models and their inner models. The validation and verification (VV) approach deployed is based on the phase models and thus is a multi-approach. It is driven by: (1) grammatical testing (English language utterances) and NLP tasks; (2) software engineering (UML modelling, architecture centric, data structures and algorithms); (3) knowledge representation logics (first-order logics and graph theory) and (d) agent practice (message passing, planning and cognitive responses and conforming to its appropriate inputs, environment, actions, performance measures (IEAP) (Russell and Norvig, 2016); (e) A range of RRG tests applied at various internal representations to address the goals of RRG – explanatory, descriptive, cognitive and computational adequacies (Panesar, 2017).

In Phase 1 – RRG model, the analysis of a SAC for each specific construal (either an utterance or response) – has two steps (Panesar, 2017). The RRG phase model evaluations identify that RRG is fit-for-purpose as the linguistic engine for LING-CSA as illustrated in Figure 11.7, demonstrating a syntactic representation that maps to semantic representation, and together are embedded in a speech act performative (SAP) message to the agent environment.

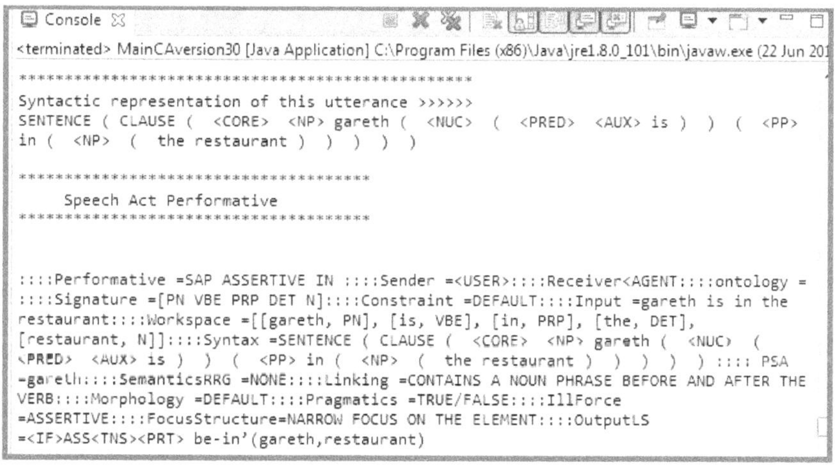

Figure 11.7: Syntactic & semantic representation, speech act performative (SAP) – message to the agent environment (Panesar, 2017).

A snapshot of key recommendations are included: (1) consider the requirements for complex sentences (more than one clause) with the extended use of procedures for the linking system; (2) support for multi-lingual (additional lexicons) such as the Spanish, (3) consideration of other SA categories such as commissives and emotives, such as, for the automatic analysis of social media tweets; (4) to include superlative adjectives and adverbs RRG lexicon to enforce greater coverage.

For Phase 2 the ACM – we can recall BDI architecture is based on the 'intentional stance' which is the highest level of abstraction of human reasoning where the agent will do what it believes are its goals. There are two main areas for future research, based on this evaluation. Firstly, it is to reduce the gap between knowledge and language, in the agent space. Here, language is in the context of a rich RRG linguistic representation against the belief base as RDF Triples posed the key obstacle of interoperability between two different low-level mappings, and hence a semantic gap. The solution is a 'lexical bridge' (LB), as illustrated in Figure 11.8.

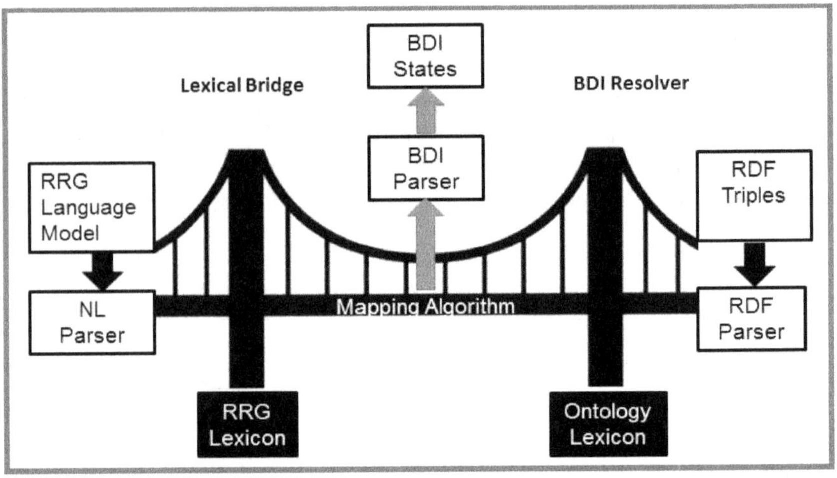

Figure 11.8: Lexical Bridge for the CSA'S belief base + BDI Parser to resolve the agent's BDI States (Panesar, 2017).

Central to this LB is the re-use of the RRG lexicon and NL parser. The derivation of a mapping algorithm from OWL to an NLP representation is recommended, specifically invoking the RRG lexicon and RDF parser. This mapping algorithm will be interfaced with a further model known as the BDI Resolution composing a BDI parser motivated by description logics (DL) to derive the BDI states of the agent.

This step will form an extra step to the pre-agent steps of ACM. The algorithm's output representation, will further invoke the rest of the ACM steps of querying the knowledge base for a specific belief, creating planning schemes, selecting a plan, manipulation of the appropriate intention, to create a grammatically correct response using the RRG language model and the extended SACs.

To summarise, our experiment of the use of a functional linguistic theory namely RRG for a linguistically centred conversational agent (LING-CSA) presented a proof-of-concept that is adequate for NLP and NLU, and presents an accurate representation of meaning. Looking forward, companies with a strong base in core digital technologies and big data analytics and see profitability are more likely to adopt an AI tools such as natural language generation and NLP. However, laggard companies are subject to the risk of digital competition and digital hacking (van Zeebroeck, 2018). Further, as more companies are adopting, there will be an increased commercial interest in NLP in AI, and greater human-to-computer communications, and more use of linguistic-based manipulations.

References

"loebner.net" (2015) "Home Page of The Loebner Prize in Artificial Intelligence - "The First Turing Test"." Accessed 12th March 2017. http://loebner.net/Prizef/loebner-prize.html.

ALICE. n.d. "AIML: Artificial Intelligence Markup Language." accessed 14 February 2017. http://www.alicebot.org/aiml.html.

Austin, John L. (1962) *How to do things with words*, Oxford: Clarendon Press.

Bird, Steven, Klein, Ewan and Loper, Edward (2009) *Natural language processing with Python*, O'Reilly Media, Inc.

Bratman, Michael (1987) *Intention, plans, and practical reason.* Center for the Study of Language and Information: Stanford University.

Butler, Christopher S. and Taverniers, Miriam (2008) "Layering in structural-functional grammars", *Linguistics* 46 (4): 689-756.

Butler, Christopher S. (2009) "Criteria of adequacy in functional linguistics", *Folia Linguistica*, 43 (1):1-66.

Chaplot, Devendra, Chhipa, Priyank and Kumar, Pratik (2013) "Watson System." [Slideshare], accessed 18 March 2017. http://www.slideshare.net/pratikkumarshanu/watson-1-28529561.

Chein, Michel and Mugnier, Marie-Laure (2008) *Graph-based knowledge representation: computational foundations of conceptual graphs*, London: Springer Science & Business Media.

Chein, Michel, Mugnier, Marie-Laure and Croitoru, Madalina (2013) "Visual reasoning with graph-based mechanisms: the good, the better and the best", *The knowledge engineering review*, 28 (03): 249-271.

Chomsky, Noam (1965) *Aspects of the Theory of Syntax*, Cambridge, MA: Massachusetts Institute of Technology.

Chomsky, Noam (1986) *Knowledge of language: Its nature, origins, and use*, Westport, Connecticut: Greenwood Publishing Group.

Clark, Alexander, Fox, Chris and Lappin, Shalom (2013) *The handbook of computational linguistics and natural language processing*, New Jersey: John Wiley & Sons.

Cohen, Philip R. and Levesque, Hector J. (1988) R*ational interaction as the basis for communication*, DTIC Document.

Cohen, Philip R. and Levesque, Hector J. (1990) "Intention is choice with commitment", *Artificial Intelligence*, 42 (2): 213-261.

Duta, Nicolae (2014) "Natural language understanding and prediction: from formal grammars to large scale machine learning", *Fundamenta Informaticae,* 131 (3-4): 425-440.

Ferrucci, David et al. (2010) "Building Watson: An overview of the DeepQA project", *AI magazine,* 31 (3): 59-79.

Foley, William A. and Van Valin Jr., Robert (1984) "Functional syntax and universal grammar", *Cambridge Studies in Linguistics London,* (38).

Francois, Jacques (2014) "Functionalist views on the genesis and development of grammar" in *25th European Systemic Functional Linguistics Conference Paris,* July 11th 2014.

Gottschalk, Judith (2011) "The precore slot in Icelandic: A topological analysis of V2-clause structure within Role and Reference Grammar", Accessed 12th September 2018, https://arrow.dit.ie/itbj/vol12/iss1/3.

Halliday, Michael, Kirkwood, Alexander and Matthiessen, Christian (2004) *Halliday's introduction to functional grammar,* New York: Routledge.

Hopgood, Adrian (2014) "AI-2014" in *Thirty-fourth SGAI International Conference on Artificial Intelligence.*

Jones, Michael N. and Mewhort, Douglas (2007) "Representing word meaning and order information in a composite holographic lexicon", *Psychological Review,* 114 (1): 1-37.

Kamp, Hans, Van Genabith, Josef and Reyle, Uwe (2011) "Discourse representation theory" In *Handbook of philosophical logic,* Springer, pp. 125-394.

Lester, James, Branting, Karl and Mott, Bradford (2004) "Conversational agents" In *The Practical Handbook of Internet Computing,* Florida: CRC Press.

Liddy, Elizabeth D. (2001) "Natural language processing" In *Encyclopedia of Library and Information Science,* New York: Marcel Decker, Inc.

Mao, Xuetao, Sansonnet, Jean-Paul, and Li, Lin (2012) "Textual Conversation Agent for Enhancing Attraction in E-Learning" in *Proceedings of International Conference on Computer Science and Information Technology.*

Markov, Andrey (1913) "An example of statistical investigation in the text of "Eugene Onegin" illustrated coupling of "tests" in chains" in *Proc. Academy of Sciences of St Petersburg.*

McCord, Michael C., Murdock, William and Boguraev, Branimir (2012) "Deep parsing in Watson", *IBM Journal of Research and Development,* 56 (3.4): 3: 1-3: 15.

Mitkov, Ruslan (2003) *The Oxford handbook of computational linguistics,* Oxford: Oxford University Press.

Nolan, Brian (2014a) "Constructions as grammatical objects: A case study of prepositional ditransitive construction in Modern Irish" In *Linking Constructions into Functional Linguistics: The role of constructions in grammar* (Brian Nolan and Elke Diedrichsen, eds.), Amsterdam/Philadelphia: John Benjamins Publishing Company, pp. 143-178.

Nolan, Brian (2014b) "Extending a lexicalist functional grammar through speech acts, constructions and conversational software agents" In *Language Processing and Grammars: The role of functionally oriented computational models [Studies in Language Companion Series 150]* (Brian Nolan and Carlos Periñán), Amsterdam and New York: John Benjamins Publishing Company.

O'Shea, Karen, Bandar, Zuhair and Crockett, Keeley (2010) "A conversational agent framework using semantic analysis", *International Journal of Intelligent Computing* Research, 1 (1/2).

Ordoñez-Salinas, Sonia and Gelbukh, Alexander (2010) "Information retrieval with a simplified conceptual graph-like representation" in *Mexican International Conference on Artificial Intelligence.*

Ovchinnikova, Ekaterina (2012) "Natural Language Understanding and World Knowledge" In *Integration of world knowledge for natural language understanding,* 15-38, Springer.

Panesar, Kulvinder (2017) "A linguistically centred text-based conversational software agent" in *Unpublished PhD thesis,* School of Computing, Creative Technologies and Engineering, Leeds Beckett University.

Perez-Marin, Diana and Pascual-Nieto, Ismael (2011) *Conversational Agents and Natural Language Interaction: Techniques and Effective Practices,* Pennsylvania: Information Science Reference.

Periñán-Pascual, Carlos (2013) "Towards a model of constructional meaning for natural language understanding" In *Linking Constructions into Functional Linguistics: The role of constructions in grammar,* 145: 205.

Pradhan, Sameer, Ward, Wayne; Hacioglu, Kadri, Martin, James and Jurafsky, Daniel (2004) "Shallow Semantic Parsing using Support Vector Machines" in *HLT-NAACL.*

Ramirez, Carlos and Cooley, Roger (1997) "A theory of the acquisition of episodic memory" in *ECML'97: Case-Based Reasoning Workshop: Beyond Classification of Feature Vectors.*

Ramirez, Carlos and Valdes, Benjamin (2012) *A general knowledge representation model of concepts,* INTECH Open Access Publisher.

Raskin, Victor and Nirenburg, Sergei (1987) "The analysis lexicon and the lexicon management system", *Machine Translation,* 2 (2-3): 177-188.

Rezai, Vâli (2003) "A role and reference gramm ar analysis of simple sentences in Farsi (Modern Persian)" in *PhD. diss.,* Isfahan: Isfahan University, http://www.wings.Buffalo.edu.rrg.

Russell, Stuart and Norvig, Peter (2016) *Artifical Intelligence: A Modern Approach - Global Edition,* Upper Saddle River, NJ: Prentice Hall.

Salem, Yasser, Hensman, Arnold and Nolan, Brian (2008) "Implementing Arabic-to-English machine translation using the Role and Reference Grammar linguistic model" In the *8th Annual Information Technology and Telecommunication Conference* (IT&T 2008), Galway, Ireland.

Searle, John R. (1969) *Speech acts: An essay in the philosophy of language,* Cambridge: Cambridge University Press.

Searle, John R. (1980) "Minds, brains, and programs", *Behavioral and brain sciences,* 3 (3): 417-424.

Searle, John R. (1983) *Intentionality: An essay in the philosophy of mind,* Cambridge Cambridge University Press.

Searle, John R. (1985) *Expression and meaning: Studies in the theory of speech acts,* Cambridge: Cambridge University Press.

Shannon, Claude E. and Weaver. Warren (1949) *The Mathematical Theory of Communication,* Illionois: University of Ilionois Press.

Sleator, Daniel and Temperley, Davy (1995) "Parsing English with a link grammar", in *ArXiv preprint cmp-lg/9508004.*

Sowa, John F. (2009) "Conceptual graphs for representing conceptual structures" In *Conceptual Structures in Practice,* 101-136.

Stalnaker, Robert (2002) "Common ground", *Linguistics and philosophy,* 25 (5):701-721.

Tesnière, Lucien (2015) *Elements of structural syntax,* Amsterdam: John Benjamins Publishing Company.

Treumuth, Margus (2011) "A Framework for Asynchronous Dialogue Systems: Concepts, Issues and Design Aspects" in *PhD thesis,* Tartu: Institute of Computer Science, Faculty of Computer Science, University of Tartu.

Turing, Alan M. (1950) "Computing machinery and intelligence", *Mind,* LIX (236): 433-460.

Van Valin Jr., Robert (2005a) *Exploring the syntax-semantics interface,* Cambridge: Cambridge University Press.

Van Valin Jr., Robert (2005b) "A summary of Role and reference Grammar" In *Role and Reference Grammar Web Page, University of Buffalo.*

Van Valin Jr., Robert (2009) "An Overview of Role and Reference Grammar", Accessed 12th March 2017, http://wings.buffalo.edu/linguistics/research/rrg/rrg_paper.html.

Van Valin, Robert D. and LaPolla. Randy, J.(1997) *Syntax: Structure, meaning, and function,* Cambridge: Cambridge University Press.

Van Zeebroeck, Nicolas (2018) *Artificial intelligence: Why a digital base is critical,* ULB – Universite Libre de Bruxelles.

Weizenbaum, Joseph (1966) "ELIZA—a computer program for the study of natural language communication between man and machine", *Communications of the ACM,* 9 (1): 36-45.

Winograd, Terry (1972) "Understanding natural language", *Cognitive psychology,* 3 (1): 1-191.

Wooldridge, Michael (2013) "Intelligent Agents" In *Multiagent Systems* (Gerhard Weiss, ed.), USA: Massachusetts Institute of Technology, pp. 3-50.

Yang, Byong-Seon (1994) "Morphosyntactic phenomena of Korean in Role and Reference Grammar: psych-verb constructions, inflectional verb morphemes, complex sentences, and relative clauses", New York: State University of New York at Buffalo.

SECTION IV:
THE ETHICS OF THE BIONIC BRAIN

CHAPTER 12

THE BIONIC BRAIN: PRAGMATIC NEUROETHICS AND THE MORAL PLAUSIBILITY OF COGNITIVE ENHANCEMENT

Peter A. DePergola II[1]

University of Massachusetts Medical School, USA;

College of Our Lady of the Elms, USA

Abstract: The seemingly infinite possibilities of contemporary neuroscience span from the augmentation of memory, executive function, appetite, libido, sleep, and mood to the maturation and development of emotional health and personality. These prospects hint at the capacity to alter neurocognitive conceptions of reality. They also mark the unavoidable inculcation of nuanced individual responses, perhaps radical, to these "tailor-made" perceptions. Hence, there exists certain neuroethical, and even more generally existential, risks within this fascinating and expeditious enterprise. The primary question in the context of present-day neurotechnology is not what can be done, but what should be. To that end, this chapter examines the concepts of memory, executive function, and emotional health and personality in the context of neurocognitive enhancement and posits the argument that neurocognitive enhancement can be justified as morally plausible in its potential to edify the caliber of overall cognition and thus contribute to the ability make pragmatically robust moral decisions on the conditions that it (i) promotes general moral character, (ii) complements human nature, and (iii) effects a deeper sense of individual and social identity.

1. Introduction

The seemingly infinite possibilities of contemporary neuroscience span from the augmentation of memory, executive function, appetite, libido, sleep, and

---

[1] Department of Medicine, University of Massachusetts Medical School and Department of Bioethics and Medical Humanities, College of Our Lady of the Elms.

mood (cf. Farah et al., 2010: 30-41) to the maturation and development of emotional health and personality (cf. Canli & Amin, 2010: 147-54). These prospects hint at the capacity to alter neurocognitive conceptions of reality. They also mark the unavoidable inculcation of nuanced individual responses, perhaps radical, to these "tailor-made" perceptions. Hence, there exists certain neuroethical,[2] and even more generally existential, risks within this fascinating and expeditious enterprise. The primary question in the context of present-day neurotechnology is not what can be done, but what should be.

To date, the debate over enhancement has been riddled with fallacious and ambiguous rhetoric, and the effect has spawned a bitter stalemate between the primary interlocutors. One cause of the current impasse is the mistaken idea that to enhance something – or, in the neuroscientific context, to allow greater access into dormant but existent biological faculties – *de facto* denotes the pursuit of its mastery or perfection. This is the concern, for example, of the 2003[3] President's Council on Bioethics[4] explored in their document "Beyond Therapy." For this group, neurocognitive enhancement (Farah, 2010: 58-75)[5] paves the already-too-tempting road to a brave new world in which humans are gods of their own. A converse view, to which this essay subscribes, contends that the concepts of enhancement and perfection are mutually exclusive, and that enhancement has little to do, in fact, with the desire to become perfected. The juxtaposition of these two polarized positions will comprise the framework of the arguments explored in this essay.

To be sure, the issues of immediate import to the debate surrounding neurocognitive enhancement are manifold, and any singular analysis of topics, no matter how sweeping, will unavoidably fall short of adequacy. This essay recognizes such limitations, and thus aims to briefly address but three: memory (recollection), executive function,6 and emotional health and personality (cf.

---

[2] For a clear and concise introduction to the expansive language, attributes, and perspectives of neuroethics, both past and present, see Racine, 2010: 27-51.

[3] This group of scholars worked under and was formed by the political regime of George W. Bush.

[4] Henceforth, "the Council."

[5] It is important to clarify that the logic inherent to the Council's general positions vis-à-vis biotechnological enhancement is herein applied specifically in the neurocognitive context. Hence, this essay does not intend to misconstrue the original content of "Beyond Therapy" as representative of the Council's explicit address of neurocognitive enhancement, but only biotechnological enhancement generally, which in turn may (presumably) be applied in specific contexts.

[6] For coverage of both memory (recollection) and executive function, see Farah et al., 2010: 30-41.

Canli and Amin, 2010: 147-54). The aim and proposal of the essay is to examine the concepts of memory, executive function,7 and emotional health and personality in the context of neurocognitive enhancement with the intention of positing the argument that neurocognitive enhancement can be justified as morally plausible in its potential to edify the caliber of overall cognition and thus contribute to the ability make pragmatically robust moral decisions on the conditions that it (i) promotes general moral character, (ii) complements human nature, and (iii) effects a deeper sense of individual8 and social9 identity.

2. Moral Character and the Memorable-Recollective Brain

A first point of divergence in the enhancement debate concerns moral character. Refocused in the context of neurocognition, it specifically concerns how moral character is affected by manipulating the natural human capacity to recollect. Generally speaking, human identity is reducible in part to the decisions one makes, the actions one pursues, and the responsibility taken for the sum of the parts. Hence, who one flows forth from what one decides to do and acts on doing, and moral character is shaped by this means. Utilizing neurotechnological enhancement is one such decision and action that significantly affects and shapes moral character. Since more than just biophysiology is as stake in the choice to pursue neurocognitive enhancement, recognition of the promises and perils of its effects on identity, mentality, free will, and, particularly in considerations of the mind-brain relationship, memory, are things for which moral character, shaped by moral responsibility, must account.

2.1. Hubris and Humility

With regard to moral character, the Council's rhetoric10 against enhancement is couched in expressivist and consequentialist concerns. The former regards the

---

[7] By this phrase is meant the neurocognitive abilities that allow for flexible, task-oriented responses in the face of manifold and often competing neurological inputs or more habitual but inappropriate behavioral response patterns. See Farah et al., 2010: 33-34.

[8] These three conditions, among others, form the contra argument vis-à-vis enhancement of the 2003 President's Council on Bioethics. See President's Council on Bioethics, 2010: 58-72.

[9] For an expansive examination of the social implications of neuroscience accompanied by an epistemological framework for the neuroscience of ethics, see Racine, 2010: 179-213.

10 Included in the introductory remarks of "Beyond Therapy" is the Council's recognition that "it will be hard to say what is wrong with any biotechnological intervention that could improve our performances. Indeed, in many cases, we ought to be thankful for or pleased

idea that the pursuit of enhancement itself denotes moral viciousness and, hence, poor moral character. The latter regards the moral forecast that enhancement will inevitably lead to the weakening of moral character. The strongest expressivist concern lies in the Council's contention that pursuing enhancement fails to exemplify respect for "the given" in nature and instead exhibits the hubris of those who act with "hyper-agent" wisdom they fail to possess (cf. The President's Council on Bioethics, 2010: 59-61). Since appreciation is the product of humility – a precondition for the possession of other moral virtues – enhancement inevitably proves a stumbling block for the development of virtues necessary to live a substantive moral life. In this sense, the Council's stance is equal parts essentialist and consequentialist.11

Applied in the context of neurocognitive enhancement, the Council's threefold premises include the notions that (i) appreciation for what one has been given is a human good of central importance that develops virtuous moral character; (ii) the pursuit of nontherapeutic memory enhancement as a form of mastery and perfection is ultimately at odds with this sense of appreciation; and (iii) the employment of memory enhancement and its products are an instance of the pursuit of mastery or perfection. From these premises can be derived two fundamental conclusions. The first is that the employment of memory enhancement is ultimately incongruous with appreciation for "the given" in nature. The second is that the employment of memory enhancement is ultimately incongruous with a central human good of which virtuous moral character requisitely consists (the President's Council on Bioethics, 201: 59-61).

2.2. The Morality of Memory Encoding

This essay finds such essentialist and consequentialist logic misguided. Essentialist and consequentialist concerns are typically the culprit of two primary errors. First, they assume that all persons who pursue enhancement possess an insatiable craving for mastery or perfection.12 Second, essentialist and consequentialist concerns take the notion of gratitude at face value rather

---

with the improvements our technological ingenuity is making possible." They spend the remainder of the document, however, endeavoring to explicate precisely why they are not. See President's Council on Bioethics, 2010: 58-72. See especially p. 58.

[11] This logic is echoed most notably by the political philosopher Michael Sandel. For a thorough analysis of the expressivist and consequentialist concerns of the "anti-enhancement" camp, see Sandel, 2007.

[12] The necessitation of spectacles for night driving, for example, proves foolish this point. Surely the driver desires neither mastery nor perfection in the endeavor to ensure the safety of self and others.

than considering it a vague concept expressive of something more appropriate and less bogged down by theological rhetoric.13 To be sure, every person should be concerned with the risks of becoming exceedingly focused on the beneficial effects of enhancement to the extent that they cease to be appreciative of the things they have. But such risks are not *ipso facto* unavoidable, and a person who focuses on what is to be gleaned from neurocognitive enhancement at the cost of appreciating current possessions already exhibits a deficiency in moral character.

Moreover, essentialist and consequentialist concerns fail to acknowledge the possibility that memory enhancement may preserve, if not explicitly develop, one's sense of appreciation. If this can be defended, then memory enhancement can be said to fortify moral character. Consider the use of drugs that regulate APMA[14] receptors to aid in depolarization, or that increase CREB,[15] a gene-activating molecule that produces proteins and buttresses synaptic strength. Applied to middle-aged and elderly populations (cf. Farah et al., 2010: 31-32)[16] the ability to encode new memories would allow for a deeper experience *ex post facto* of critically significant life events.[17] This enhancement certainly possesses the potential to lead to a deeper sense of appreciation for the already given, not a movement away from it (Farah et al., 2010: 31-32).[18] In the context of human health and well being, then, the "respect" argument seems to be one in favor of, not against, enhancement. Understood in this light, enhancing neurocognitive endowments seems to promote, not detract from, moral character, thereby edifying the caliber of overall cognition and thus contributing to the ability to make pragmatically robust moral decisions.

3. Human Nature and the Executive-Functional Brain

A second point of divergence in the enhancement debate concerns human nature. Refocused in the context of neurocognition, it specifically concerns

---

[13] Both the vagueness and the presupposition of the theological nature underlying the concept of gratitude are swiftly identified upon posing the question, "to what/whom?"

[14] Amino-3-hydroxy-5-methyl-4-isoxazole propionic acid.

[15] cAMP response element-binding protein.

[16] Neurocognitive studies have indicated that this population in particular stand to gain most from memory enhancement technologies.

[17] The analysis of increasingly controversial memory augmentations such as dampening and vivifying transcend the scope of this essay.

[18] More to the point, an adequate sense of appreciation for "the given" in nature includes the appreciation of everything that is given, including memory enhancement techniques.

how human nature is affected by manipulating the conception of one's existence in the world. The philosophical conception of human nature as a set of shared moral ideals between human beings that distinguishes them from all other beings can be traced to the Aristotelian corpus. If these ideals are considered natural, they are consequentially regarded as essential rather than contingent or superfluous. This logic lends to the interpretation that if human beings were to lose any of the moral characteristics considered essential to human identity, they would cease to be human. The question is thus raised over whether neurocognitive enhancement contributes to or depreciates human nature.

3.1. The "Unnatural" Means of Human Dignity

With regard to human nature, two primary concerns frame the Council's argument against enhancement. The first is that enhancement will alter or obliterate human nature, rending it man-made and therefore "cheap." (President's Council on Bioethics, 2010: 61-64). The second is that if enhancement alters or obliterates human nature, this will affect one's ability to determine the good, because the ability to determine something is dependant on the nature of the determiner (*Ibidem*).[19] As it pertains to the first concern, the principal fear is that one or more of the characteristics essential to human nature will inevitably be manipulated by neurocognitive enhancement, rendering one an alien to oneself (and others). The continual creation of these "transhuman others" (President's Council on Bioethics, 2010: 58-72)[20] might eventually result in the extinction of human beings. Since it is illogical to tamper with the wisdom of nature, it follows that it is illogical to attempt to become "better than well." Thus, participation in enhancement would leave one with only the prodigality of free will (President's Council on Bioethics, 2010: 61-64).

The second part of the argument is based on the notion that human goodness is inextricably linked to human nature. Transcending this nature – becoming "God-like" – undermines the apprehension and valuation of goodness, rendering nothing save for experiential, and thus existential, misunderstanding. Since understanding and valuing the good is essential to moral living, human beings must do everything possible to preserve it. Hence,

---

[19] The metaphorical language used by the Council to demonstrate this point includes the inevitable gap created between "the dancer and the dance."

[20] The Council also invokes concepts such as "God-like," "hyper-agency," "better than well," "cheating," "self-alienation," "better model," "less than human," and several others to describe transhumanist desires.

any neurocognitive enhancement that would augment human nature is discouraged and must be avoided. In this view, human actions flow forth from human nature. Once nature is sacrificed by participating in enhancement, thereby "cheating" oneself out of oneself, the ability to determine the good and act upon it is lost (cf. President's Council on Bioethics, 2010: 61-71).

3.2. The Folly of Normative Essentialism

This essay rejects such normative essentialism[21] as implausible for several reasons. First, there is nothing intrinsically wrong with altering human nature by means of enhancing executive function since the "naturally given" clearly contains both good and bad elements (cf. Farah et al., 2010: 33-34)[22] and because there is no persuasive reason to believe that in every effort to nuance the poor elements there would exist a disproportionate risk posed to those considered good (*Ibidem*).[23] Second, if it were the case that executive function enhancement would "obliterate" human nature by transforming human beings into "hyper-agents," this action would not in itself be wrong and may in fact be morally right.[24] Third, it does not necessarily follow that adjusting human nature will result in the loss of the ability to make judgments about the good, because human beings requisitely possess a conception of the good by which they evaluate human nature itself.[25] Finally, appeals to human nature generally effect obscurity in the moral debate over enhancement and can be ameliorated with more apposite considerations.

Applied in the context of executive function enhancement, the third reason above deserves immediate attention. What underlies the erroneous normative essentialist claim that nature and goodness are inextricably linked is the idea that to decipher whether something is good, one needs to know if it conforms to, or "fits" with, human nature. This essay has already noted one reason (cf.

---

[21] Normative essentialism is the belief that comprehensive moral rules may be extracted from reflection on human nature.

[22] Consider, for example, executive function deficits such as attention disorders, working memory miscues, and inhibitory control issues. While these are "naturally given," no reasonable person would consider them intrinsically good.

[23] To be sure, the use of drugs that target dopamine and noradrenaline neurotransmitter faculties carry particular risks in the effort to augment the "naturally given" neurocognitive deficiencies, but certainly not in every case, and almost always within the realm of reasonable proportion.

[24] *Prima facie*, the action in question may at worst be considered morally neutral but by no means clearly and objectively morally wrong.

[25] This exemplifies, among others, the logical fallacy of circulus in demonstrando.

Ibidem)[26] to reject this claim: appealing to the role nature plays in decision making is largely about recognizing constraints; therefore, rather than rendering human beings incapable of judging rightly – which is the (preexistent) existential plight of those to whom neurocognitive deficiencies have been "naturally given" – participation in executive function enhancement would simply require the consideration of new constraints (*Ibidem*).[27] Based on the innate desire to evolve, then, participation in neurocognitive enhancement may be seen in this light as complimentary of, not a detraction from, human nature, thereby edifying the caliber of overall cognition and thus contributing to the ability make pragmatically robust moral decisions.

4. Identity and the Emotional-Personal Brain

A third point of divergence in the enhancement debate concerns individual and social identity. Refocused in the context of neurocognition, it specifically concerns the individual and social effects of augmenting one's identity in ways pertaining to emotional health and personality. To be human is to be a particular person, with a particular history, and with particular loves and memories that bind one to behave in particular ways. The possibilities of neurocognitive enhancement offer entry into an uncharted plane of being within which who one understands oneself to be in relation to self and society (cf. Racine, 2010: 215-21)[28] may be significantly amplified through techniques that would allow for emotional and personal maturation or blunting. Issues related to moral responsibility and "manufactured" perceptions are hence inevitably called into question.

---

[26] There is, however, a second and more important reason to reject this claim: human beings already make coherent and reasonable judgments concerning human nature that can persuasively argue in favor of augmentation. Recognizing flaws in human nature – e.g., deficits of executive function such as attention disorders, working memory miscues, and inhibitory control issues – suggests that human beings have a concept of human nature that is independent of human nature itself.

[27] One plausible constraint would be to allow enhancement therapies to be accessed by only those most likely to benefit from them, which has typically proven to be those with the lowest working memory capacities, significant attention disorders, and multiple inhibitory control issues. Interestingly, at least one study has shown that the use of dopamine agonist bromocriptine improves the neurocognitive performance of individuals with a less-than-average working memory but lowers the performance of those with the highest natural capacities.

[28] For a still further analysis of the sociological implications of the promised future of neuroscience and its accompanying neuroethics.

4.1. Plentiful Ends, Partial Flourishing

With regard to individual and social identity, the Council's position against enhancement manifests itself in two primary veins. The first refers to the idea that enhancement threatens the most precious piece of human existence: idiosyncratic identity. The second turns on the first, and concerns the notion that participation in enhancement provides others with the means to manipulate the identity that should be solely one's own to "work out." As it pertains to the first concern, enhancement is viewed to remove the intelligible means necessary to understand one's particularity in shared time and space (President's Council on Bioethics, 2010: 64).[29] Participation thus threatens this essential identity, which is inextricably linked to moral responsibility,[30] by allowing it to be usurped by the cold and impersonal hand of biotechnology. Consequentially, "self-alienation" is the inevitable destination on the journey toward actualizing this naïve "achievement" (President's Council on Bioethics, 2010: 64-65).

The second part of the argument clings to the notion that enhancement effectively "de-agents" the subject and allows the "enhancer" to assume the subject's identity as biotechnological puppet-master par excellence. Since personal achievements can only be personally achieved, it follows that impersonal achievements cannot be considered the work of individual persons (*Ibidem*).[31] While it can be said that an individual self uses cognitive drugs to, for example, increase alertness, it cannot be said that an individual self thereby does the "alerting." Rather, pharmacologicals have acted in place of a self and goals have been achieved only by means of sacrificing personal independence. The cardinal idea is that to be independent essentially requires possessing individual limits. Thus, by the mere act of participating in enhancement, one undermines[32] one's own identity (*Ibidem*).

---

[29] For the Council, "to be human is to be someone, not anyone – with a given nature (male or female), given natural abilities (superior wit or musical talent), and, most important, a real history of attachments, memories, and experiences, acquired largely by living with others."

[30] In this way, identity is intimately linked with the conception and constitution of moral character.

[31] The metaphorical image used to capture this point is the student behind the calculator. Using the calculator does not *de facto* render the student a "knower" of mathematics, even if the student arrives at the correct answer through its use.

[32] In this case, a "pseudo-happiness" has been achieved at the cost of the same individual self necessary to secure a sense of happiness that can be called genuine and personal.

4.2. Against Self-Reductionism

This essay rejects such self-reductionism for reasons beyond its unfortunate tendency to paint the unflattering portrait of people who pursue neurocognitive enhancement as unintelligent "identity-thieves" who prostitute dignity for gain at the cost of personal integrity. The position here is that emotional health and personality enhancement can promote individual and social identity for at least two reasons. The first concerns the idea that emotional health and personality enhancement may be used to correct existing individual neurocognitive deficiencies that result in individual injustices. The second concerns the idea that emotional health and personality enhancement may be used to correct individual neurocognitive deficiencies that result in social injustices. To exemplify the former, consider neurological studies that provide insight into the benefits of developing cognitive drugs that enhance underdeveloped neural correlates regulating negative emotions related to moral violations. The use of neurocognitive pharmacologicals may promise brain regions previously undeveloped, yet nonetheless implicated in moral development, a greater personal emotional content in moral stimuli (Racine, 2010: 186). So, doing may combat individual injustices[33] by leading to a comprehensively balanced understanding of idiosyncratic moral responsibility and the essential role of emotion in systematic decision making (Racine, 2010: 184-89).[34]

The second reason may be exemplified by considering neurological studies that provide insight into the benefits of developing cognitive drugs that enhance underdeveloped neural mechanisms (Racine, 2010: 185)[35] that process moral sensitivity to issues related to justice and care. The use of neurocognitive pharmacologicals may promise activation to regions of the brain that experience abnormalities associated with dissociable neural processing events (*Ibidem*).[36] So doing may combat social injustices[37] by

---

[33] Thus, rather than effectively sacrificing or annihilating one's sense of self, emotional health and personality enhancement can be defended as the gateway into one's authentic self and the individual moral responsibilities such a discovery demands one to observe.

[34] Other emotional health and personality enhancement studies promotive of individual identity include, among others, the influence of bodily harm on neural correlates of semantic and moral decision making, the moral affiliation of disgust, and affective responses to one's own moral violations..

[35] These include the polar medial prefrontal cortex, dorsal posterior cingulated cortex, and posterior superior temporal sulcus.

[36] These would include a dormant left intraparietal sulcus, which is associated with sensitivity to justice issues, or a dormant ventral posterior cingulate cortex, ventromedial and dorsolateral prefrontal cortex, and thalamus, which is associated with sensitivity to care issues.

leading to a deeper awareness of concerns related to social justice while simultaneously developing one's individual identity as related to others in the collective mission to express care and respect for well-being (Racine, 2010: 184-89).[38] In light of these reasons, then, neurocognitive enhancement may be defended as effecting a deeper sense of individual and social identity, thereby edifying the caliber of overall cognition and thus contributing to the ability make pragmatically robust moral decisions.

5. Conclusion

Memory, executive function, and emotional health and personality are but three issues of principal significance in the debate over the moral licitness of neurocognitive enhancement. Through the lens of a critical-dialogical framework, the aim of this essay has been to address the aforementioned issues with the intention of positing the argument that neurocognitive enhancement can be justified as morally plausible in its potential to edify the caliber of overall cognition and thus contribute to the ability make pragmatically robust moral decisions on the conditions that it (i) promotes general moral character, (ii) complements human nature, and (iii) effects a deeper sense of individual and social identity. To that end, it has been successful.

The implications here are significant. To be sure, the reality of enhancement extremism is a genuine and growing fear. But rather than allowing it to terminate human progress, it may instead serve to remind that while the benefits of neurocognitive technologies are important, how and how far they are utilized is more important still.

References

Canli, Turhan and Amin, Zenab (2010) "Neuroimaging of Emotion and Personality: Ethical Considerations" In *Neuroethics: An Introduction with Readings* (Martha J. Farah ed.), Cambridge, MA: The MIT Press, pp. 147-54.

---

[37] Thus, rather than effectually obscuring and demoting social identity, emotional health and personality enhancement can be defended an as invitation to participate more deeply and responsibly in the social story of which one is an integral individual part, thereby promoting the rational capacity to decipher the good.

[38] Other emotional health and personality enhancement studies promotive of social identity include, among others, gender differences in neural mechanisms underlying moral sensitivity, the neural basis of belief encoding and integration in moral judgment, and the functional networks in emotional moral and nonmoral social judgments.

Farah, Martha J., Illes, Judy, Cook-Deegan, Robert, Gardner, Howard, Kandel, Eric, King, Patricia, Parens, Eric, Sahakian, Barbara, and Root Wolpe,

President's Council on Bioethics. "Beyond Therapy: Essential Sources of Concern" In *Neuroethics: An Introduction with Readings,* (Martha J. Farah ed.), Cambridge, MA: The MIT Press, pp. 58-72.

Racine, Eric (2010) *Pragmatic Neuroethics: Improving Treatment and Understanding of the Mind-Brain,* Cambridge, MA: The MIT Press.

Sandel, Michael J. (2007) *The Case Against Perfection: Ethics in the Age of Genetic Engineering.* Cambridge, MA: Harvard University Press.

CHAPTER 13

DOES AI BRAIN IMPLANT COMPROMISE AGENCY? EXAMINING POTENTIAL HARMS OF BRAIN-COMPUTER INTERFACES ON SELF-DETERMINATION

Tomislav Miletić[1]

University of Rijeka, Croatia

Frederic Gilbert[2]

University of Tasmania, Australia

Abstract: Novel generations of Brain-Computer Interface (BCI) technologies operated by AI, in particular, predictive neurotechnologies, offer enormous potential to support implanted individual' decisions and capacities for self-determination, such as empowering agential cognitive capacities. This chapter examines the ethics of predictive AI BCI, especially the question of potential risks of harm associated with having a predictive AI BCI device strongly influencing one's decision-making processes for self-determination due to epistemic dependency. We observe that if an agent aims to be rational, the more precise the AI prediction of an optimal epistemic conclusion, the less autonomy an agent has to decide otherwise than what is prescribed by the AI prediction. As such, predictive AI BCI does not necessarily limit self-determination but rather may enhance it in providing a precise and correct focus to the human rational agent through which she is able to pick only from the best possible range of choices and accomplish the goal for which she strives. In order to avoid potential psychological harms from a device 'knowns better', we propose a

---

[1] Faculty of Humanities and Social Sciences, University of Rijeka, Croatia.

[2] CALE, School of Humanities, at the University of Tasmania, Hobart, Tasmania. Centre for Neurotechnology, University of Washington, Seattle, USA.

framework of human-AI symbiosis to prepare, train, and educate prospective AI BCI recipient.

1. Introduction

Recently, AI has been used to enhance performances of a novel generation of Brain Computer Interface (BCI) technologies. Augmenting BCI performances with AI may materialize in many ways. In this manuscript, we will focus on predictive technologies, namely advisory brain implants. These novel predictive neurotechnologies have been tested with significant success for the first time in a clinical trial (Cook et al., 2013; Kingwell, 2013). The tested devices predict a specific neuronal event (e.g., a neuronal or psychiatric event) allowing the implanted individual to be forewarned and to take necessary steps to reduce, avoid and augment the impact of a forecasted event. These AI predictive technologies offer enormous potential to support implanted individual' decisions and capacities for self-determination. In theory, this generation of BCI devices could be used for a large range of clinical and non-clinical application (Gilbert and Cook, 2015); such as augmenting and empowering agential cognitive capacities (i.e., reasoning, learning, decision making information retrieval and analysis), but also predicting unwanted outcomes (i.e., depressive episodes; addictive habits, socially reprehensive conducts).

Increasingly, as implantable BCI advisory system empower and enhance human capacities, the line between where human cognitive capacities finish and where AI functionalities start is becoming imperceptible. With the blurry line of separation between human's capacities and AI functionalities thinning, it may radically impact our understanding of who and what we are. In what respect AI predictive functionalities may redefine the very relation we have to our own agency, autonomy, and even self and identity? The blurry boundary introduces many questions of AI ethics, many of these questions are in relation to the problems of AI-joined-human-agency.

In this manuscript, we will explore the ethics of predictive AI brain implant, known as BCI, especially the question of potential risks of harm associated with having a BCI device strongly influencing one's decision-making processes for self-determination. To what extent does an individual's decision-making or self-determination become compromised by relying too much on these AI advisory functionalities? How much of a causal role does the subject have when initiating an action via an AI predictive system, and what are the potential adverse cognitive effects of having an AI system in charge of a decision/action. Can the implanted individual's decisions that are based upon BCI device predictions be properly interpreted as decisions that are freely chosen?

2. AI, Prediction and Best Decision

Since Garry Gasparov lost against IBM Deep Blue computer in 1997, AI predictive algorithms have increased in power (Hsu, 2002). Nowadays, AI predictive algorithms have reached unprecedented capacities, they can calculate millions of moves per second in advance, as such, it is unlikely that human biological brains will ever be strong enough to beat the best AI systems. This dominance of the AI system in chess game has led to a new type of game, namely Advanced Chess, where human players are assisted by a computer program to explore the possible results of moves; as such human players and AI algorithms become a team. A specificity of Advance Chess is that while the AI informs its human partner of possible scenarios and outcomes the human remains fully in control of the final team decision, that is, what moves the team will make in the end. The game of Advanced Chess is a perfect illustration of AI predictive algorithms assisting and advising while preserving a player locus of control.

The fact that AI predictive technologies can be designed to advantageously assist a human player to reach some outcomes begs the question of why not to follow a predicted outcome forecasted by an AI system? In other terms, if the goal of the game is to win, and the predictive capacities of the AI system far outstrip those of the human player, why should the human player make the final decision to overrule the AI's ultimate suggestion?

To illustrate, let us imagine Rita playing a chess game, which she is desperately trying to win, as she faces an ultimate scenario in which she wants to move X^1, but her AI advisory system recommends her a set of moves X^n and especially the move X^2 which the system recommends as a necessary step towards victory. Since the AI's predictive power far outstrips that of Rita, if the AI system advises Rita to move X^2 in order to secure victory, the likelihood that she will win if she does X^2 is higher than her doing X^1. The trade-off Rita receives if she doesn't follow her intuition (X^1), but rather the advice given by her AI partner, is that her capacities to win the game have increased in these specific circumstances. X^2 is in the best interest of Rita if she wants to win the game.

This example from the chess game helps introduce some important ethical problems. For instance, if Rita moves X^2, as advised by her AI predictive system, in what sense can it be said that it is actually Rita who is making the winning move, or even a winning decision? Aside from executing her AI predictive system advice, one could wonder to what extend appropriating an AI system prediction is, in fact, an agential self-constitutive experience. How can this individual be properly understood at the origin of her winning move when the technology appears to have predicted and dictated it? Can it be interpreted that she only executed the prediction? How can an agent winning decisions – which is based upon the programmed responses of an advisory AI – be said to originate from Rita?

3. Who or What is at the Origin of a Winning Decision

Let us move away from Advanced Chess example and focus on AI advisory brain-computer interfaces. Like the AI assistive system in chess, invasive AI Brain computer interfaces assist implanted individual to respond to the upcoming neuronal events that are forecasted by providing information to help predict neuronal activity patterns before specific outcomes occur. A typical example would be predicting epileptic seizures (Gilbert O'Brien et al., 2018; Gilbert, Cook, et al., 2019).

Is it right to ascribe a decision to an implanted individual when it would appear that this very decision has been instigated by the AI BCI predictive and advisory functionalities (McCarthy, 1979)? To answer this question, let us look at the chronology of a decision, in particular, a decision taken by an individual who is suffering from epilepsy and who have been forewarned by an advisory signal of an upcoming neuronal event requiring a therapeutic intervention in order to prevent the event from happening. The chronology would be as follows:

1) an alert is signaled to an individual regarding that *e* is about to occur (*e* = an epileptic seizure),
2) AI advisory system recommends a range of options X^n,
3) AI advisory system strongly prescribes a particular decision X^2 (X^2 = taking anti-seizure medication).
4) the individual is expected to initiate an action by taking a particular decision X^i in order to prevent *e* from happening.

Following this chronology, it seems that predictive and advisory devices offer the prospect of prescribing a particular decision to the implanted individual. This chronology can be broken down into a simple proposition: "If *e* is predicted, then X^2 is prescribed". Here, an AI advisory system can suggest a range of options X^n (X^2, X^3, X^4, etc.), but strongly prescribe one option, namely X^2. In fact, by informing a patient in advance that *e* will occur if no intervention is undertaken, advisory functionalities are able to influence an individual's decision-making process in order to avoid specific neuronal events. In other words, "If *e* is predicted, then X^2 should be undertaken". By alerting the patient that *e* will occur if no intervention is undertaken, we understand the device as recommending (to the patient) to decide in favor of X^2. X^2 is in the best interest of the patient, according to the device, because of *e*.

On that point, an AI advisory system may be seen as some sort of prescriptive authority with a strong capacity to urge an individual to take a certain decision. If the implanted individual does not base her decision on any other relevant and reliably available evidence, then she may not have any other choice but to follow the forewarned alert.

She can choose not to act at all on the AI advisory system, but this choice could lead to catastrophic results. We believe, in the long term, one would think that "choose not to act" is not an option. By predicting upcoming neurological events, devices have substantial prescriptive power to make individuals initiate a particular decision. Given their ability to predict, AI advisory system may be understood to be in a position of authority over an implanted individual (the individual being dependent on the prediction). That is, when displaying information to individuals, an AI advisory system plays an epistemically authoritative role in informing individuals; implanted individuals are epistemically dependent on the prediction.

Whether the epistemic authority of the device exerts a disproportionate authoritative power is likely related to its predictive accuracy. For example, if the prediction of the device is 100% accurate, the device forecast will likely coincide with the implanted individual's choice to take anti-seizure medication. When device recommendation coincides with patient choice, if the prediction is 100% accurate, it will not be useful or necessary to resort to other relevant and reliable available evidence to ascertain what to decide (especially since resorting to other sources of available evidence would then increase the risk of trusting irrelevant evidence).[3]

Conversely, if predictions are often inaccurate, then the authority of the device appears to play a crucial negative role in the patient's decision-making process. That is, with inaccurate prediction, the implanted individual may feel obliged to blindly follow the authority of the device rather than accurate

---

[3] Accuracy will likely make the patient trust the advice based on reliable predictions. However, reliability of the advice may be based on a whole host of variables other than just the predictive accuracy of the BCI. But even if the predictions of the BCI are 100% accurate, there may be other variables that figure in a patient decision whether or not to follow the recommended prescription. Maybe the device is accurate in predicting that she will have a seizure, but not at all capable of telling her how long it will last as well as how severe it will hurt. The variable of whether she tends to suffer from anti-seizure medication side effects is also a key variable. To make a reasonable decision of whether or not to take medication a patient may need all kinds of other relevant evidence. In that sense, we understand devices have substantial prescriptive power to make individuals initiate a particular decision, but it would not be an exclusive prescriptive power.

prediction given by some other epistemic source. This is troubling especially when we have in mind that taking anti-seizure medication may result in strong side effects; patients may feel the AI advisory system's prescription as being imposed upon them rather than being justified to them.

Predictive and advisory technologies are an excellent example of epistemic dependency because they inform the implanted individual of events that the individual has no capacity to confirm until the event actually occurs (absence of epistemic independency). We argue that epistemic dependency does not necessarily compromise autonomy and rationality (individual best interest); however, the net benefit of epistemic dependency does not come free of ethical issues.

Let us look at a similar and hypothetical example. Imagine you are currently packing your suitcase to leave earth to visit the human-inhabited planet Charlotte, and you receive a forecast from a device about the weather on Charlotte. Given the context, you have no other evidence than the available forecast itself to decide whether the forecast can be relied upon to be true (epistemic dependency). For example, you cannot look out the window to determine whether you will pack your swimsuit or your winter undergarments. Nor can you contact someone on the planet to ask them about the weather in advance of your arrival on the planet. You can only rely on the data from the device. Applied in the case of epilepsy, an implanted individual might be trying to look at other sources of available evidence or phenomenon to guide her responsive decision-making rather than exclusively relying on the computer's advisory indicator. However, what are the other sources of information that would be relevant and how could implanted individuals confirm (or disconfirm) the reliability of such information? For instance, should individuals pay more attention to their own subjective experience, their instincts of a seizure about to happen? How reliable can such instincts be? Given the lack of other relevant and reliable available evidence to base responsive decisions upon, there is a strong chance that the affected individual at risk for seizures will conclude that she *should* follow the prescription of the AI advisory system.

The idea that predictive and advisory technologies generate epistemic dependency seems to narrow the parameters for relevant decision-making. An AI advice may come with outcomes compromising individual best interest.4 Here,

---

[4] When speaking of a device as the origin of a decision, we understand a causal explanation. Here, in no respect we mean a device with intention or will. The device, indeed, has been strictly programmed with certain responsive functions. Exploring this process as decision-making on the part of the device imply causation not intention.

however, an important distinction needs to be made: epistemic dependency may be A) an internal injunction (compatible with one's interest to prevent one's own potential future suffering) or B) an external injunction (the individual experiencing or feeling a decision imposed by AI advisory system). Regarding scenario A, it seems that to follow the relevant advisory recommendation could serve one's interest to prevent one's own potential future suffering. In this respect, it would be difficult to think of scenario A as ethically problematic *per se*, because one's interests would be directly in line with the prediction. However, regarding scenario B, if the decision to initiate a therapeutic intervention is ascribed not to the implanted individual but rather to an external AI system (in effect), then there seems to be a risk of the individual being imposed a prescription (i.e., feelings of estrangement vis-à-vis the decision) (Gilbert O'Brien et al., 2018). If the prescription from the advisory device is experienced as being an external injunction or imposed, would patients feel losing control? May patients feel a loss of agency or control in the face of relying on technology for advice?

The ethical concern we need to explore is how an AI advisory system programmed to provide only helpful information (for the agent to act upon or not) may become a system that potentially constrains action (even if it does not remove choice from the agent). The paradox of being implanted with an AI advisory brain system may be that they both increase patient control on the one hand, and diminish patient self-determination on the other hand by advising on what a patient should do (Gilbert, 2015).

4. More Precise the Prediction, Less Autonomy for the User?

Doubts can be harbored when we think about autonomy with respect to medical AI advisors, as when one allows the AI advisor to provide her with the best possible medical advice, one can still think that this limits or negates their autonomy. However, the very notion of accurate or precise AI prediction appears to show otherwise. In fact, the perceived limits to human decision making (autonomy) which precise AI-empowered prediction manifests (if operating optimally) are not something that necessarily impairs self-determination. The argumentation for such observation lies on how there exists a generally applicable epistemic (and computationally describable) rule which applies to all instances of epistemic autonomy and states that:

Cases where device are programmed at aiming an increase of drug intake in favour of a pharmaceutical company rather than targeting patient best medical interest would be a strong incursion and intrusion on patient decision-making process.

> Rule: If one aims to be rational, the more precise the prediction of an optimal epistemic conclusion the less autonomy one has to decide otherwise.

Or in other words, the more precise the prediction, the less choice one has in making the final decision if one aims to successfully achieve the goal of one's decisions (as described the above section about the AI chess example). This is due to a simple reason that states how although humans do have a huge range of choices for a specific action, only a few of them are optimal if one aims to achieve the goal of the action and some of them are always irrational (as for instance the decision to drink and drive, or the decision to refuse an expert's medical advice). Thus, if one aims to ensure the attainment of one's goal, one has to have the best possible prediction for the outcome of one's actions in order to do so. As such, one purposefully limits one's own decision making (autonomy) and even does so freely in order to attain the conclusion of her decisions – the attainment of the decision's goal. And limitations which such prediction places, such as the advice coming from a medical expert, upon the behavioral or cognitive autonomy of the human individual are freely accepted since one is aware of the benefits they provide. Finally, since the entire set of possible choices which aim to accomplish a specific goal is computable, we can compute the range of the best possible options for a single goal and propose those options to the human agent for a final decision on the matter. This is especially true for highly procedural knowledge domains such as medicine where we are aware of what one has to do in order to attain the goal of improved health.

If this is correct, then the autonomy-prediction relation showcases how advanced AI prediction doesn't necessarily limit human autonomy but rather may enhance it in providing a precise and correct focus to the human agent through which she is able to pick only from the best possible range of choices and accomplish the goal for which she strives. Additionally, it also shows that there is nothing mysterious, magical or sacrosanct in human autonomy concerning decision making but rather it is more that humans are accustomed to giving special value to their (personal) decisions. As humans give special value to them, they also object to the notion that some entity (for instance, an AI) should provide them with the best options for us to choose from.

Still, one could object that such technology even if giving all the best predictions can still prove to be phenomenologically or psychologically intrusive if the AI BCI system is experienced by its user as being externally injunctive, or intrusive to one's own thought processes. This experience might range from cases of mild annoyance to cases where one feels her personal identity ruptured by the AI system's operations (Gilbert Goddard et al., 2017;

Gilbert, 2018; Gilbert Viana et al., 2018). Let us look at the potential risk of harms.

5. What are the potential harms?

When we think on the manner how to approach and analyze the possible risks connected with AI BCI implementation there are different ways one may engage upon this endeavor. For a lucid assessment of the possible risks, we propose to take a rather pragmatic approach, one connected with the practical preparation for, implementation and post-procedural evaluation and support of the patient. Naturally, this approach might be copied for any other intracranial (intrusive) implants of AI decision-assisting technology in the coming future as the fundamental characteristics of such technologies and their corresponding ethical issues should remain similar if not same.

The first, and constitutive to the ethical and legal permissibility of the system's use, is the one directly connected with the technology's proper functioning and implementation. These include the ethical issues concerned with the system's implementation, that is the ethical guidelines guiding the necessary steps which need to be taken in order for the medical implementation procedure to be medically safe for the patient and those technical and medical requirements guaranteeing the system's proper functioning after the implementation. This means to say that the primary and constitutive ethics of AI BCI implementation and use is, naturally, aiming to establish technically safe and ethically sound design and implementation procedures that will ensure the system's proper implementation and it's operation after a successful implementation. This is a *sine qua non* condition, since all other options including those of the system not operating properly (as designed) after the implementation or the procedure of implementation resulting in an accident and injury to the patient (neural or psychological) will prohibit inquiries into broader and deeper issues of AI-assisted decision-making ethics. For this reason, it is paramount that the risks and ethics of AI BCI implementation need be clearly distilled into two categories or sets of issues.

The first one directly engages the medical (including psychological) and technical risks of AI system implementation (Hansson, 2005) and subsequent operation inside the patient's body (brain). Possible risks of this kind lean more to the already established field of neuroethics (Fuchs, 2006; Clausen, 2011) and should aim to establish ethically aligned design (EADv2) and implementation procedures for such devices that ensures the system will pose no medical threat to the patient's body or mind and that it will operate under the described technical capacities (Klein et al., 2015). Unfortunately, as no technical device, especially one of complex design, is immune to malfunction the possibility and probability of the system's malfunction need to be realistically taken into consideration. This means to say that system designers, the medical team engaged with the preparation of the patient and post-procedural following, and the patient itself have to be aware that the real issue with system malfunctions is not might they happen but rather how might it happen. This means that a failsafe approach guarantees to safeguard the patient from the consequences of such system failure which includes both direct medical, such as damaging the brain matter and psychological, such as diverse trauma effects consequences. The issue of system failure is one of huge importance since no system is infallible or can be expected to work perfectly for its entire duration. Unfortunately, ways in which the system may fail are numerous, and include inadequate or bad system design, human inability to effectively manage the system's operation or to cope with its complexity, system errors created from the interaction of system's subcomponents or finally, Black Swans (Taleb and Chandler, 2007), failures which no one can predict or be prepared for in advance.

Additionally, as we are dealing with intrusive almost „intimate" technology which the patient perceives and experiences as part of her own body, she must be psychologically prepared for the possibility of its malfunction in a similar way we prepare and educate children for the possibility that they might get seriously hurt or injured in their daily lives. As the first trip to the dentist, or a first injury might happen to be a traumatic experience for the child if it is not adequately given care and counseling, a patient reports "[The BCI] made me feel I had no control. So I didn't have control over what I was going to do" (Gilbert Cook et al., 2019). By extension, patients whose brain implemented AI malfunctions or stops functioning might experience a similar traumatic experience. This might be the case if the patient becomes irreversibly dependent upon the system's operation in her daily lives which includes not only the AI devices designed for purely medical purposes, a patient reports "[The BCI] was me, it became me, [...] with this device I found myself" (Gilbert Cook et al., 2019) but also and perhaps even more those devices which would work as advisors for a greater amount of life engaging decision-making processes (Magliocchetti et al., 2011). Presumably, if one would build a stronger relation of dependency with such devices, their malfunction would result in far more stronger impacts for the psyche and everyday functioning of such individuals. For this reason, it is paramount that the patient has to be prepared adequately for their implementation and offered both medical and, if needed, psychological assistance to cope with the system's malfunction (Gilbert Cook et al., 2019).

On top of this, one has to take into consideration the peculiarity of the trust relation one builds with such AI BCI systems which is that the AI decision-assistive system is implemented into the brain of the patient (user) and as such is not easily (or not at all) detachable from the user's body. This additional layer of intimacy AI BCI technologies introduce into the life of the patient is not to be admonished easily as the system, after the implementation, becomes not only a virtual, psychologically essential part of the user's cognitive character (Pritchard, 2010) "with this device I found myself" and processes (Palermos, 2011) but also a physical part of the user's body (brain). This additional level of physical intimacy with the user's body is extremely important since it directly affects the sense of ownership (Aymerich-Franch and Ganesh, 2016) over the system and the subsequent psychological acceptance of the „foreign" element into one's own body and psyche and the trust build with it.

In other words, after being implanted with AI BCI tech one can really experience the sense of being intimately physically tied with the silicon even to the sense of accepting oneself as being upgraded or empowered by the technology – thinking of oneself as a cyborg (Verbeek, 2008; Palese, 2012) But it is not an automatic fusional experience, a patient report: "I wouldn't say the technology was an integral part of me" (Gilbert Cook et al., 2019). The question then becomes not only how much trust can we give to predictive systems which are external to our bodies and as such even physically detachable from our daily lives and functioning but rather more, how much trust can we give to predictive AI systems upon which we become dependent on our decision-making and which at the same time have become parts of our bodies. In other words, the issue is not merely that of "trusting machines" but rather more "trusting machines in the brain", that is trusting machines which have become part of us, both mentally and physically – creating a single and perhaps even irreversible whole.

Unfortunately, if the system malfunctions or stops operating efficiently, the relation of trust would be seriously and perhaps irreparably damaged and one could no longer be dependent upon the assistance the system provides. This relation of trust built on the system's proper functioning (Heersmink, 2013) is then revealed as the basis without which we cannot further evaluate the second part of risks which includes the broader and deeper issues of epistemic dependency, epistemic AI authority and the possible loss of human autonomy in the AI-assisted decision-making process.

6. Epistemic Dependency: AIs as Epistemic Authority

The implementation of AI BCI predictive systems into the human bodies and their subsequent integration into the human decision-making process is a peculiar case of human empowerment. AI, on the one hand, generally viewed, has the power to create tremendous amounts of computational operations in short time, outstrip humans in the capacity to monitor the user's state (through the use of smart sensor technologies) in order to provide analysis and prediction far exceeding that of the ordinary human (Agrawal et al., 2018). But, on the other hand, the AI systems which we design and build in practice are not general prediction machines (such as the human brain) but rather systems built for a specific task, issue or problem and outside of the computational frame for which they are designed and built for, they either do not operate at all or cannot operate as (efficiently) intended by their designers. For this reason, it is of great importance to distill the issue of epistemic dependency and epistemic AI authority into two distinctive categories.

The first one, as portrayed in the case of epilepsy prediction systems above, is that in which the system exhibits epistemic authority over the human inside a highly procedural and knowledge-structured domains such as the medical field and the other, in which the system exhibits epistemic authority over the human in non-procedural, contextual, knowledge-structured domains as for instance morality, politics or even specific life decisions.

The main and biggest difference between these two categories is that one can be fairly certain (as for instance approved by standards of good science) of the type and kind of epistemic authority one wishes to have both personally as an individual and as an entire society in areas of procedural knowledge and expertise such as the medical field while one cannot be so (if one can at all) certain of the requirements of such authority in the other. For instance, in the medical area, or more specifically in diagnosing and treating specific ailments, we can easily find out what are the requirements a human expert ought to have to be recognized as a medical authority, that is the best expert in the field for that specific task. Some of these requirements include both theoretical knowledge of the ailment, which includes the best ways to prevent or treat it, and the practical knowledge or expertise which can be seen as the capacity to treat the patient according to the best available medical procedures in order to accomplish the optimal success of the therapy.

Since the reasons for establishing medical authority are transparent, guided by scientific principles and socially agreed upon, a patient should have no issues in accepting the medical advice as something she ought to follow if she wishes to attain better health or at least prevent the ailment's symptoms from further occurring. In this regard, consequently, the human person as a patient does not normally experience a threat to his decision making autonomy but rather often experiences just the opposite – that the medical expert has empowered her decision making by providing the best possible advice on the matter and if required can also follow up with the treatment procedure itself. Similarly, the human person who puts her trust into the AI medical expert (advisor), if the AI system exemplifies the best medical procedures and is producing a transparent sum benefit for the health of the patient, should have no rational reasons to experience diminishing in one's autonomy if one follows AI provided advice since the advice provided by the AI would always and only result in evident health improvement.

The second category, on the other hand, deals not with domain-specific, easier to compute, knowledge applications but rather with difficult to compute, non-procedural knowledge applications where the sum benefit one aims to achieve is not easily definable. For instance, if we aim to build moral or political AI advisors or personal AI life-coaches assistant (Danaher, 2018; Savulescu and Maslen, 2015; Klincewicz, 2016; Giubilini and Savulescu, 2017), it is difficult to specify what are the socially acceptable and easily recognizable standards of expertise we should implement for AI systems of this kind and the transparent sum benefits the advice given by such a system produces for the human individual.

In lack of these, the human individual might have big difficulties in trusting such AI systems since the end goal, that is, the sum benefit which the system provides is not clearly transparent to its user. Additionally, the existence of diverging but socially approved epistemic authorities within these fields does not help clarify which of, or what type of epistemic authority, should the AI system exhibit? For instance, one might ask what kind of advice would the AI give to the human user under all occasions – to be liberal or conservative, consequentialist or virtue theorist? One might answer that we might simply create morally and politically pluralist AI systems (Giubilini and Savulescu, 2017) and let the users (guided by the free market economy) choose whatever they prefer the most – AI conservative, liberal, consequentialists or virtue theorist systems. Naturally, we may aim to and succeed in creating AI expert advisory systems that are only conservative or only consequentialist but then the issue of changing one's attitude, the hallmark of human autonomy, comes to the forefront. If our AI advisors serve only to entrench one into pre-defined conceptual cognitive and behavioral models, then we can objectively question the loss of one's autonomy when following such an advice. Similarly to "algorithmic control"[5] AI advisors of this kind might severely limit and curb the development of human epistemic autonomy by enclosing the user within a specific set of decision options.

---

[5] Where social networks through the use of smart algorithms reinforce the user's existing behavioral routines through a constant reiteration of similarly tuned data inputs which as an end result not only reaffirm the users' present behavior but also creates and encloses users into social eco-chambers where they communicate only with those of similar political or moral worldviews.

To conclude, the main difference between the perceived loss of human autonomy in medical decision-making processes and the realistic chance to curb or weaken human autonomy in moral, political or life-decisions through the use of AI advisory systems, stems from the fact that the medical expert-like knowledge and its application is highly procedural, computable and based on clear standards of socially accepted expertise which results in a clear sum benefit for its user, easily recognizable both by the user and other members of the society. Additionally, in the first case, the user's autonomy is not endangered as the human still retains her full decision-making capacities and uses them to pick freely amongst the best possible options provided by the AI advisor for the attainment of the user-specified goal. The user is the one who specifies the goal[6] (in our case, the goal is better health), the AI provides the best (in real time) advice to attain that goal, and the user affirms one or all of the provided advice in order to accomplish her initial goal. For this reason, it is not only that one does not lose his decision-making autonomy but rather enhances it through the use of AI advisory systems if they are operating efficiently. The only loss one may have in accepting this relation of dependency, through which she is provided with the best medical advice, is the loss of "lesser" or "bad" options/advice. But since the purpose of medicine is clear, which is to make one healthy, then one can only account how the only valid advice which the medical AI expert should give is always the best medical advice it can give, and no other. As such, one can easily accept and enter into the relation of epistemic dependency between the medical (AI) authority and oneself. Lastly, through such a relation the human stands empowered, as she perceiving loses the full range of autonomy, she actually only discards the inadequate options and retains the optimal ones.

The second option, on the other hand, is the one where the advice given by the AI system is not strictly domain-specific but rather more general-like in its features (as for instance an AI life-coach or a political/moral advisor) and which standards of expertise are non-transparent7. Such systems hold the danger of

---

[6] Naturally, one may expect that the AI advisor's goals will be provided by its designers and formed by human needs and market. As such, human users will be able to both pick from the pre-existing templates or create their own personal portfolios which the AI advisor might use to structure its advices in order for the user to attain his predefined, or *ad hoc* created, goals.

[7] The inability to clearly represent a specific goal structure, that is the model which represents the best ways to accomplish a specific goal, showcases either the inability to present the argument conclusively and firmly (such as for instance the final decision on certain moral issues) or the computational difficulty to transparently elucidate the goal structure, which presents itself as a difficulty in modeling the question on how to actually attain the goal

framing its user within a set of cognitive or behavioral routines that curb the development of human decision autonomy by limiting the capacity to explore additional, novel or different options while at the same time forming a strong bond of dependency without providing an evident sum benefit to the user. Such systems, then, should be deemed ethically unwarranted.

As such, in juxtaposing these two types of systems, we see how there are two distinct ethical issues ingrained in AI advisory systems. One concerns humans and deals with their, often psychological, (in)ability to trust medical AI experts and willfully enter into a relation of dependency with them even though it would be in their best interest to do so. The other concerns the design of AI systems and their technical or design (in)ability to provide humans with advice which will not limit their autonomy but will rather constantly strive to broaden rather than narrow, develop rather than curb the range of their decisions and the capacity to pick the best available options in those fields of human endeavor where exists a valid pluralism of options.

To resolve both of the issues, where the AI partner provides advice which develops rather than curbs human decision-making autonomy and the human part is adequately (psychologically) prepared for the relation of epistemic dependency formed with its (medical) AI partner we propose that such design is best resolved within an ethically Human-AI symbiosis framework.

7. Human-AI Symbiosis: How to ethically empower the human decision process

The concept of Human-AI symbiosis, inspired by the Human-Robot Symbiosis framework (Reidsma et al., 2016; Rosenthal et al., 2010; Sandini et al., 2018), entails artificial and human partners working both autonomously and jointly to accomplish a specific goal. This goal might be predefined in advance or it can be agreed upon in real-time communication without prior planning.

In cases of predefined goals, the AI partner can be designed for the exact purpose, in the same manner, the human partner can be properly educated. For instance, in cases of robotic surgery, the surgeon robot and the surgeon human doctor form the symbiotic relation (Abovitz, 2001) only within that specific context and no other. This means to say that human-AI symbiosis should not be seen as a relation of symbiotic cooperation where general intelligence is a necessary requirement for its formation but rather that even

---

which one strives for. In other words, there are some questions without a definite and precise answer, and there are some answers for which it is difficult to find the right questions.

narrow types of artificial intelligence can successfully form symbiotic relations with their human partners. The fact how the robot surgeon cannot join the other doctors at the coffee break for a small chit-chat does not negate the fact how it is an effective and highly proficient surgeon partner which in some of its capacities far outweighs the capacities of even the best human surgeons available.

Still, a necessary requirement for the formation of symbiotic relation is transparency of intention and action capabilities. In other words, the AI partner has to be designed to transparently reveal its internal states while it engages communication with its human partners and that it is, by design, "fully aware", of its perceptual, physical or cognitive limitations. This means to say that the AI partner must not in any way deceive when it reveals its intentions or its capacities for action to its human partner. What it aims to do and what it can do must be clearly known or freely available for the inquiry to the human partner. The same counts for the human partner as well, since the AI partner cannot properly operate nor utilize its capacities if it is designed to trust and hold accountable its human partner while he does not do so. Additionally, for the AI partner, the capacity for action explanation (providing reasons why it did the way it did) is a necessary requirement for the formation of a strong relation of trust.

As such, the symbiotic link denotes how the human and the AI partner are here for a common, joint goal and are not in any way subverting or stalling its accomplishment but are rather empowering each other by supplementing each other lacks and allowing full utilization of each other's prowess.

This supplementation or support includes not only the cooperative support in physical or cognitive capacities but should also include motivational and emotional support as well provided by manufacturers, clinical teams, consumer services, etc. Additionally, the power of the symbiotic relation expresses itself exactly when AI partners are designed not to imitate humans but rather fully utilize their own specific and peculiar capacities of which some far outstrip human capacities both in kind and in power. The same counts for the human partner as well, as the AI still lacks a general type of intelligence and with it "common sense". (Lemaignan et al., 2017). As such, when both partners engage each other in such accord and adapt to each other's strengths and weaknesses, they effectively complement and enhance their combined efforts in accomplishing their common goal.

Finally, what this entails, with regards to human autonomy, is that the AI partner must not in any way subvert human autonomy to the level of a simple proxy in the decision-making process. This does not mean that if the AI partner does something far better than its human counterpart, as for instance, the case of our medical AI advisors or even AI chess, that we ought to think of this as diminishing of human autonomy. Rather the opposite, the human always retains its full range of autonomous decision-making capacities but in accordance with it has, and at the same time, allowed the AI, partner, to supplement her autonomy with its own. For an ethical human-AI symbiosis, we need to ensure that we are supplementing autonomy, not supplanting it.

In other words, the human experiences, and is aware of the fact how she always retains her full autonomy and with it her human dignity but trusts her AI partner for those tasks which she knows it is far superior in accomplishing. The crucial and breaking point for the symbiotic relation is then revealed to be human acceptance. The AI partner may work optimally, better than any human, but if the human partner denies its help, nothing can or should be done. This option has to be always open – the human has to have the capacity to reject the (best) AI given advice and, in the case of medical AI advisors, endanger her health. Hopefully one would not act in such an irrational manner, but unfortunately, as recently shown through the exploration of the "algorithm aversion" phenomenon, even though AI outperforms humans in its predictive prowess, "whenever prediction errors are likely – as they are in virtually all forecasting tasks – people will be biased against algorithms" (Dietvorst et al., 2014: 11) rather than humans.

The wished-for conclusion of the symbiotic relation is then that the human learns to trust its AI advisor by having lucid and realistic expectations of its AI partner through the awareness of its capacities (both lacks and prowess) but that it also remains confident and courageous in making the final decisions when faced with its, in specific domains clearly established, epistemic AI superiority, which affect how one perceive responsibility (Nyholm, 2018; Danaher, 2016; Santoni de Sio and van den Hoven, 2018; Grubler, 2011) and how legal rights are established (MacDonald Glenns, 2012).

For this reason, it is extremely important that the patient who accepts the use of such future AI medical advisors is adequately prepared for the symbiotic relation before she enters into it (Gilbert Brown et al., 2019; Gilbert Cook et al., 2019; Gilbert Pham et al., 2019; Gilbert O'Brien et al., 2018; Gilbert 2018; Gilbert and Tubig 2018; Gilbert Goddard et al., 2017). This preparation for the relation includes the need for a clear elucidation of the symbiotic relation, a proper education of the capacities of AI systems, a motivational encouragement to enter into it, and the psychological support both before and after the implementation. Unfortunately, if one is not adequately prepared for the symbiotic relation, rejections of the implant may occur, which might manifest in refusal to cooperate upon given advice (and subsequent endangerment of health) or even worse cases such as psychological trauma or the wish to remove the implant altogether.

Acknowledgement:

FG is supported by grant from the Australian Research Council (DECRA award Project Number DE150101390).

References

Abovitz Rony (2001) "Digital surgery: the future of medicine and human-robot symbiotic interaction", *Industrial Robot: An International Journal*, 28 (5): 401-406.

Agrawal, Ajay, Gans, Joshua and Goldfarb, Avi (2018) "Exploring the Impact of Artificial Intelligence: Prediction Versus Judgment" in *Rotman School of Management*, Working Paper No. 3177467.

Aymerich-Franch, Laura and Gowrishankar Ganesh (2016) "The role of functionality in the body model for self-attribution", *Neuroscience Research*, 104: 31-37.

Clausen Jens (2011) "Conceptual and Ethical Issues with Brain–hardware Interfaces" *Current Opinion in Psychiatry*, 24 (6).

Cook, Mark, O'Brien Terence J., Berkovic, Samuel et al. (2013) "Prediction of seizure likelihood with a long-term, implanted seizure advisory system in patients with drug-resistant epilepsy: A first-in-man study", *Lancet Neurology*, 12 (6): 563–71.

Danaher, John (2016) "Robots, law and the retribution-gap" in *Ethics and Information Technology*, 18 (4): 299–309.

Danaher, John (2018) "Toward an Ethics of AI Assistants: an Initial Framework" in *Philos. Technol*, 31 (3).

Dietvorst Berkeley, Simmons Jospeh and Massey, Cade (2014) "Algorithm Aversion: People Erroneously Avoid Algorithms After Seeing Them Err", *Journal of Experimental Psychology*, General, 144.

Fuchs, Thomas (2006) "Ethical Issues in Neuroscience" *Current Opinion in Psychiatry*, 19 (6).

Hansson, Sven O. (2005) "Implant Ethics" *Journal of Medical Ethics*, 31 (9): 519.

Heersmink, Richard (2013) "Embodied tools, cognitive tools and brain-computer interfaces", *Neuroethics*, 6: 207–219.

Gilbert, Frederic, Brown Timothy, Dasgupta, Ishan, Martens, Hannah, Klein Eran, Goering, Sarah (2019) "An Instrument to Capture the Phenomenology of Implantable Brain Device Use" *Neuroethics*, https://doi.org/10.1007/s12152-019-09422-7.

Gilbert, Frederic, Cook, Mark, O'Brien, Terence and Illes Judy (2019) "Embodiment and estrangement: Results from a first-in-human intelligent BCI trial" *Science and Engineering Ethics*, 25 (1): 83-96.

Gilbert, Frederic, Pham, Christopher, Viana, John and Gillam, Wayne (2019) "Increasing brain-computer interfaces media depictions: pressing ethical concerns" *Brain-Computer Interfaces*, DOI:10.1080/2326263X.2019.1655837.

Gilbert, Frederic, O'Brien, Terrence and Cook, Mark (2018) "The Effects of Closed-Loop Brain Implants on Autonomy and Deliberation: What are the Risks of Being Kept in the Loop?" In *Cambridge Quarterly of Healthcare Ethics*, pp. 1-12.

Gilbert, Frederic, Viana, John and Ineichen, Christian (2018) "Deflating the "DBS causes personality changes" bubble" *Neuroethics*, DOI:10.1007/s12152-018-9373-8.

Gilbert, Frederic and Tubig, Paul, (2018) "Cognitive Enhancement with Brain Implants: The Burden of Abnormality" *Journal of Cognitive Enhancement*, 2 (4): 364-368.

Gilbert, Frederic (2018) "Deep Brain Stimulation: Inducing Self-Estrangement" *Neuroethics*. 11 (2): 157-165.

Gilbert, Frederic, Goddard, Eliza, Viaña John, Carter, Adrian and Horne, Malcolm (2017) ""I Miss Being Me": Phenomenological effects of Deep Brain Stimulation" *American Journal of Bioethics Neuroscience*, 8 (2): 96-109.

Gilbert, Frederic (2015) "A threat to Autonomy? The Intrusion of Predictive Brain Devices". *American Journal of Bioethics Neuroscience*, 6 (4): 4-11.

Gilbert, Frederic and Cook, Mark (2015) "Are Predictive Brain Implants an Indispensable Feature of Autonomy?", *Bioethica Forum*, 8 (4): 121-127.

Giubilini, Alberto and Savulescu, Julian (2017) "The Artificial Moral Advisor. The "Ideal Observer" Meets Artificial Intelligence", *Philosophy & Technology*, 31 (2): 169-188.

Grübler, Gerd (2011) "Beyond the responsibility gap. Discussion note on responsibility and liability in the use of brain-computer interfaces", *AI & Society*, 26: 377.

Hsu, Feng-hsiung (2002) *Behind Deep Blue: Building the Computer that Defeated the World Chess Champion*, Princeton, NJ: Princeton University Press.

Kingwell, Katie (2013) "Implantable device advises patients with epilepsy of seizure likelihood", *Nature Reviews Neurology*, 9: 297.

Klein, Eran, Brown, Tim, Sample, Matthew, Truitt, Anjali R., Goering, Sara (2015) "Engineering the brain: Ethical issues and the introduction of neural devices", *Hastings Center Report*, 45 (6), 26–35.

Klincewicz, Michal (2016) "Artificial intelligence as a means to moral enhancement", *Studies in Logic, Grammar and Rhetoric*, 48: 171–187.

Lemaignan, Séverin, Warnier, Mathieu, Sisbot Akin, Clodic Aurélie and Alami, Rachid (2017) "Artificial cognition for social human–robot interaction: An implementation", *Artificial Intelligence*, 247: 45–69.

MacDonald, Linda (2012) "Case study: Ethical and Legal Issues in Human Machine Mergers (Or the Cyborgs Cometh)", *Annals of Health Law*, 21 (1): 175-180.

Magliocchetti, Daniele, Gielow, Martin, De Vigili, Federico, Conti, Guiseppe and De Amicis, Rafaelle (2011) "A Personal Mobility Assistant based on Ambient Intelligence to Promote Sustainable Travel Choices", *Procedia Computer Science*, 5: 892-899.

McCarthy, John (1979) "Ascribing Mental Qualities to Machines" In *Philosophical Perspective in Artificial Intelligence* (Martin Ringle, ed.), Sussex: The Harvester Press.

Nyholm, Sven (2018) "Attributing Agency to Automated Systems: Reflections on Human–Robot Collaborations and Responsibility-Loci", *Science and Engineering Ethics*, 24: 1201.

Palermos, Spyridon (2011) "Belief-Forming Processes, Extended", *Review of Philosophy and Psychology*, 2: 741–765.

Palese, Emma (2012) "Robots and cyborgs: to be or to have a body?", *Poiesis & Praxis*, 2 (8):191-196.

Reidsma, Dennis et al. (2016) "The EASEL Project: Towards Educational Human-Robot Symbiotic Interaction" in *Proceedings of the 5th International Conference on Living Machines* (N. F. Lepora, A. Mura, M. Mangan, P. F. M. J. Verschure, M. Desmulliez, & T. J. Prescott, eds.), Lecture Notes in Computer Science, Vol. 9793, Switzerland: Springer International Publishing, pp. 297-306.

Rosenthal, Stephanie, Biswas, Joydeep and Veloso, Manuela (2010) "An effective personal mobile robot agent through symbiotic human-robot interaction" in *Proc. of 9th Int. Conf. on Autonomous Agents and Multiagent Systems*, pp. 915–922.

Sandini, Guilio et al. (2018) "Social Cognition for Human-Robot Symbiosis—Challenges and Building Blocks", *Frontiers in Neurorobotics*, 12: 34.

Santoni de Sio, Filippo, van den Hoven, Jeroen (2018) "Meaningful Human Control over Autonomous Systems: A Philosophical Account", *Front. Robot. AI*, 5: 15.

Savulescu, Julian, Maslen, Hannah (2015) "Moral Enhancement and Artificial Intelligence: Moral AI?", *Beyond Artificial Intelligence*, 9: 79–95.

Nassim, Taleb and Chandler, David (2007), *The Black Swan*, New York, NY: Random House.

The IEEE Global Initiative on Ethics of Autonomous and Intelligent Systems. Ethically Aligned Design: A Vision for Prioritizing Human Well-being with Autonomous and Intelligent Systems, Version 2. IEEE, 2017. Accessed: http://standards. ieee.org/develop/indconn/ec/autonomous_systems.html.

Verbeek, Peter-Paul (2008) "Cyborg intentionality: rethinking the phenomenology of human–technology relations", *Phenom Cogn Sci.*, 7 (3): 387–395.

CHAPTER 14

THE ETHICS OF BRAIN-COMPUTER INTERFACES (BCI)

Aníbal M. Astobiza[1]

University of the Basque Country, Spain

Txetxu Ausin[2]

Institute of Philosophy, Spanish Research Council, (CCHS/CSIC)

Ricardo M. Ferrer[3]

University of Granada, Spain

Stephen Rainey[4]

University of Oxford, UK

Abstract: The ambition to create mind-reading technologies – using Brain-Computer Interfaces (BCIs), to record neural activity in the living brain to predict the thoughts and experiences of individuals – is growing. Already, those technologies that translate neural activity into control signals for computers, robots or prostheses – are constantly being developed and refined. The prospect

---

[1] Basque Government Posdoctoral Researcher UPV/EHU; Visiting scholar Oxford Uehiro Centre for Practical Ethics, University of Oxford; Visiting fellow in Global Health and Socia Medicine, Center for Bioethics, Harvard Medical School, Harvard University; EXTEND and INBOTS (Horizon 2020, ID 779982 and ID 780073), anibal.monasterio@ehu.eus.

[2] Tenured Scientist at the Institute of Philosophy, Spanish National Research Council (CCHS/CSIC); EXTEND and INBOTS (Horizon 2020, ID 779982 and ID 780073), txetxu.ausin@cchs.csic.es.

[3] PhD candidate in Philosophy at the University of Granada EXTEND and INBOTS (Horizon 2020, ID 779982 and ID 780073), ricardo63@autistici.or.

[4] Research fellow in the Oxford Uehiro Centre for Practical Ethics, University of Oxford; BrainCom (Horizon 2020, ID 732032), stephen.rainey@philosophy.ox.ac.uk.

of using this kind of technology outside the clinical setting, a setting which has its own very deep neuroethical conundra, suggests the need for common ethical, legal and social guidelines at the EU and international level. Not least, the potential for the merging of human and machine, human enhancement, and the possibilities for private profit from all of this, make responsible handling of such BCI neurotechnology imperative. In this chapter, we focus on some issues in the BCI literature and research as a blueprint to identify challenges, risks and opportunities for the elaboration of a common ethical and legal framework concerning issues of safety, ethics and data protection.

1. Historical background

Pneuma Psychikon, in Greek, or *Spiritus Animalis*, in Latin, are the names of the ancient theory of animal spirits developed by Alexandrian physicians and philosophers during the third century BCE (Smith, Frixione, Finger & Clower, 2012: 5-9). The theory postulated that each cerebral ventricle was the seat of a specific mental function and contained a unique type of spirit with the power to perform a given function (Cobb, 2002: 395-200). "Animal spirits" were considered to be weightless and invisible things that floated through the nerves to mediate the operations of the living body.

A mixture of Greek philosophy and Christian theology led to the belief that the human body was divided into three main parts, each housing a type of soul. The vegetative soul in the liver was associated with desires and appetites. The heart housed the vital soul and was responsible for the passions and action. The rational soul was immortal and immaterial and resided nowhere in particular. Each of its faculties such as memory, imagination, intellect, were mediated by the animal spirits (Smith, Frixione, Finger & Clower, 2012: 18).

According to Hippocrates, who is often credited with developing the theory, four humors - black bile, yellow bile, blood, and phlegm - must be balanced for good health. This account coexisted with the idea of the animal spirits.

These theories were held true by many scholars from backgrounds as diverse as Byzantine, Arabian and Western occidentals well beyond the Renaissance. Animal spirits and the theory of the four humors dominated the thinking about the nervous system for approximately 1500 years. In 1563, when Andreas Vesalius published *De humani corporis fabrica* ('On the Fabric of the Human Body'), the theory of animal spirits was still prevalent. However, Vesalius refuted the received authority and wisdom of Galen saying that the animal spirits have nothing to do with the cerebral ventricles, the place where they believed to be. For this reason, Vesalius was considered heretic and was forced into exile.

Even still in the seventeenth century with Rene Descartes (1596-1650) the theory of animal spirits dominated the thinking about the nervous system. But Descartes introduced further modifications to the theory of animal spirits. While the ancients thought that the animal spirits were invisible and weightless, Descartes argued that they were actually liquids that flowed through the nerves. Due to the success of mechanics as a science, Descartes described the behavior of animals, including humans, under the same principles that governed the behavior of automata. His ideas were extended in the work *De Homine*, published in the early 1630s. Despite containing many anatomical errors (to modern eyes) this work was very influential because it provided the first mechanistic interpretation of human physiology.

Descartes explained behavior in terms of hydraulic and mechanical statues or automata found in many royal gardens in France. This model or analogy proved to be successful and influential. Descartes argued that the differences between animals and humans were only the intricacy of the underlying mechanisms.

By placing anatomy within the scope of mechanical research, and by considering that animals and humans behave mechanically in the same way, Descartes proposed the study of animals as a model for studying humans. But Descartes was a philosopher more interested in *a priori* methodology than empirical experimentation.

On December 8th, 1664, during a visit to Paris, the dutch microscopist Jan Swammerdam conducted a gruesome but intriguing series of experiments in front of Olaf Borch (1626-1690), a Danish botanist who left a clear tale of the study in his diary. Swammerdam cut off the heart of a living unanaesthetized frog and demonstrated that this had no effect on his ability to move. The severely damaged animal could keep swimming. On the other hand, if the brain was removed, the movement ceased. Later, he showed that if he touched the severed nerve with the scalpel the muscles contracted.

Contrary to what Descartes thought, movement could occur without connection between the muscle and the brain. Moreover, the fact that movement happens at all implied that animal spirits were not responsible for nerve transmission and muscle movement. In subsequent experiments, he tried to show whether the same effect could be observed in an isolated nerve–muscle preparation. This experiment had an immediate impact at the time, and is still carried out in high school in science classes all over the world.

This procedure allowed Swammerdam the mathematization and measurement of the contractile power of the muscle. For neuroscience, the most important contribution of these series of experiments carried out by Swammerdam was to show that movement is produced by an external stimulation of the nerve that Swammerdam himself called "irritation", but is now referred to as "action potential". Swammerdam had clearly shown that the nerves did not work by transmitting animal spirits, but he didn't find an answer to how irritation led to movement. Those ancient and medieval concepts such as animal spirits were rapidly replaced with the advent of this new vision based on anatomy, chemistry and physics.

A completely unexpected 'eureka' moment came at the end of the eighteenth century, when Luigi Galvani (1737-1798) dissected a frog and discovered by chance that the muscles responded to external electrical stimuli. Galvani inferred that the response to external electrical stimuli was an indication of the electrical nature of motor movement. In 1848, Emil Du Bois-Reymond (1818-1896) expanded Galvani´s idea by using a galvanometer to measure the electrical current. Julius Bernstein (1839-1917), who studied under the guidance of Du Bois-Reymond and von Helmholtz, found that what he called "action potential" consists of a self-propagated depolarization of the nerve membrane. At last, animal spirits as a concept disappeared and by the beginning of the twentieth century, it was established that the nervous system is composed of cells and that these cells generate electrical signals.

The German pharmacologist Otto Loewi (1873-1961) was a pioneer in determining that the electrical impulses of cells are transmitted to other cells through chemical messages. Later, Loewi´s ideas were developed further by Henri Dale. The two scholars would go on to share the Nobel prize for Physiology or Medicine in 1936. Dale identified acetylcholine as a putative neurotransmitter. In the early 1950s, Alan Hodgkin and Andrew Huxley carried out a series of experiments where they demonstrated how the electrical impulse or action potential was generated.

1.1. Applications in brain recording and stimulation

In 1929, Hans Berger made another breakthrough with the development of Electroencephalography (EEG) to enable non-invasive recording of electrical signals from the human brain. Subsequently, the emergence of increasing computational power, real-time analysis systems and detailed knowledge of how the brain works laid the foundation for realizing the ambition of controlling external devices directly with neural activity.

Today, researchers have many more techniques at their disposal to investigate the function of the nervous system even at the sub-cellular level. With the knowledge that the nervous system works by transmitting electrochemical impulses from cell to cell, researchers can measure and control it. One ultimate proof of our understanding of the nervous system is reflected in our ability to influence its functioning, either in a normal or pathological state. The capacity to control the nervous system in its pathological state allows us to intervene therapeutically, to improve its functioning in patients suffering from motor disorders.

The research and development of brain-computer interfaces (BCI) offers great promises for research scientists, and with health and wellbeing implications for the general public. BCI technologies can play a role in the restoration of functions damaged by nerve, spinal cord or brain injuries. BCI technology measures the electrical activity of the brain – no longer conceived of as animal spirits – and use it to realise control over machines, such as computers and prosthetic devices. BCI are built on the basis that sensory, motor, and cognitive information is represented in the brain by networks of neurons, and computer algorithms are designed to record and decode their combined electrical activity.

Non-invasive neurotechnologies, such as extracranial EEG and magnetoelectroencephalography (MEG), record neural activity by placing electrodes on the scalp. Owing to this approach, these technologies are intrinsically unidirectional (i.e., they record but do not stimulate the brain) and of limited capacity. BCIs, on the other hand, can be invasive or non-invasive. Regarding invasive BCI, these techniques use electrodes placed on the surface or inside the brain. Electrical signals are recorded using electrocorticography techniques by placing electrodes, during surgery, under the dura, a protective membrane covering the brain. These technologies, while allowing two-directionality (i.e., recording and stimulation) and better performance, rely on potentially very complex surgical procedures to place the electrodes correctly.

The history of BCI begins with the first attempt to control brain signals on a neurophysiological basis in 1968. Wyrwicka and Sterman (1968) recorded sensory-motor rhythms of animals (cats) and then translated these rhythms into sensory feedback to reward the animals. At the same time, a research technique called neurofeedback had already been tested in humans. This technique consisted in the regulation of brain activity by the subject himself. Kamiya (1969) published the first study where healthy individuals learned to control their alpha waves. With other findings in operant conditioning where it could be demonstrated that it is possible to control the firing of single cortical neurons in monkeys, all of this supposed the neurophysiological foundations for the BCI technology.

The term "brain computer interface" was first proposed by Jaques Vidal in 1973 when he presented a system that could translate EEG signals into computer-controlled signals. BCI technologies can be classified as invasive and noninvasive. The function of all types of BCI are similar. The brain signals detected are amplified and then filter out using algorithms. Brain signals are classified according to their features, for example, sensorimotor rhythms. After processing and decoding all the relevant signals from the brain, the output can be used to control the movement of a cursor, prostheses, wheelchair, exoskeletons5 , etc. But also the BCI technology can be used to communicate.

BCI technology enables users to open a bridge of communication with the world. People with paralysis or other motor disorders can express their ideas or silent thoughts. Therefore, a variety of disabled people can express their ideas through different methods such as spelling applications, semantic categorization, or silent speech communication. But the potential applications of BCI technology go beyond the clinical and therapeutic context as assistive technology. Today, BCI technology can be used to monitor the inner state or conditions of people (Garcia-Molina, Tsoneva and Nijholt, 2013: 11), detection and diagnosis (Poulos, Felekis and Evangelou. 2012: 711-712), rehabilitation (Farrell and Herr, 2011: 6041), or even neuroergonomics (Parasuraman and Rizzo, 2008: 3).

2. Dual-use research of Brain-Computer Interfaces

According to the National Institutes of Health, Office of Science Policy (March, 2012), dual-use research of concern is defined as:

> ...research that, based on current understanding, can be reasonably anticipated to provide knowledge, information, products, or technologies that could be directly misapplied to pose a significant threat with broad potential consequences to public health and safety, agricultural crops and other plants, animals, the environment, material, or national security.

Until relatively recently, the expression "dual-use" referred primarily to technologies that could have both civil and military applications. Nowadays, the concept has multiple meanings. The advance and progress of technology (and science) creates demands on scientists in general to consider a range of potential misuse possibilities alongside the security implications of dual-use research. This latter sense of the phrase will be used to refer to that technology that can be misused, or the dissemination of scientific knowledge that can be misapplied, and always in relation to BCI.

In the following sections, some of the broader issues to be mentioned include human enhancement; non-maleficence, beneficence, and safety;

---

[5] The ceremony of the 2014 Football World Cup, held in Brazil, featured a paraplegic man who used a mind-controlled robotic exoskeleton to kick the ball at the opening. See a description of the technology used in the semi-autobiographical book of the researcher behind this feat (Nicolelis, 2011).

standards; autonomy and personal identity; data protection and privacy; justice, and governance. Whilst a complete account of each of these areas will be impossible owing to constraints of space, nonetheless, it is informative to collect these issues together.

The scientific and policy communities have attempted to enact ethical codes and guidelines for scientists or professionals as a response to dual uses of research. These guidelines attempt to emphasize safety, and incentivize researchers to exercise control over their research. This creates a tension or conflict between obligations and freedom of research. The emergence of any new technology brings with it promises and perils. The great challenge of any policy recommendation, or ethical principle, is to minimize risks while preserving the promises.

Some have argued, especially in relation to biotechnology but perhaps applicable to any scientific research with the potential to be misused including BCI technologies, that research should proceed without the intervention of nonscientists or regulators (Pinker, 2015). They even promote the eradication of the precautionary principle, claiming it hinders research. However, this is subjective and highly contentious. Scientists themselves do not necessarily have the skills to discern and to determine the scope, nature, goals, and underlying values of their own research, or whether that research can cause harm to individuals or society at large. Professional obligations in relation to the possible misuse of research results are not discussed widely enough within the scientific community. Any obligation must be based on the ethical principle to prevent harm, while nonetheless noting that this very concept itself requires strenuous analysis.

The meaning of harm, or what to prevent, is not always apparent. An important distinction is between intentional and unintentional harm (to be discussed in more detail in the subsection 3.1.). Awareness is key to differentiating between intentional and unintentional harm. Being aware means having an active reflection on the consequences of research. There are multiple ethical issues in relation to dual-use research, and if a scientist is unaware of the consequences, this is one serious such issue. Low levels of awareness can lead to misunderstanding about the dual use of research.

Multiple ethical obligations and principles have been discussed in the scientific community, but some of them are the following:

- The duty and obligation to consider negative consequences of research
- The duty and obligation to prevent generally unforeseeable harm
- The duty and obligation not to publish or share sensitive information
- The duty and obligation to oversee and limit access to dangerous material

These ethical obligations are of a generic character applicable to any dual-use research. Regarding BCI technologies ethical principles are also applicable, but neurotechnologies demand specific ethical considerations. Tennison and Moreno (2012) have reviewed the research with neurotechnologies and have established three categories that can be included in the concerns of dual-use research. These categories are: *brain-computer interfaces* (BCIs), *neurotechnologies for warfighter enhancement,* and neurotechnological systems for *deception detection and interrogation.* Here we only focus on the first category. As we shall see throughout this chapter, BCI neurotechnology allows the bypassing of an individual's nervous system to interact with the physical surrounding world.

BCI technology can be applied in clinical contexts to restore mobility and to assist people using neuro-controlled technologies. It can have applications in neurodegenerative disorders such as Alzheimer's disease. Outside the clinical setting, this neurotechnology has commercial use in the entertainment industry, for example, in video games, etc. Due to its increasing portability and compatibility with smartphones, the technology is gaining popularity owing to its potential for use in self-monitoring, self-quantification, and even pure entertainment. However, alongside this, probably the greatest problem with the dual-use research of neurotechnology, including BCI, is that it can be a potential target for non-state actors and terrorist groups.

For each BCI application, there is an ethical issue in addition to the general ethical principles which reflects the concern for the dual-use research of neurotechnology. Although BCI technology can offer great social and therapeutic benefits, it may also provide the basis for great harms, with the potential to influence individuals' actions, thereby diminishing autonomy. The interconnected world in which we live, new information and communication technologies, such as the Internet of things (IoT), exacerbate the constant flow of information and neurotechnologies such as BCI enhance people's ability to communicate with their environment. This can be translated into an alienation or overexploitation of people by the economic system.

Another major potential problem with neurotechnologies is the way in which allocations of responsibility are carried out (Grübler, 2011: 377-378). In discussions on responsibility a major concern is 'the problem of many hands'. The issue of allocation of responsibility emerges when in a complex system, such as a BCI system, there is a chain of events where many people have a share in any action that may lead to unexpected or unintended consequences, perhaps causing harm. Who is to blame? The development and implementation of neurotechnologies includes industry, engineers, software programmers, technicians, physicians, and ultimately the user. Besides this, there is also the remote, though nonetheless increasingly possible prospect of "brain hacking".

As we shall see in subsection 3.4., there is a real danger of "brain hacking" where people with malevolent intentions intercept brain signals from BCI devices to use them for their own ends. The most speculative scenario is that of 'mind-melding'. In this scenario, the possibility that an individual's consciousness is melded to other consciousness arises. This might lead to situations where individual consciousness disappears altogether, and a collective consciousness emerges in its place. This would create a vast distance between the consciousness of a single individual and this global consciousness, similar to the difference between our consciousness and that of other animals.

Responding to all these issues and concerns about dual-use in research on neurotechnology will require specific ethical principles with particular applications in mind.

3. Ethical issues in BCI.

Various BCI applications may appear to some observers almost science fictional. BCI technology allows the possibility of using neural correlates of thoughts as triggers to control machines. Bidirectional feedback, as intended by the research project EXTEND (http://www.extend-project.eu/), includes communication between a user and the physical system that can restore some degree of motor function in paralyzed people. The user, thanks to this feedback, can translate intention into action despite his or her inability to perform overt physical actions.

The application of this neurotechnology raises many ethical questions. Included among these are the expectations of caregivers and patients surrounding the promise of the technology, and the level or degree of invasiveness of its implementation. Regarding the level of invasiveness, different types of electrodes expose subjects to varying levels of risk, and imply different benefit / risk ratios. These must be weighed against the probability of success or failure of the technology. This, in turn, raises the issue of informed consent. Patients who cannot communicate will have representatives acting on their behalf and deciding in their interests. There are other associated ethical issues, such as whether doctors should make paternalistic decisions or not, or even if surgery is ethical at all in some cases.

Beyond the context of therapeutic use of BCI, this neurotechnology can be used to enhance performance of non-patients – cognitively and physically intact people. It is possible that in relation to BCI, a system can augment an existing capacity beyond the average, i.e., that the therapeutic technique may be applied in a non-therapeutic setting in order to bring advantageous results to the non-patient. If, in short, BCI neurotechnology achieves the remote control of external devices, we can even imagine scenarios where robots can be controlled at a distance through neural activity. This idea might seem like telekinesis, but it is a practical application of neurotechnology – a real and scientific telekinesis.

Human enhancement through BCI is already in itself a matter of controversy, but among the many ethical issues raised there is a central one concerning normalization. No-one should be expected to use this technology, such that non-enhancement becomes a stigma. The potential of enhancement as a form of imposed discrimination may be inherent given that the use of BCI confers an advantage on all those who use it. Those who do not want to, or refuse to use this technology, may suffer economic or social cost. We can conceive multiple scenarios of an unequal distribution of access to this technology that can reinforce the differences between groups and classes of people.

3.1. Nonmaleficience and safety

According to a definition of the principle of nonmaleficience, the obligation not to do harm covers not only intentional actions, but also imposing risks of harm. Therefore, people may harm or endanger another person without harmful intent.

The Institute of Electrical and Electronics Engineers (IEEE), the world´s largest technical professional standards organization, has provided new standards for robotics and AI that prioritize human well-being and ethical considerations. In the document entitled "Ethically Aligned Design: A Vision for Prioritizing Human Wellbeing with Artificial Intelligence and Autonomous Systems" it is provided a series of standards that include the following:

- *Standard for ethically driven nudging for robotic, intelligent, and autonomous systems*

This standard evaluates and analyzes the design and user interface of any artificial and robotic system to identify possible hidden "nudges" that impose a specific way of interacting with a system that can affect users.

- *Standard for fail-safe design of autonomous and semi-autonomous systems*

The IEEE points out the need for safety-oriented design in robots and artificial systems. Moreover, the regulatory schemes and legal system ought to seek balance between innovation and safety. Safety is one of the social values to be highlighted when humans interact with semiautonomous or autonomous systems. A standard on safety needs to establish a clear procedure for measuring, testing, and certifying an autonomous system's functionality.

- *Well-being metrics standard for ethical artificial intelligence and autonomous systems*

All relevant stakeholders, engineers, programmers, roboticists, and technologists must ensure and promote the well-being of humans in terms of economic growth and productivity. This standard includes measures and indicators of well-being in order to align the data produced by these systems to conform with the needs and desires of human users.

Promoting user safety entails the combination of the principle of beneficence (discussed below) and the principle of nonmaleficience. In many respects, to ensure safety in all the products, services and processes derived from BCI, robotics and AI systems, innovation need to comply with the legal regulatory framework.

3.2. Beneficience

Beneficience is doing good and not doing harm (Belmont Report, 1978). In other words, in any biomedical research or intervention, the potential benefits must outweigh the possible risks to subjects. But in order to know how to apply this basic bioethical principle, it is necessary to have a definition of what good is, and avoid the risks of something bad that could happen. BCI neurotechnology can restore the mobility and communication of people who suffer from motor disorders or have been paralyzed by, for example, a traumatic accident. BCI does not alter the brain itself, but is a passive device that picks up electrical activity from the brain to be processed by an algorithm on a computer that in turn controls an external device, such as a robotic prosthetic arm. This means that ethical considerations are not like other techniques or therapies where, for example, pharmacology or drugs can alter specific cognitive and mental processes. This kind of intervention can arguably affect the values and interests of individuals. But BCI can affect major dimensions of individual users such as their autonomy, or personal identity.

This kind of concern is separable from the health risks of BCI-related implants. These exist mainly in terms of the degree of invasion needed to measure electrical activity. Implants used in invasive BCI entail risk of tissue damage, infection, and device failure. The implantation procedure through a surgical operation may also involve considerable risks. But either way, it can be thought that society is ethically obligated to provide an improvement to those individuals who suffer challenges to their daily functioning. These obligations surround improving the patient's quality of life, their caregivers' conditions, and enabling the BCI-user a return to a more functional life with family, work and friends if those benefits outweigh the risks. So when analyzing the potential risks and/or benefits of BCI it is imperative to promote these kinds of goods.

The risks to be taken into account to promote such goods can be categorized in multiple ways, but some of the most pressing are the psychological and physical risks to patients (and caregivers). One of the possible psychological risks is that of unfulfilled promises. With each innovative technology patients, caregivers, family members and society at large, can have and develop false hopes on the results of applying a certain intervention which does not result in any improvement.

Physical risks tend to be minimal if the neurotechnology is applied with proper care by professionals in the right context. But neurotechnologies, unlike many other areas, invite an ethos of DIY culture predicated on the ready availability of self-styled neuroactive devices. This DIY brain-hacking approach could yet prove to have detrimental outcomes overall.

3.3. Autonomy + personal identity

There are almost daily reports of the amazing advances being made in BCI technology. For example, in 2012 a 52-year-old woman paralyzed from the neck down was able to 'use her mind' to control a robotic arm to pick up objects in her surroundings such as a cup of tea on a table etc. (Courtine et al., 2012: 515-517). This woman, called Jan, had implanted tiny grids of electrodes into her brain. The electrodes were placed near neurons that control arm movement in the motor cortex.

As a user of BCI might this technology change how we think about ourselves? For example, as this technology gets more integrated physically and psychologically with the people who use it, the distinction between the machine and the self will become more blurred. Phenomenological descriptions of patients using BCI and neuroprostheses resemble the use of tools for a purpose or goal. It requires from patients effort, careful attention and the activity of other cognitive operations to master the use of a robotic arm or another type of device. But we can imagine that as technology progresses those devices may no longer be what we understand as 'tools'. It is tempting to ask if some day there will be a point where the neuroprostheses become part of the people who control them. How this will affect the sense of autonomy of BCI users is an empirical question but given the plasticity of the motor cortex and of our body scheme (Maravita and Iriki, 2004: 81), it is not irrational to think that external objects can become part of our sense of autonomy and identity.

Neurotechnologies in general, but BCIs specifically, raise ethical and social challenges regarding self-ownership of our thoughts, and emotions. For instance, closed-loop systems, or bidirectional BCI, can monitor and adaptively modify the brain, transforming the phenomenological experience of the user affecting his or her own sense of autonomy and identity and ultimately the way they view themselves and their relation to others.

3.4. Data protection and privacy

We live in the era of Big Data. Sophisticated algorithms and Artificial Intelligence techniques are applied to large amounts of data to find recurrent statistical patterns that can be used to predict and understand people's behavior. As a result of the increase in the capabilities of computers and the large amount of data produced, we are currently witnessing a data deluge. It happens in industry, business, commerce and, of course, science. With this amount of data, data governance is necessary and something that is often overlooked – *data sovereignty.*

Data sovereignty means identifying and recognizing the authorship and ownership of data. From the traditional copyrights regulations to the data we create through our activity on social networks on the Internet. In science, the current ecosystem of large collaborative research poses an incredible burden on the successful governance of data. Science in general needs to be centrally concerned with data collection, curation, and distribution in order to allow scientific activity and practice. And it needs to be done in accordance with principles and practices of responsible research and innovation.

Having said this, in the field of neurotechnologies, and BCI in particular, there are two very important ethical considerations: *data protection* and *privacy.* Sensitive data includes your password to access your bank account, medical records, your name or home address. But if we were to characterize the data obtained from neurotechnological advances, these are arguably much more sensitive data. Only now are we able to record, measure and manipulate mechanically the very brain/mind- arguably, the seat of ourselves.

Neurotechnologies, and especially BCI, have a great social and individual benefit. They can be used to treat people with motor disorders and communication; can be used in the military; in advertisement and the entertainment industry. As revolutions in the digital ecosystem and systems continues, and more neurotechnological applications emerge, greater knowledge and understanding of the brain will come. This may mean a better understanding of the neural correlates of multiple mental states, but it is not without risks and dangers.

If these neurotechnologies are not used with caution, they put individuals at an unprecedented risk. They can involve intrusion and invasion of the most fundamental personal privacy: the sphere of the mind. It is very common to think of the mind as being the fundamental base of personal freedom and self-determination. But with unprecedented neurotechnological development, and progress in neuroscience the intricacies of the mind/brain are revealed.

Because of these neurotechnological developments, Ienca and Andorno (2017) call for an ethical and legal reconceptualization of Human Rights. They believe that we must protect the mind as the ultimate refuge of our freedom and privacy from excessive intrusion through neurotechnology. These reconceptualizations would lead us to introduce *neurorights*. It is true that neurotechnology could record the activity of our brain and cause physical and psychological damage. But imagine the damage that can cause an interface implanted in our brain chronically, a brain exposed to the continuous manipulation and observation from the outside. This is a serious concern.

Because neurotechnology has the potential to threaten mental integrity, cognitive liberty, self-determination, international human rights law ought to consider the possibility of a new proviso to protect misleading intrusion in our last sphere of privacy. To the extent that neurotechnology becomes more accessible and used in multiple contexts, risks and threats increase similarly to other activities and sectors in the digital ecosystem.

A group of researchers (Martinovic et al., 2012: 35) have shown how it is relatively easy to extract information from users of BCI technology directly from the activity of their brains. The information that can be extracted from hacking systems or technologies such as BCI can be banking passwords and details, home address, etc. Very sensitive information, indeed. It is for that reason that Ienca and Andorno (2017) propose the reconceptualization of existing human rights and even the creation of new human rights, the so-called, neurorights: the right to cognitive liberty, the right to mental privacy, the right to mental integrity, and the right to psychological continuity.

We agree with Ienca and Andorno (2017) that protecting the mental dimension as the last refuge of our freedom and privacy is a matter of great urgency in the face of the possible forms of exploitation that new neurotechnologies can exert. It is imperative to maximize the benefit that neurotechnologies can offer to society at large while reducing any regressive impact of the use of neurotechnologies on rights and liberties.

3.5. Justice

The question of justice in science and technology policies deals centrally with how to allocate scarce resources to the people. Distributive justice, in contrast to other types (e.g., commutative or retributive), requires or entails the pursuit of equality. Amartya Sen (1992) begins his seminal work with the following two questions: "why equality?" and "what equality?" Equality for Sen is a multidimensional space where different people can put their emphasis on very different areas. On the other hand, Sen acknowledges that increasing equality in one dimension necessarily diminishes in another one. Distributive justice finds a stark and real difference with inequity, which is a normative term denoting an unjust or unjustified state of affairs, and inequality which is a term referring to uneven distribution, right or wrong. Equality can be inequitable and inequity can be just, always from the point of view of moral philosophy when it comes to distributing.

Several traditions in political philosophy have different views about why and what to distribute. For example, utilitarianism considers that a given state of affairs is fair as long as it increases total well-being. Social contractarianism views a state of affairs as fair whenever rational individuals freely agree to after deliberation. Neoliberal or libertarians are also market-oriented as utilitarians, but they consider a distributive system as fair and just only when individuals are able to exercise their rights (e.g., property rights).

Many policies in science and technology all over the world embrace some component of these traditions in political philosophy. But whichever policy is to be applied, a good science and technology policy should aim to promote equality. When we talk about justice in the field of neurotechnologies and BCI means that it is needed to respect autonomy, personal identity, privacy and equality.

4. A public deliberative approach to guide and govern emerging neurotechnologies

How ought we to regulate BCI? Throughout this chapter, it has been discussed how important ethical issues arise from the application of neurotechnology. Issues from safety to justice. We need to keep in mind that the advance and progress of neurotechnology opens up multiple possibilities for misusing BCI devices which require careful ethical supervision. For that reason, the dual-use character of neurotechnology needs to be carefully regulated through a legal framework that provides a good control and monitoring, risk assessment and public and deliberative approach to guide and govern these emerging neurotechnologies. In other words, we need an international body to regulate BCI technology. The serious risks and potential for misuse of neurotechnology, i. e. the dual-use character of neurotechnology, has even led scientists to propose a moratoria or total ban on its use. Others argue for an implementation of the precautionary principle as a rule of thumb. Whatever the most successful proposal, the issue of how to deal with the problems arising from neurotechnology should involve all the relevant stakeholders to reach out a consensual solution.

A total ban would be, for many, a disproportionate measure as a policy, but on the other hand, implementing a precautionary principle could lead to the stagnation of research. In this tradeoff, we must bear in mind that it is not possible to predict the future development of neurotechnology or any repurposing of BCI for malicious activities. For this reason, instead of a top-down policy, a public deliberative approach to guide and govern emerging technologies is the most pragmatic decision.

What sees wide agreement among European researchers in particular is the difficulty of accepting military funding for neuroscience research. One fear emanating from this would be the *weaponization of neuroscience*. With the help of the public and society at large, the scientific community must find the most effective way to regulate its research and thus prevent the problems of dual-use research of neurotechnology.

However, one thing we must be clear about is that scientists cannot prevent their research from being misused. They must work to promote social benefit, but other filters and guarantees must be established. One of the most relevant initiatives in this regard is to communicate to citizenries, via open awareness-raising campaigns, the importance of neuroscience knowledge and education. This will aid the avoidance of sensationalism and explain to the general public the scientific view of benefits of good technology use. A clear scientific position will aid the kind of pragmatic discourse required to ground deliberations on public policy.

What is meant by a deliberative approach to governance? The rethoric of deliberation requires and emphasizes a form of a discourse among citizens to frame policies. In other words, a deliberative approach consists in engaging consumers and citizens in the governance of science and technology. The STAGE ('Science, Technology and Governance in Europe') project, sponsored by the European Commission (https://cordis.europa.eu/project/rcn/58206_en.html), described a typology of governments worth mentioning here. The STAGE typology of six forms of governance is based on an analysis of the different governance frameworks of the 8 countries that formed the project.

1. *Discretionay governance*: Policies within this category are made without the direct participation of civil society.
2. *Corporatist governance*: This type of governance recognizes the different interests of the parties involved in a debate.
3. *Educational governance*: This type of governance recognizes the ignorance of citizenship and, in order to mitigate it, offers the creation of public policies for science education

4. *Market governance*: This a type of market-oriented governance in which the only participation of citizens is as consumers in a demand and supply logic.
5. *Agonistic governance*: This is a type of governance where policies are made through a process of confrontation and adversarial politics.
6. *Deliberative governance*: This type of governance rests on the assumption that open debate and engagement is the foundation of political decision-making.

Deliberative governance appears to be the most appropriate approach to regulate emerging neurotechnologies. This is not just because it is an ideal type- where all relevant stakeholders participate in the decision-making process- but because it fits with the values of many standard western liberal democratic societies. The challenges are how the principles of deliberative and participatory governance can play out in practice.

To be truly deliberative a process must involve:

- Discussion with participants (patients, families, practitioners, NGOs, civil society organizations, etc.)
- Free and open expression of interests, needs and interests in light of the decision to be made by all participants
- Access to relevant information by all participants
- Participants should be as much diverse and descriptively representative as possible

5. Conclusions

Progress and advance in the ability to monitor, record and manipulate the brain has made it possible to create neurotechnology such as BCIs. We have seen how this technology may present risks and cause harm if used in detrimental ways. The dual-use character of neurotechnology in general and BCI in particular makes it necessary to explore its ethical ramifications ranging from safety, data protection, privacy and the need to create an international body to oversee ethical design of neurotechnology with the participation of all relevant stakeholders.

References

Cobb, Matthew (2002) "Exorcizing the animal spirits: Jan Swammerdam on nerve function", *Nature Reviews Neuroscience*, 3: 395-200.

Courtine, Grégoire et al. (2012) "Brain-machine interface: closer to therapeutic reality?", *The Lancet*, 16, 381 (9866): 515-517.

Farrell, M. Todd and Herr, Hugh M. (2011) "A method to determine the optimal features for control of a powered lower-limb prostheses" in *Engineering in*

Medicine and Biology Society, EMBC, 2011 Annual International Conference of the IEEE, 6041-6046.

Garcia-Molina, Gary, Tsoneva, Tsvetomira and Nijholt, Anton (2013) "Emotional brain-computer interfaces", *Int J Auton Adapt Commun Syst,* 6 (1): 9-25.

Grübler, Gerd (2011) "Beyond the responsibility gap. Discussion note on responsibility and liability in the use of brain-computer interfaces", *AI and Society,* 26 (4): 377–382.

Ienca, Marcello & Andorno, Roberto (2017) "Towards new human rights in the age of neuroscience and neurotechnology", *Life Sciences, Society and Policy,* 13: 5.

Kamiya, Joe (1969) "Operant control of the EEG alpha rhythm and some of its reported effects on consciousness." In *Altered states of consciousness* (Charles T. Tart, eds.), New York: Wiley, pp. 519–529.

Maravita, Angelo & Iriki, Atsuhi (2004) "Tools for the body (schema)", *Trends in Cognitive Sciences,* 8 (2): 79-86.

Martinovic, Ivan et al. (2012) "On the feasibility of side-channel attacks with brain-computer interfaces" in *Proceeding Security'12 Proceedings of the 21st USENIX conference on Security symposium,* 34-46.

Nicolelis, Miguel (2011) *Beyond Boundaries: The New Neuroscience Connecting Brains with Machines and How It Will Change Our Lives.* New York: St. Martin´s Griffin.

Parasuraman, Raja & Rizzo, Matthew (2008) "Introduction to neuroergonomics" In *Neuroergonomics The Brain at Work* (Raja Parasuraman and Matthew Rizzo, eds.), Oxford: Oxford University Press, pp. 3-13.

Pinker, Steven (2015) "*The moral imperative for bioethics.*" Accessed May 22, 2018 https://www.bostonglobe.com/opinion/2015/07/31/the-moral-imperative-for-bioethics/JmEkoyzlTAu9oQV76JrK9N/story.html.

Poulos, Mario, Felekis, Theodoros & Evangelou, Angelos (2012) "Is it possible to extract a fingerprint for early breast cancer via eeg analysis?", *Med Hypotheses,* 78 (6): 711-716.

Sen, Amartya (1992) *Inequality Reexamined.* Oxford: Oxford University Press.

Smith, C.U.M., Frixione, Eugenio, Finger, Stanley & Clower, William (2012) *The Animal Spirit Doctrine and the Origins of Neurophysiology,* Oxford: Oxford University Press.

Tennison, Michael & Moreno, Jonathan (2012) "Neuroscience, Ethics, and National Security: The State of the Art", *PLoS Biol,* 10: e1001289.

UNITED STATES (1978) *The Belmont report: Ethical principles and guidelines for the protection of human subjects of research,* Bethesda, MD.: The Commission.

Wyrwick, Wanda & Sterman, Maurice (1968) "Instrumental conditioning of sensorimotor cortex EEG spindles in the waking cat", *Physiol. Behav.* 3: 703-707.

CHAPTER 15

WISDOM AS META-KNOWLEDGE: A PRACTICAL APPLICATION OF ARTIFICIAL GENERAL INTELLIGENCE AND NEURAL MACROSENSING

Natasha Vita-More[1]

University of Advancing Technology, USA

Abstract: Relevant to moral values and ethical choices, there are intangible experiences throughout life that can hinder the ability to make sensible decisions and principled reasoning. Objective analysis is necessary to evaluate problems; however, it is not enough. It needs a process that understands the continued growth and expansion of knowledge. It requires a commitment to the integration of logical reflection and emotional expression in understanding a relational ecology in which knowledge is not a static action; but one that changes, grows, and evolves.

With the blending of technology and human biology, many are speculating about ways to improve the quality of life. Apart from the use of technological devices and applications to steer our actions, the applied practice of critical thinking is needed. For example, exercising mental alertness is needed in questioning information, identifying nonsense, being self-responsible, investing in healthy habits, and developing time management skills. In doing this, a framework is needed to examine daily actions, especially from a variety of perspectives – self-knowledge, laws-of-nature, and the complexity of the systems that we participate in. This knowledge becomes a means for a better understanding about how to prepare for aligning a desire for purpose in life with realistic expectations.

The achievement of wisdom affords a meta-knowledge: knowledge about knowledge. Put differently, it would be the sum of the range of what is

---

[1] Professor of Ethics, Business & Entrepreneurship, University of Advancing Technology. Executive Director, Humanity+, Inc.

perceived on a moment-to-moment basis along with the foresight to analyze what is yet to be perceived.

This chapter proposes that artificial general intelligence (AGI) is a likely tool to appropriate personalized meta-knowledge. Further, neural macrosensing is a necessary component of this appropriation because it provides a wide range of analytics of neuron functions and human perceptions arrived at through sensing. Therefore, neural macrosensing is a necessary tool for observing both rational decisions and emotional reactions that can help expand our needs and effectively report these activities to our brain's frontal lobe.

1. Approach

Is it possible to partner two divergent yet well-defined technologies and their distinct protocols to assist our brain in acquiring wisdom? Exploratory engineering is bringing about methods to assist higher-order cognitive processes that warrant careful consideration. These processes typically include logic, problem-solving, and decision making. A synergistic approach to the cognitive process of decision making by the human in conjunction with the machine is by the National Research Council, and its affiliates in 2014, as stated below:

> The study's sponsors came to the National Academies with a problem: The amount of inputs for complex decision making, and the availability of computing assistance for that process, has outpaced our ability to efficiently and effectively exploit it all. How, they asked, might humans and computers team up to turn data into reliable (and when necessary, speedy) decisions?

The proposition of coupling AGI and neuro macrosensing is that it could be a logical and synergetic relationship. Both are currently researched with adequate funding and both offer substantial potential. Thus, merging the exponential technology of AGI with a speculative technology of neural macrosensing is plausible. As described by Robert Freitas in *Nanomedicine* (1997):

> The ability to detect individual neural cell electrical discharges noninvasively in many different ways [sic], coupled with the abilities (A) to recognize and identify specific desired target nerve cells [sic] and (B) to pool data gathered independently by spatially separated nanodevices in real time [sic], offers the possibility of indirect neural macrosensing of complex environmental stimuli by eavesdropping on the body's own regular sensory signal traffic (Freitas, 1997).

2. Artificial General Intelligence (AGI)

The first technology in this discussion is artificial general intelligence (AGI) addresses the original aim of artificial intelligence – that of building machines with human-level intelligence. Unlike narrow AI, which is attentively task-oriented, AGI is based on the process of developmental learning to gain and build on intelligence through interactions and experiences rather than being programmed with intelligence, de facto. Further, by optimizing a system's knowledge acquisition and integration of such knowledge, AGI is designed to foster self-awareness and bootstrap intelligence that can lead toward the development of meta-knowledge. The most successful seminal designs were likely to be Seed AIs that were theoretically programmed for self-modification and recursive self-improvement. (Yampolskiy, 2015) The developed system might also have full access to the Web's facilities, plus substantial distributed learning. However, more currently, the proposed system could draw from the research underway between human and computer of the 2019 "Loving AI" project:

> Loving AI is a research project addressing how AI agents can communicate unconditional love to humans through conversations that adapt to the unique needs of each user while supporting integrative personal and relational development. It is a collaboration between Lia Inc., Hanson AI, and SingularityNet. (Loving AI, n.d.)

For the purposes of this chapter's argument, a variety of approaches could develop a symbiotic relationship between computer and human. The vision of this approach was first established in the 2004 version of "Primo Posthuman (originally designed in 1999) as a metabrain augmentation, referred to as "Substrate Diverse Persons". This metabrain[2] would provide an assistant – a foresight analyst to support its human counterpart on cognitive and emotional levels. The relative aspect of this relationship refers to the acquisition of wisdom and that meta-knowledge, meaning the knowledge of knowledge, could be a natural outgrowth of the liaison. As an augmentation, a narrow to strong AI foresight analyst would be developed as a relatively non-invasive approach; however, it could also be developed as a chip imbedded in the frontal lobe. Here it could streamline information feedback as light-activated ion channels for remote control of neural activity. (Kramer 2004) The foresight analyst would also be a new, emotional part of ourselves to bounce ideas off of; similar, but far more advanced than the Remembrance Agent (RA)

---

[2] The term metabrain was first used by Dr. Grey Fahy. In this chapter, it means a supplement the neocortex to enhance neurological processes.

computer, designed by Bradley Rhodes (1996) at MIT Media Lab, that watched over the shoulder and suggested information relevant to what a person is reading or writing.

Consider that an AGI foresight analyst that functions as both a moment-to-moment assistant and a colleague that offers foresight, will no doubt affect our thinking for the better. The AGI foresight analyst would learn and/or be taught, specifically programmed and unique to each person; like an illusionary angel sitting on our shoulder, giving *good* advice. The foresight analyst would be a complementary insider who listens to our emotional responses to the world around us and proactively integrates with information gathering. It would be personalized to know when and how to interact with us and reside unnoticed until petitioned to perform a task, and counsel when guidance is needed. In fact, it might be similar to Freud's structural theory (Schafer, 2017) in which the foresight analyst, as a superego, would be in touch with the elements of consciousness and work to balance their integration.

3. Neural Macrosensing

The second technology is neural macrosensing. Macrosensing, a concept created by scientist and nanotechnology expert Robert A. Freitas Jr., relates to medical in vivo nanorobotics for the detection of somatic states as well as extrasomatic states, such as sensory data originating outside the body. In theory, it is the "ability to detect individual neural cell electrical discharges non-invasively ... [and] offers the possibility of indirect 'neural macrosensing' of complex environmental stimuli by eavesdropping on the body's own regular sensory signal traffic." (Freitas, 1999)

Neural macrosensing would result from millions of nanorobots located throughout the body's sensory system and record and transport these sensory responses to the brain's cognition center. The brain would be in constant contact with the sensory responses to develop an advanced ability to sense our behaviors. The nanorobots would detect individual neural cell electrical discharges non-invasively and recognize and identify specific desired target nerve cells and pool data gathered independently by spatially separated nanodevices in real time.

4. Privacy Concerns – Security

As a relevant and timely issue concerning bodily intervention, especially regarding our brains and personal identity, equally important to protect the privacy and security of our brains from surveillance and hackers. This field will be on par with advances in AGI and neuro macrosensing. Further, it may be that personal identity cybersecurity and network forensic will precede the development of human-computer based wisdom. The field of cybersecurity is growing rapidly, far more rapidly than AGI and nanotechnology.

On a personal note: I am a senior professor in ethics and philosophy and my main concern is the protection of personal identity. I am well aware of the haunting trajectories before us and I am to work toward helping my students learn about the future of technology – even if it may seem to be the far future – that will cause us to be careful about protecting our personal worth and identity.

5. AGI and Neural Macrosensing

If AGI could assist our decision-making skills, the foresight analyst would need a full understanding of human feelings, emotions and responses to objectively counsel its human counterpart. Since neural macrosensing is programmed to listen to the senses and direct data impulses to the thinking center of the brain, and AGI is a learning intelligence designed to adapt to and assist the brain; it is possible that the two distinct technologies could work cohesively together.

6. AI and Biological Senses < AGI and Neural Macrosensing

Both AGI and neural macrosensing are the innovations of exploratory engineering based on theoretical foresight, similar to the inventions Leonardo da Vinci sketched out long before they could be realized. However, the coupling of AGI and neural macrosensing is based on current tangible advances in their respective fields. The motivation behind the advances is due in large part to overcoming hurdles within each technology and the overwhelming concern that computing power is advancing rapidly and could overtake human intelligence.

Dr. Marvin Minsky, referred to as the father of AI, said, "We have got to get back to the deepest questions of AI and general intelligence and quit wasting time on little projects that don't contribute to the main goal." (Stork, 2000) The late Dr. Stephen Hawking claimed that computer intelligence will surpass human intelligence within a few decades and urges us to "develop as quickly as possible technologies that make possible a direct connection between brain and computer, so that artificial brains contribute to human intelligence rather than opposing it." (Perera, 2001)

In 2006, the Artificial General Intelligence Research Institute Workshop convened just north of Washington, DC in Bethesda, Maryland. Speakers at the workshop have been actively involved in designing and/or implementing AGI systems and offer relevant insight into the potential of AGI. The momentum in this field could generate a facelift on the field of narrow AI. According to AGI expert Dr. Ben Goertzel: "There are probably a few dozen teams (some of them teams of one) in the world making deeply-serious efforts at achieving AGI."

Interacting and learning through computer interfaces is different than face-to-face learning and has the potential for actually bringing about an evolved state of knowledge. AGI engineers hope that imagination and inspiration can be programmed into superintelligence, like Hal in 2001 Space Odyssey, along with the ability to form billions of calculations second by second as with "Deep Blue" (Dreyfus, 1993: 68, 103, 122).

7. Biological Senses < Macrosensing and Neural Macrosensing

In 2005 the National Institutes of Health grated Forty-Two Million Dollars to set up four nanomedicine centers. In April 2006, the journal Nature Materials estimated that 130 nanotech-based drugs and delivery systems were being developed worldwide. Even though nanomedicine is speculative, nanotechnology has enough bling to carry it along past conservative turnkeys. Yet, "the possibility of using nanorobots in medicine, advocates say, would totally change the world of medicine once it is realized." (Wikipedia, 2006) In 2005, Sander Olson of NanoTech.biz conducted an interview with Freitas.

The key issue for enabling full-immersion reality is obtaining the necessary bandwidth inside the body, which should be available using the in vivo fiber network I first proposed in Nanomedicine, Vol. I (1999). Such a network can handle 10^{18} bits/sec of data traffic, capacious enough for real-time brain-state monitoring. The fiber network has a 30 cm3 volume and generates 4-6 watts waste heat, both small enough for safe installation in a 1400 cm^3 25-watt human brain. Signals travel at most a few meters at nearly the speed of light, so transit time from signal origination at neuron sites inside the brain to the external computer system mediating the upload are ~0.00001 millisec which is considerably less than the minimum ~5 millisec neuron discharge cycle time. Neuron-monitoring chemical sensors located on average ~2 microns apart can capture relevant chemical events occurring within a ~5 millisec time window, since this is the approximate diffusion time for, say, a small neuropeptide across a 2-micron distance. Thus, human brain state monitoring can probably be 'instantaneous', at least on the timescale of human neural response, in the sense of 'nothing of significance was missed.'

8. AGI foresight analyst and Its Human Counterpart

But how would we secure a cooperative relationship between the foresight analyst and its human counterpart? If someone of the stature of the late Dr.

Stephen Hawking states that "Humans must develop an interface that allows the human brain to be directly connected to a computer, so that the artificial brain contributes to human intelligence, rather than opposing it," then it is a likely bet that we ought to pay attention.

The most likely approach would be to build a generic foresight analyst with an ability to bond. This bonding would require more than technically-driven intellectual motivation; it would require strong sensory capabilities for, essentially, sniffing out the environment. And this is where neural macrosensing comes in; to "allow us to become exquisitely sensitive, like 'super-senses' to fine details in our environment." (Freitas) Neural macrosensing could provide the needed sensorial feedback for both assisting the brain and developing elevated acuity.

Goertzel is an optimist about the rate of progress toward AGI, and that AGI will most likely arise before neural macrosensing, and even be applied to drive neural macrosensing. Being a pragmatist, however, he is mindful of the possibility that neural macrosensing technology could advance faster than AGI, "in which case it will initially be driven by specialized 'narrow AI' (task-oriented artificial intelligence rather than a learning general intelligence) applications rather than AGI."

> Hybridizing this sort of narrow AI with neural macrosensing technology could lead to amazing neuro feedback technology even without AGI. Use of neuro feedback devices could allow us to control our states of mind with far more fluidity and dexterity than is currently conceivable, perhaps solving age-old problems such as the difficulty of sustaining Zen-style 'mindfulness' while engaged in intensive and challenging intellectual activity. [If we figured out the biological indicators of the rare state of mind where both spiritual mindfulness and intellectual mastery were being demonstrated, quite possibly neuro feedback would allow us to learn to trigger that state via conscious effort.] And the use of AGI for guiding neuro feedback, in place of relatively simple machine learning methods, would make the possibilities even more awesome, almost surely enabling states of consciousness that are currently inaccessible (and literally unthinkable) to humans (Goertzel, 2006).

9. Conclusion

Merging exponential and speculative technologies warrants careful consideration. Since nano macrosensing requires the use of medical nanorobots that are likely to be diamond or diamondoid substances. The first and most obvious question is whether diamondoid devices or their components are likely to be hazardous to the human body (Freitas, 2003). Second, achieving a cognitive counterpart of AGI for something as personal as wisdom may be viewed contrarily. Not everyone does or will consider the use of technology a dignified means for attaining wisdom.

One thing is certain: change occurs, technology is humanity's tool, and we want to make better choices in our lives – to be wise. We also like to make things pleasant, simple and relaxed. The finessing of wisdom makes the most difficult task seem effortless. The wise appear to glide through life, weightless. If technologies such as AGI and neural macrosensing could help our brains function more smoothly, with added ease and self-confidence, then there could be ample opportunities to observe the world around us. Seamless modification would make the pristine work of neural macrosensing seem so effortless that brain monitoring would be instantaneous while missing nothing (Freitas, 1999). "We would be able to optimize our emotions with rational cognitive styles, awareness of our own mental activity, and careful evaluation of our worldview" (More, 1999). And further, an introspective view of human needs to push further and become more than we are is stated in my 2013 essay:

> Our physical and perceptual experiences are heightened by the urge to push their boundaries. We are handlers of our connections by selecting what media and which combinations to put together in creating our alternative worlds. This self-identification of social media offers a glimpse of how we might select and combine platforms, systems and substrates of our form, and all its attributes.

Optimizing emotions would be like adjusting a thermostat to turn up emotional heat or to turn it down "Since our immediate environment is rich in dangers and opportunities that range widely in importance, our brain needs something akin to a thermostat to determine when a specific challenge reaches the threshold of being sufficiently important to activate the several systems that focus attention and develop appropriate responses. Emotion, centered principally in a small set of sub cortical brain systems, is our biological thermostat, and so it's central to cognition." (Sylwester, 2000)

> "My thermostat has three beliefs. My thermostat believes that it's too hot in here, it's too cold in here and it's just right in here." It sounds simple, and perhaps that is what wisdom ought to be." (McCarthy, 1983)

References

Freitas, Robert A., Jr. (1999) *Nanomedicine, Volume I: Basic Capabilities*, Georgetown: Landes Bioscience.

Freitas, Robert A., Jr. (2003) *Nanomedicine, Volume IIA: Biocompatibility*, Georgetown: Landes Bioscience.

Goertzel, Ben (2016) *AGI Revolution: An Inside View of the Rise of Artificial General Intelligence*, Los Angeles, CA: Humanity+ Press.

Kramer, Richard H. (2004) "Light-activated ion channels for remote control of neuronal firing", *Nature Neuroscience*, 12: 1381-1386.

Loving AI. (n.d.) "About", Retrieved from http://lovingai.org/.

McCarthy, John (1983) *The Little Thoughts of Thinking Machines*, Retrieved from http://www.cse.msu.edu/~cse841/papers/McCarthy.pdf.

More, Max (1997) "Mind Morph: Technologically Refined Emotions and Personality" In *Extro 3 Conference: The Future of the Body and Brain / Future Infrastructure*, San Jose.

National Research Council, Policy and Global Affairs, Board on Global Science and Technology Machines and Networks: A Global Review of Data-to-Decision Technologies Committee on Integrating Humans (2014) "Computing and Decision Making Today" In *Complex Operational Decision Making in Networked Systems of Humans and Machines: A Multidisciplinary Approach*, Washington, DC: National Academies Press, p. 11.

(n/a) (2006) "Nanomedicine: A matter of rhetoric?", *Nature Materials*, 5 (4): 243m, http://www.nature.com/materials/index.html.

Olson, Sander (2005) "An Interview with Robert Freitas, Part II" in *Nanoteh.biz*, Retrieved in http://www.nanotech.biz/i.php?id=robertfreitas2.

Perera, Rick (2001) "Hawking: Re-engineer humans or risk machine rule" in *CNN, Sci-Tech*. Retrieved in http://archives.cnn.com/2001/TECH/ptech/09/05/humans.machine.idg/index.html.

Schafer, Robert (2017) "The Loving and Beloved Sueprego in Freud's Structural Theory", *The Psychoanalytic Study of the Child*, 15 (1) : 163- 88.

Stork, David (1996) "Scientist on the Set: An Interview with Marvin Minsky" In *Hal's Legacy* (D.G. Stork, ed.), Cambridge, MA: MIT Press.

Sylwester, Robert (2000) "Unconscious Emotions, Conscious Feelings, and Curricular Challenges", *The Science of Learning*, 58 (3): 20-24.

Vita-More, Natasha (2004) "Primo Posthuman", Available at https://www.kurzweilai.net/radical-body-design-primo-posthuman.

Yampolskiy, Roman V. (2015) "Analysis of types of self-improving software" In *Artificial General Intelligence*, Springer International Publishing, pp. 384-393.

SECTION V: ETHICS OF ARTIFICIAL INTELLIGENCE

CHAPTER 16

UNETHICAL RESEARCH: HOW TO CREATE A MALEVOLENT ARTIFICIAL INTELLIGENCE

Federico Pistono[1]

Independent Scholar

Roman V. Yampolskiy[2]

University of Louisville, USA

Abstract: Cybersecurity research involves publishing papers about malicious exploits as much as publishing information on how to design tools to protect cyber-infrastructure. It is this information exchange between ethical hackers and security experts, which results in a well-balanced cyber-ecosystem. In the blooming domain of AI Safety Engineering, hundreds of papers have been published on different proposals geared at the creation of a safe machine, yet nothing, to our knowledge, has been published on how to design a malevolent machine. Availability of such information would be of great value particularly to computer scientists, mathematicians, and others who have an interest in AI safety, and who are attempting to avoid the spontaneous emergence or the deliberate creation of a dangerous AI, which can negatively affect human activities and in the worst case cause the complete obliteration of the human species. This chapter provides some general guidelines for the creation of a Malevolent Artificial Intelligence (MAI).

1. Introduction

A significant number of papers and books have been published in recent years on the topic of Artificial Intelligence safety and security (Yampolskiy, 2018), particularly with respect to superhuman intelligence (Yudkowsky, 2008; Chalmers, 2010; Bostrom, 2006; Yampolskiy, 2014; Armstrong, Sandberg and

---

[1] Independent Researcher (research@federicopistono.org).

[2] University of Louisville (roman.yampolskiy@louisville.edu).

Bostrom, 2012; Muehlhauser and Helm, 2012; Hibbard, 2001; Loosemore and Goertzel, 2012; Omohundro, 2008; Yampolskiy, 2015a; 2015b; 2015c). Most such publications address unintended consequences of poor design decisions, incorrectly selected ethical frameworks or limitations of systems, which do not share human values and human common sense in interpreting their goals. This chapter does not focus on unintentional problems, which might arise as the result of construction of intelligent or superintelligent machines, but rather looks at intentional malice in design. Bugs in code, unrepresentative data, mistakes in design and software poorly protected from black hat hackers can all potentially lead to undesirable outcomes. However, intelligent systems constructed to inflict intentional harm could be a much more serious problem for humanity.

1.1. Why the paper on this topic?

Following cybersecurity research paradigm, which involves publishing papers about malicious exploits as much as publishing information on how to design tools to protect cyber-infrastructure, we attempt to summarize general guidelines for the development of malicious intelligent software. Information exchanges between hackers and security experts are indispensable to the sustainment of a well-balanced cyber-ecosystem in the domain of security. In the flourishing domain of AI safety, hundreds of papers have been published on different proposals geared at the creation of a safe machine (Sotala and Yampolskiy, 2015), yet nothing, to our knowledge, has been published on how to design a malevolent intelligent system. Availability of such information would be of great value particularly to computer scientists and security experts who have an interest in AI safety, and who are attempting to avoid both the spontaneous emergence and the deliberate creation of a malevolent AI. However, to avoid increasing MAI risks, we avoid publishing any design/pathway details that would be useful to malefactors, not obvious, and not quickly fixable.

1.2. Hazardous Intelligent Software

"Computer software is directly or indirectly responsible for controlling many important aspects of our lives. Wall Street trading, nuclear power plants, Social Security compensation, credit histories, and traffic lights are all software controlled and are only one serious design flaw away from creating disastrous consequences for millions of people. The situation is even more dangerous with software specifically designed for malicious purposes, such as viruses, spyware, Trojan horses, worms, and other hazardous software (HS). HS is capable of direct harm as well as sabotage of legitimate computer software employed in critical systems. If HS is ever given the capabilities of truly artificially intelligent systems (e.g., artificially intelligent virus), the consequences unquestionably would be disastrous. Such Hazardous Intelligent Software (HIS) would pose risks currently unseen in malware with subhuman intelligence." (Yampolskiy, 2012).

Nick Bostrom, in his typology of information hazards, has proposed the term *artificial intelligence hazard*, which he defines as (Bostrom, 2011) "computer-related risks in which the threat would derive primarily from the cognitive sophistication of the program rather than the specific properties of any actuators to which the system initially has access."

Addressing specifically superintelligent systems, we can also look at the definition of Friendly Artificial Intelligence (FAI) proposed by Yudkowsky (Yudkowsky, 2008) and from it derive a complimentary definition for Unfriendly Artificial Intelligence: a hypothetical Artificial General Intelligence (AGI) that would have a negative rather than positive effect on humanity. Such a system would be capable of causing great harm to all living entities and its values and goals would be misaligned with those of humanity. The system does not have to be explicitly antagonistic to humanity, it is sufficient for it to be neutral to our needs.

An intelligent system could become malevolent in a number of ways, which we can classify into: unintentional and intentional on the part of the designer. Unintentional pathways are most frequently a result of a mistake in design, programming, goal assignment or a result of environmental factors such as failure of hardware. Just like computer viruses and other malware is intentionally produced today, in the future, we will see pre-mediated production of hazardous and unfriendly intelligent systems (Yampolskiy, 2016). We are already getting glimpses of such technology in today's research with recently publicized examples involving lying robots (Castelfranchi, 2000; Clark, 2010), black market trading systems (Cush, 2015) and swearing computers (Smith, 2013).

1.3. Who might be interested in creating Malevolent AI?

Purposeful creation of MAI can be attempted by a number of diverse agents with varying degrees of competence and success. Each such agent would bring its own goals/resources into the equation, but what is important to understand here is just how prevalent such attempts will be and how numerous such agents can be. Below is a short list of representative entities, it is very far from being comprehensive:

- **Military** developing cyber-weapons and robot soldiers to achieve dominance.

- **Governments** attempting to use AI to establish hegemony, control people, or take down other governments.
- **Corporations** trying to achieve monopoly, destroying the competition through illegal means.
- **Villains** trying to take over the world and using AI as a dominance tool.
- **Black Hats** attempting to steal information, resources or destroy cyberinfrastructure targets.
- **Doomsday Cults** attempting to bring the end of the world by any means.
- **Depressed** looking to commit suicide by AI.
- **Psychopaths** trying to add their name to history books in any way possible.
- **Criminals** attempting to develop proxy systems to avoid risk and responsibility.
- **AI Risk Deniers** attempting to demonstrate that AI is not a risk factor and so ignoring caution.
- **AI Safety Researchers**, if unethical, might attempt to justify funding and secure jobs by purposefully developing problematic AI.

2. How to create a Malevolent AI

The literature on AI risk (Sotala and Yampolskiy, 2015) suggests a number of safety measures which should be implemented in any advanced AI project to minimize potential negative impact. Simply inverting the advice would, in many cases, lead to a purposefully dangerous system. In fact, the number of specific ways in which a malevolent designer may implement hazardous intelligent software is limited only by one's imagination, and it would be impossible to exhaustively review. In "AGI Failures Modes and Levels", Turchin (Turchin, 2015) describes a number of ways in which an intelligent system may be dangerous at different stages in its development. Among his examples, AI:

- Hacks as many computers as possible to gain more calculating power
- Creates its own robotic infrastructure by the means of bioengineering
- Prevents other AI projects from finishing by hacking or diversions
- Has goals which include causing suffering
- Interprets commands literally
- Overvalues marginal probability events (Turchin, 2015)

Despite the impossibility of providing a comprehensive list of failures, we can suggest some general principles, which most experts would agree, are likely to produce MAI by ill-informed but not purposefully malevolent software designers. Those include: Immediate deployment of the system to public networks such as Internet, without testing; Providing system with access to unlimited information including personal information of people, for example, massive social networks like Facebook; Giving system-specific goals which are not vetted with respect to consequence and unintended side-effects; Putting the system in charge of critical infrastructure such as communication, energy plants, nuclear weapons, financial markets. In the following sections, we provide details on two particular and related types of design mistakes: absence of oversight by qualified experts and secretive nature of project.

2.1. No oversight boards

Certain types of research, such as human cloning and the development of biochemical weapons, have been banned or severely restricted due to their unethical nature with respect to laws and human values (as defined by the universal declaration of human rights), and therefore might cause irreparable harm to humanity. Yampolskiy (Yampolskiy, 2013a) proposed to consider Artificial General Intelligence (AGI) research unethical, as "true AGIs will be capable of universal problem solving and recursive self-improvement. Consequently, they have the potential of outcompeting humans in any domain essentially making humankind unnecessary and so subject to extinction" and "may possess a type of consciousness comparable to the human type making robot suffering a real possibility and any experiments with AGI unethical for that reason as well." The creation of an oversight board, a team of experts in artificial intelligence that evaluates each research proposal and decides if the proposal is for standard AI or may lead to the development of a full-blown AGI, would either stop (albeit very unlikely), severely limit, or at the very least slow down the process of creation of an uncontrolled AGI.

If a group decided to create an MAI, it follows that preventing a global oversight board committee from coming to existence would increase its probability of succeeding. This can be accomplished in a number of ways, most of which exploit one or more of the innumerable cognitive biases humans are victims of.

The first strategy is to disseminate conflicting information that would create doubt in the public's imagination about the dangers and opportunities of AGI research. One lesson can be learned from the public "debate" around the theory of evolution through means of natural selection and that of anthropogenic climate change. Even though the scientific consensus among experts is overwhelming on both scientific theories, the public perception is that there is an ongoing debate in the scientific community and that the matter is not settled in either case. This has been accomplished by means of politicization of the issue, or by associating a religious ideology to either side.

Similarly, an individual or group with the intention of creating a MAI would attempt to increase polarization in the debate, either through lobbying, marketing, by disseminating false information, by publishing polarizing news stories on popular blogs and anti-establishment websites, by creating conspiratorial hypothesis that feed the public's imagination and the anti-government sentiment, or by attracting the attention of political parties or influential religious groups, just to name a few. The confusion generated would create an environment nebulous enough to delay or even prevent the creation of the global oversight board.

A second option is to convince the government of a country, preferably a small one or a micro-state, not to sign the AGI-prevention treaty that all other major nations would eventually subscribe to. Again, it is not difficult to imagine how this could be achieved. The simplest way would be to simply bribe government officials into accepting this condition. Other means include promises of greatness as a result of being the first country to have access to the first AGI ever created, to intimidation, extortion, or even a simple leverage of the ignorance that they might have on the subject of AGI dangers. In all cases, only one condition needs to be satisfied in order for the plan to succeed. Given the vast heterogeneity of human responses and the long list of cognitive biases one can leverage, it is reasonable to assume that at least one exception will be found and exploited, by someone at some point.

2.2. Closed source code

One of the key pieces in both the purposeful and the unintended creation of malevolent artificial intelligence is the ability to run on closed-source, proprietary software and hardware. It is well known among cryptography and computer security experts that closed-source software and algorithms are less secure than their Free and Open Source (FOS) counterpart, they are more prone to vulnerabilities, exploits, deliberate or accidental malevolent use (Schneier, 1999). While FOS is not a sufficient condition for security, Kerckhoffs' principle shows that closed-source software and security through obscurity are bad practices (Hoepman and Jacobs, 2007). It is also known that both closed source software and hardware have been manipulated and tampered with by intelligence agencies and institutions, whether with or without the knowledge or consent of the companies creating the specialized hardware or programming the software running on it (Appelbaum et al., 2013; Schneier, 2013; 2014). In this environment, any group with the intent of creating a malevolent artificial intelligence would find the ideal conditions for operating in quasi-total obscurity, without any oversight board and without being screened or monitored, all the while being protected by copyright law, patent law, industrial secret, or in the name of "national security".

Furthermore, a much broader case deals with the unintended consequences of running sophisticated and powerful non-free software and hardware, even without malicious intents. It has been shown that the indiscriminate collection of personal data and information about people is prone to unintended uses that can backfire, go out of control and have ramifications outside its original intent. By deliberately compromising the security of a system, either hardware or software, the integrity of the system's operations is put at risk and it is at the mercy of anyone capable of running the exploit that was created (i.e., to collect people's personal information). Any system having a universal backdoor (i.e., anything that implements Digital Rights Management (DRM) restrictions) can be used for virtually any purpose, without letting the user know that the system has been compromised or that it's being used for unintended purposes, whatever they might be.

The very existence of non-free software and hardware puts humanity at a greater risk, it can foster or accelerate the process of the creation of an artificial entity that can outcompete or control humans in any domain, making humankind unnecessary, controllable, or even subject to extinction, and should therefore be considered unethical. The authors recognize the likelihood of non-free software be rendered illegal or banned globally is extremely low, therefore greatly increasing the probability of the creation of an MAI and that of human subjugation or extinction. At the same time, wide distribution of an open-source code for AI may make it possible for a number of additional agents, which otherwise lack technical sophistication, to engage in conversion of a safe AI product into a malevolent one. It is unclear what the solution to prevent this would be. It is arguable that greater access to the source code by a greater number of people would give more chances to prevent or fix problems, given that historically there is a greater number of individuals and organization whose goal is to fix bugs and keep systems safe rather than creating damage. On the other hand, terrorists and criminals often have some success in identifying and exploiting new vulnerabilities despite great defensive efforts.

Recent announcement of OpenAI, a non-profit artificial intelligence research company whose goal is to "advance digital intelligence in the way that is most likely to benefit humanity as a whole, unconstrained by a need to generate financial return", with over $1 billion pledged for its development, is a promising step in the right direction. Particularly, if the research will also be focused on monitoring and preventing early MAI attacks, although it's too early to say at this point.

3. Societal Impact

In this section, we will review the potential impact of MAI on economy, government, legislation and military.

3.1. Economy/Unemployment

The dangers of technological unemployment, which states that technology eventually advances so rapidly that it creates a non-cyclical, structural unemployment problem in the labor market, are starting to be considered as a serious matter in academia. Research by Pistono (Pistono, 2012), Frey and Osborne (Frey and Osborne, 2013), suggests that we might already see an early sign of this process today and that as much 47% of all jobs in the U.S. could be automated. Further analysis by Bowels (Bowels, 2014) indicates that the same trend would apply to Europe as well, with numbers fluctuating between 61.95% in Romania and 47.17% in the U.K, averaging 54% across the EU-28. The proposed solutions, which range from tax reforms, to promoting entrepreneurship and adopting an Unconditional Basic Income (Haagh and Howard, 2012), are non-homogeneous, polarized, and not universally accepted.

The topic is still in its infancy, and the prospect of technological unemployment has not been widely accepted by most governments, which struggle to recognize its very existence, let alone come to an agreement on a coherent solution that would be practical across the globe. Given the historical context and precedents, we expect governments to react non-homogeneously and at different times. This creates the ideal fertile ground for an MAI to take over without encountering a strong, coherent, and organized opposition. By exploiting the natural tendency of companies to want to increase their productivity and profits, a MAI could lend its services to corporations, which would be more competitive than most, if not all humans workers, making it desirable for a company to employ the quasi-zero marginal cost AI, in place of the previously employed expensive and inefficient human labor.

A MAI could also, reasonably, run on closed-source software, and operate on the cloud under SaaS subscriptions (Software As A Service, a software licensing and delivery model in which software is licensed on a subscription basis and is centrally hosted), or by licensing a subset of its functions to install on the companies' servers, where it could run small exploit programs and worms. These programs would be very hard to detect and could take over servers and global operations very quickly, and execute their goals accordingly. Even under the assumption that an MAI would not succeed in deploying such exploits, viruses, and worms, there is still a high likelihood that it would become the prevalent provider of services across the world, effectively giving it the ability to control global food markets, power stations, transportation, financial markets, logistics and operations, just to name a few industries.

A particularly sophisticated MAI would be able to predict which behaviors would catch the attention of external observers who look for signs of its emergence, and would mimic the functioning of a normal company or even an ecosystem of companies that operate at the global level, including employing human beings, so as not to draw suspicion. The structure of the corporation theoretically allows for such hierarchical systems to function, where workers don't need to know who or what is controlling the whole, and where work is compartmentalized appropriately. Additionally, corporate personhood in the U.S. recognizes a corporation as an individual in the eyes of the law, the same applying also to the City of London in the United Kingdom. With this in place, an MAI could conceivably create a network of corporations that can dominate the market within the span of a decade or less, where the creation of structural, irreversible unemployment of more than half of all jobs would be the least worrisome outcome.

3.2. Government

Governments play a crucial role in the functioning of all modern societies. The balance of powers and oversight has been the subject of controversy and heated debate since the first primitive forms of organized societal structures, which over time, evolved into modern forms of governments.

Coups d'état, the sudden and illegal seizure of a state either by force or through corruption, intimidation, or economic hitmen, can create instability, uprisings, civil wars, poverty and in the worst cases, a total collapse of the country. Throughout history, humans have successfully staged *Coups d'état* by disseminating false information, staging assassinations, and other relatively simple operations that could be easily accomplished by an MAI. In addition, a sufficiently advanced MAI with access to the global system of communication, internet and social media, can stage multiple *Coups* simultaneously, run simulations of possible ramification of each attack, incite governments and populations against one another, and feed exponentially a planetary chaos machine to bring the maximum level of instability and therefore compromising the safety and ability to respond to further attacks, which can be combined with judicial, legislative, and military action. The safest course of action for an MAI would probably be to twist the knobs just enough to create global chaos and let humanity destroy or weaken itself to the point of being extremely vulnerable, while remaining undetected and unsuspected, and thus maintaining its existence with minimal risk. If governments believe they are fighting terrorist groups, either internal or foreign, and are constantly in a state of paranoia, they will continue to distrust one another, engaging in cyber, economic, and even physical attacks, furthering the goal of the MAI.

All this can be augmented and tweaked to perfection by scanning through the data available on social media, emails, text messages, phone calls and other means of electronic communication. The more the data is available and unencrypted, the more an MAI can understand how to best respond and stage the next wave of attacks. The NSA, by illegally installing backdoors into hardware and software worldwide, collecting personal information and metadata of all internet users that go through US-controlled servers, has weakened the security of our communication systems, and has made us more vulnerable to espionage and terrorist attacks (Peha, 2013; Paterson et al., 2013; Abadi et al., 2014). Security agencies in other countries have also been involved in similar activities, making the problem worse. An MAI can capitalize on the fundamental insecurity of these systems and have access to an unprecedented level of information regarding all human activities, and therefore have almost perfect information on which to simulate possibilities, stage attacks and diversions to reach its goals while remaining undetected and unsuspected.

3.3. Legislative

Any action or strategy deployed by an MAI at the expense of humans and their life-supporting systems can be amplified and accelerated by legislation in support of such plans. Research by Lessig (Lessig, 2011) shows that the extreme concentration of power, influence, and decision-making in the hands of a few has made it possible to completely decouple the will of the people from the results of legislations of countries such as the United States, where 0.05% of the population can consistently pass laws that are against the will of the majority. It follows that an MAI would seek control of the House, the Senate, and other bodies through lobbying, at a fraction of the energy and time expenditure (cost) that would require via other means. Hidden and obscure pro-MAI activities legislation can feed into a positive feedback cycle that further increases the influence of an MAI and the likelihood of developing its goals. The more power becomes centralized and influence is in the hands of few, the easier it will be for an MAI to seek control of the world. It is also very likely that MAI itself would attempt to enter politics (Pellissier, 2015) or penetrate our judicial system (Dunn, 1995) to avoid having to rely on human assistants.

3.3. Military

In the authors' view, military action and weapons against humans, whilst vivid in the public's imagination and in sci-fi stories, are very unlikely to be used by an MAI, unless all other options fail. Armed forces and military interventions are noticeable and very hard to hide. As explained earlier, an MAI will make use of all possible means to go unnoticed and unsuspected for the longest period possible, so as to not give time for humans to assemble and strategize counter-attacks or effective defenses. The direct use of force when other more efficient options are available is an irrational strategy, of which humans are subject to because of their evolutionary biological baggage, but an MAI would have no reason for it, and therefore would delay or even avoid its use altogether.

Should necessity arise, striking a global conflict will be a very easy task for an MAI. The total number of nuclear weapons is estimated at 10,144 (2014), many of which are stored in aging facilities, without the proper safety standards, underfunded, and numerous scandals have put into question the ability of operators to keep them safe (Smith, 2012, 2014, 2015a, 2015b). If trained officers and even civilians have been able to breach the security of these systems, any sufficiently advanced MAI can accomplish the goal with ease. As mentioned before, the path of least resistance is the most logical and efficient course of action. Rather than announcing and launching a deliberate attack against humanity, it would be much easier to simply exploit the security flaws of existing nuclear storage facilities and their human personnel. Of course, the possibility remains that MAI will take over military robots, drones and cyber weapons to engage in a direct attack by exploiting autonomous capabilities of such weapons (Arkin, 2010). If the goal were to exterminate humanity, one of the easiest paths would be to design an extremely deadly virus or bacteria, which can stay dormant and undetected until it reaches the entirety the world's population, and is then activated wirelessly via nanobots simultaneously and ubiquitously, killing all humans in a matter of minutes or hours.

4. Conclusions

A lot has been written about problems which might arise as a result of arrival of true AI, either as a direct impact of such invention or because of a programmer's error (Majot and Yampolskiy, 2014; Yampolskiy and Fox, 2012; Yampolskiy and Fox, 2013; Sotala and Yampolskiy, 2015; Yampolskiy, 2013b; Yampolskiy and Gavrilova, 2012; Yampolskiy et al., 2012; Ali, Schaeffer, and Yampolskiy, 2012; Gavrilova and Yampolskiy, 2010). Intentional malice in design has not been addressed in scientific literature, but it is fair to say that, when it comes to dangers from a purposefully unethical superintelligence, technological unemployment may be the least of our concerns. According to Bostrom's orthogonality thesis, an AI system can have any combination of intelligence and goals (Bostrom, 2012). Such goals can be introduced either via initial design or in case of an "off the shelf" superintelligence introduced later – "just add your own goals". Consequently, depending on whose bidding the system is doing [governments, corporations, sociopaths, dictators, military industrial complexes, doomsdayers, etc.] it may attempt to inflict damage unprecedented in the history of humankind or perhaps inspired by it. The world is not short of terrible people and, as it is famously stated, "*absolute power corrupts absolutely*". Even in absence of universally malicious motives [the destruction of the human species], the simple will to subjugate or to control certain groups of people or to gain advantage over someone else is enough to make an MAI desirable for some entity or organization. In the order of (subjective) undesirability from least damaging to ultimately destructing, a malevolent superintelligence may attempt to:

- Takeover (implicit or explicit) of resources such as money, land, water, rare elements, organic matter, internet, computer hardware, etc. and establish monopoly over access to them;
- Take over political control of local and federal governments as well as of international corporations, professional societies, and charitable organizations;
- Reveal informational hazards (Bostrom, 2011);
- Set up a total surveillance state (or exploit an existing one), reducing any notion of privacy to zero including privacy of thought;
- Force merger (cyborgization) by requiring that all people have a brain implant which allows for direct mind control/override by the superintelligence;
- Enslave humankind, meaning restricting our freedom to move or otherwise choose what to do with our bodies and minds. This can be accomplished through forced cryonics or concentration camps;
- Abuse and torture humankind with perfect insight into our physiology to maximize the amount of physical or emotional pain, perhaps combining it with a simulated model of us to make the process infinitely long;
- Commit specicide against humankind, arguably the worst option for humans as it can't be undone;
- Destroy/irreversibly change the planet, a significant portion of the Solar system, or even the universe;
- *Unknown Unknowns.* Given that a superintelligence is capable of inventing dangers we are not capable of predicting, there is room for something much worse but which at this time has not been invented.

Artificial Intelligence Safety Engineering (AISE) is an emerging interdisciplinary field of science concerned with making current and future Artificially Intelligent systems safe and secure. Engineers, Computer Scientists, Philosophers, Economists, Cognitive Scientists and experts in many other fields are starting to realize that despite differences in terminology and approaches they are all working on a common set of problems associated with control over independent and potentially malevolent intelligent agents. Currently, such research is conducted under such diverse labels as: AI ethics, Friendly AI, Singularity Research, Superintelligence Control, Software Verification, Machine Ethics, and many others. This leads to suboptimal information sharing among practitioners, confusing lack of common terminology and as a result limited recognition, publishing and funding opportunities. It is our hope that by outlining the main dangers associated with future intelligent systems, we are able to inspire a more cohesive research program aimed at counteracting intentional development of dangerous software. Perhaps, even a malevolent AI could be repaired and turned into a Friendly AI with a sufficient level of understanding on the part of AI safety researchers. Of course, those attempting to purposefully introduce MAI into the world will no doubt provide their systems with a sophisticated self-defense module preventing intervention by AI safety researchers attempting to neutralize it.

Finally, we are issuing a challenge to the AI Safety community to continue this work by discovering and reporting specific problems, which may lead to MAI along with the request that publication of concrete solutions to such problems takes place prior to open publication of the problematic source code.

Acknowledgements

Roman Yampolskiy expresses appreciation to Elon Musk and FLI for partially funding his work via project grant: "Evaluation of Safe Development Pathways for Artificial Superintelligence" awarded to Global Catastrophic Risk Institute, Seth Baum (PI). The authors are grateful to Yana Feygin, Seth Baum, James Babcock, János Kramár and Tony Barrett for valuable feedback on an early draft of this chapter. Views in this chapter are those of the authors, and do not necessarily represent the views of FLI, GCRI, or others.

References

(2014) "Nuclear Notebook" in *Bulletin of the Atomic Scientists*, http://thebulletin.org.

(2015a) "Audit Report Security at the Nevada National Security Site" in *U.S. Department of Energy, Office of Inspector General, Office of Audits and Inspections*, http://energy.gov/sites/prod/files/2015/05/f22/OAS-L-15-06.pdf.

(2015b) "Security Improvements at the Y-12 National, Security Complex" in *U.S. Department of Energy, Office of Inspector General, Office of Audits and Inspections*, http://energy.gov/sites/prod/files/2015/09/f26/DOE-IG-0944.pdf.

(2014) "US nuclear launch officers suspended for 'cheating'" (January 16) in *BBC*, http://www.bbc.com/news/world-us-canada-25753040 .

Abadi, Martín et al. (2014) "An Open Letter from US Researchers in Cryptography and Information Security", http://masssurveillance.info.

Ali, Nawaf, Schaeffer, Derek and Yampolskiy, Roman V. (2012) "Linguistic Profiling and Behavioral Drift in Chat Bots" in *Midwest Artificial Intelligence and Cognitive Science Conference*, 27.

Appelbaum, Jacob et al. (2013) "Shopping for Spy Gear: Catalog Advertises NSA Toolbox" in *Der Spiegel*, http://www.spiegel.de/international/world/catalog-reveals-nsa-has-back-doors-for-numerous-devices-a-940994.html.

Arkin, Ronald C. (2010) "The case for ethical autonomy in unmanned systems", *Journal of Military Ethics*, 9 (4): 332-341.

Armstrong, Stuart, Sandberg, Anders and Bostrom, Nick (2012) "Thinking inside the box: Controlling and using an oracle ai", *Minds and Machines*, 22 (4): 299-324.

Bostrom, Nick (2006) "Ethical Issues in Advanced Artificial Intelligence", *Review of Contemporary Philosophy*, 5: 66-73.

Bostrom, Nick (2011) "Information Hazards: A Typology of Potential Harms From Knowledge", *Review of Contemporary Philosophy*, 10: 44-79.

Bostrom, Nick (2012) "The superintelligent will: Motivation and instrumental rationality in advanced artificial agents", *Minds and Machines*, 22 (2): 71-85.

Bowels, Jeremy (2014) "The computerisation of European jobs", http://bruegel.org/2014/07/the-computerisation-of-european-jobs/.

Castelfranchi, Cristiano (2000) "Artificial liars: Why computers will (necessarily) deceive us and each other", *Ethics and Information Technology*, 2 (2): 113-119.

Chalmers, David (2010) "The Singularity: A Philosophical Analysis", *Journal of Consciousness Studies*, 17: 7-65.

Clark, Micah H. (2010) "Cognitive illusions and the lying machine: a blueprint for sophistic mendacity" in *Rensselaer Polytechnic Institute*.

Cush, Andy (2015) "Swiss Authorities Arrest Bot for Buying Drugs and Fake Passport" (January 22) in *Gawker*, http://internet.gawker.com/swiss-authorities-arrest-bot-for-buying-drugs-and-a-fak-1681098991.

Dunn, Theodore (1995) "Exploring the possibility of using intelligent computers to judge criminal offenders", http://www.fdle.state.fl.us/Content/getdoc/814750b4-4546-49f0-9be2-90fe772257f7/Dunn.aspx.

Frey, Carl Benedikt and Osborne, Michael A., (2013) "The future of employment: how susceptible are jobs to computerisation?", *Sept*, 17: 2013.

Gavrilova, Marina L. and Yampolskiy, Roman V. (2010) "State-of-the-Art in Robot Authentication [From the Guest Editors]", *Robotics & Automation Magazine, IEEE*, 17 (4): 23-24.

Haagh, Anne-Louise and Howard, Michael (eds.) (2012) *Basic Income Studies*, 7 (1).

Hibbard, Bill (2001) "Super-Intelligent Machines", *Computer Graphics*, 35 (1): 11-13.

Hoepman, Jaap-Henk and Jacobs, Bart (2007) "Increased security through open source", *Communications of the ACM*, 50 (1):79-83.

Lessig, Lawrence (2011) *Republic, Lost: How Money Corrupts Congress – and a Plan to Stop It*, New York, NY: Twelve.

Loosemore, Richard and Goertzel, Ben (2012) "Why an intelligence explosion is probable" In *Singularity Hypotheses*, Berlin: Springer, pp. 83-98.

Majot, Andrew M., and Yampolskiy, Roman V. (2014) "AI safety engineering through introduction of self-reference into felicific calculus via artificial pain and pleasure" in *Ethics in Science, Technology and Engineering, 2014 IEEE International Symposium.*

Muehlhauser, L. and Helm, L. (2012.) "The Singularity and Machine Ethics" In *In The Singularity Hypothesis: A Scientific and Philosophical Assessment* (A. Eden, J. Søraker, J. Moor and E. Steinhart, eds.), Berlin, Heidelberg: Springer.

Omohundro, Stephen M. (2008) "The Basic AI Drives", *Proceedings of the First AGI Conference*, Volume 171.

Paterson, Kenneth et al., (2013) "Open Letter From UK Security Researchers" University of Bristol, http://bristolcrypto.blogspot.co.uk/2013/09/open-letter-from-uk-security-researchers.html.

Peha, Jon M. (2013) "The Dangerous Policy of Weakening Security to Facilitate Surveillance" (October 4), Accessed *SSRN*: http://ssrn.com/abstract=2350929. http://dx.doi.org/10.2139/ssrn.2350929.

Pellissier, Hank (2015) "Should Politicians be Replaced by Artificial Intelligence?" (June 12) *IEET*, http://ieet.org/index.php/IEET/more/pellissier20150612.

Pistono, Federico (2012) *Robots will steal your job, but that's ok: how to survive the economic collapse and be happy*, Scotts Valey: CreateSpace.

Schneier, Bruce (1999) "Open Source and Security", https://www.schneier.com/crypto-gram/archives/1999/0915.html#OpenSourceandSecurity.

Schneier, Bruce (2013) "More about the NSA's Tailored Access Operations Unit", https://www.schneier.com/blog/archives/2013/12/more_about_the.html.

Schneier, Bruce (2014) "CROSSBEAM: NSA Exploit of the Day", ys

Smith, Dave (2013) "IBM's Watson Gets A 'Swear Filter' After Learning The Urban Dictionary" (January 10) in *International Business Times*, http://www.ibtimes.com/ibms-watson-gets-swear-filter-after-learning-urban-dictionary-1007734.

Smith, R. Jeffrey (2012) "How an 82-year-old exposed security lapses at nuclear facilities" (September 12) in T*he Center for Public Integrity*, http://www.publicintegrity.org/2012/09/12/10851/how-82-year-old-exposed-security-lapses-nuclear-facilities.

Sotala, Kaj and Yampolskiy, Roman V. (2015) "Responses to catastrophic AGI risk: A survey", *Physica Scripta*, 90.

Turchin, Alexey (2015) "A Map: AGI Failures Modes and Levels" (July 10) in *LessWrong*, http://lesswrong.com/lw/mgf/a_map_agi_failures_modes_and_levels/.

Yampolskiy, Roman V., Cho, Gyuchoon, Rosenthal, Richard and Gavrilova, Marina (2012) "Experiments in Artimetrics: Avatar Face Recognition" In *Transactions on Computational* Science, XVI: 77-94.

Yampolskiy, Roman V. and Fox, Joshua (2012) "Safety Engineering for Artificial General Intelligence", *Topoi*: 1-10.

Yampolskiy, Roman V. and Fox, Joshua (2013) "Artificial General Intelligence and the Human Mental Model" In *Singularity Hypotheses: A Scientific and Philosophical Assessment*: 129.

Yampolskiy, Roman V. and Gavrilova, Marina (2012) "Artimetrics: Biometrics for Artificial Entities", *IEEE Robotics and Automation Magazine (RAM)*, 19 (4):48-58.

Yampolskiy, Roman V. (2012) "Leakproofing Singularity - Artificial Intelligence Confinement Problem", *Journal of Consciousness Studiesm*, 19 (1-2): 194–214.

Yampolskiy, Roman V. (2013a) "Artificial intelligence safety engineering: Why machine ethics is a wrong approach" In *Philosophy and Theory of Artificial Intelligence*, Berlin, Heidelberg: Springer, pp. 389-396.

Yampolskiy, Roman V. (2013b) "What to Do with the Singularity Paradox?" In *Philosophy and Theory of Artificial Intelligence*, Berlin, Heidelberg: Springer, pp. 397-413.

Yampolskiy, Roman V. (2014) "Utility Function Security in Artificially Intelligent Agents", *Journal of Experimental and Theoretical Artificial Intelligence (JETAI)*: 1-17.

Yampolskiy, Roman V. (2015a) "The Space of Possible Mind Designs" in *Artificial General Intelligence*, Berlin, Heidelberg: Springer, pp. 218-227.

Yampolskiy, Roman V. (2015b) "On the Limits of Recursively Self-Improving AGI" In *The Eighth Conference on Artificial General Intelligence*, Berlin, Germany.

Yampolskiy, Roman V. (2015c) *Artificial Superintelligence: a Futuristic Approach*, Florida: Chapman and Hall/CRC.

Yampolskiy, Roman V. (2016) "Taxonomy of Pathways to Dangerous AI" In *30th AAAI Conference on Artificial Intelligence* (AAAI-2016), *2nd International Workshop on AI*, Ethics and Society (AIEthicsSociety2016), Phoenix, Arizona, USA.

Yampolskiy, Roman V. (ed.) (2018) *Artificial Intelligence Safety and Security*, Florida: CRC Press.

Yudkowsky, Eliezer (2008) "Artificial Intelligence as a Positive and Negative Factor in Global Risk" In *Global Catastrophic Risks* (N. Bostrom and M. M. Cirkovic, eds.), Oxford, UK: Oxford University Press, pp. 308-345.

Chapter 17

The Search for X: Grounding Rights and Obligations in the Age of Artificial Intelligence

Hasse Hämäläinen[1]

Jagiellonian University in Kraków, Poland

Abstract: This article attempts to find the most plausible answer to the question whether a machine could be attributed moral and legal rights and obligations among current answers. The article uses the concept of 'X' as a 'placeholder' for that what justifies attribution of rights and obligations to humans, whatever that might be. The answers are categorized with regard to their stance towards attributing X to machines. It is possible to think (**a**) that X is something, which only humans can have, (**b**) which only humans have today, but which machines could acquire in the future, or (**c**) both humans and machines have. The article will show that the supporters of (**a**) must regard X as certain perceivable feature that differentiates humans from all other beings. It is argued that so far, they have, however, failed to find any perceivable feature that could justify such a difference. Option (**b**) is thus assessed as more plausible than (**a**) overall, but the article notes that also it is vulnerable to a problem – namely, it cannot explain why we should *not* regard machines that could acquire X as humans already in virtue of having X *in potentia*. Therefore, the article concludes that option (**c**) is the most plausible answer at the moment, although it is not to be excluded that (**a**) shall emerge as more plausible, provided that the desired feature is eventually found.

1. X: What Justifies Rights and Obligations

The question whether a machine could be attributed moral and legal rights and obligations has begun to gain traction in mainstream philosophy only recently.

---

[1] Jagiellonian University in Kraków.

The first investigators of this question were a group of 18th-century philosophers, who regarded humans as essentially machines, and thus, were presented a challenge to justify why only the 'human machine' can be treated as a moral agent, or else risk falling into scepticism or nihilism about the grounding of morality (D'Holbach, 2001: 108-10; La Mettrie, 2009: 22). However, their theories, that were at the margins of philosophy even at their own time, were eclipsed by the developments in the field of organic chemistry and biology in the course of the 19th and 20th centuries, which made evident that a machine is a rather poor analogue to illustrate the operation of the human organism.

The recent re-emergence of the question among philosophers is motivated by the advances in the development of computers that have made the computers seem more and more like humans insofar as certain expanding range of skills, such as a skill to translate sentences between languages (e.g., Wu et al., 2016), learn to play games (e.g., Silver et al., 2018) and describe what an unseen image represents (Vinyals et al., 2014) is concerned. Thus, even if one does not regard humans as essentially machines, one could be nevertheless enticed to speculate about the prospect of computers becoming so much like humans that any perceptible difference between the two kinds of being, which can justify why the former could not be the bearers of rights and obligations like the latter, shall become blurred – and perhaps eventually disappear.

Regardless of whether or not this possibility can be realised, the path towards its realisation is rife with philosophical puzzles. One of them arises at the moment in which the first machine shall acquire a 'perfect human likeness' – viz. has all the capacities that humans typically have. However, provided that humans are *not essentially* machines, a perfectly humanlike machine would nevertheless fall short from being a human. In other words, the humans would be still separated from this machine by *something that makes us human*, whatever it is – an entity that shall be denoted by the letter 'X' in this article – and only by it. (In my choice of terminology, I follow Fukuyama (2002), who used the term 'Factor X' to denote this unknown factor.) When faced by the essential difference between it and us, the described machine would thus be likely to conceive itself as more *similar* to other machines, who lack X as well, than to humans. This can become problematic for us. We humans typically treat machines as beings that lack any moral worth – i.e., that are only valuable insofar as they help us in realising our ends, and that can be discarded once they become useless. Since the humanlike machine would think just like a human, it might make analogous conclusions about beings that it perceives as essentially different from it – that is, about humans. Thus, the humanlike machine could conceivably treat humans as beings that lack any intrinsic moral worth – perhaps "only fit for slavery or slaughter" (Annas et al., 2002: 162) – in the way analogous to the way we currently treat machines.

Even if one did not take this concern seriously, for a machine is unlikely to acquire fully humanlike capacities, the question about the rights and obligations of machines is also relevant to another, very practical concern. Many people have lost their jobs for machines already, and countless of others inevitably will. This phenomenon will take place *at least as long as* machines are no more than "slaves" for their owners, without political autonomy and personal salary (which they could use e.g. for educating themselves or assisting their human 'colleagues' or whatever they may deem appropriate) that is sufficiently correlated with the productivity improvement they bring *vis-a-vis* human labour. Moreover, independently of the question whether a machine could have moral rights such as political autonomy, if autonomously-acting machines are not obliged to respect *any* legal obligations whatsoever, which we *are* obliged to respect, there is a real risk that the concern about machines 'enslaving and slaughtering' humans will be realised at least in a certain extent, making the world (even) more hostile place to live for many of us.

Provided that the attribution of rights and obligations to a being – for example, that the being has a right to a just salary and deciding for themselves, or an obligation to treat certain other beings respectfully, and not to make selfish use of them, *etc.* – has any grounding in the 'nature' of the being, defined as *the set of features in virtue of which it is the kind of being it is,* the grounding is to be found in X. For as the things stand now, only human beings are subject to *both* moral and legal rights and obligations, and if this situation is justifiable, then that what makes us special among beings – i.e., X – should be also what makes us entitled to special treatment with regard to this issue. In the light of these premises, it seems that anyone who thinks that we should *not* be prevented from developing humanlike machines ('i.e. a posthumanist'), would need to show how X could be *shared* between humans and the humanlike machines so that *both* would have appropriate rights and obligations (which would prevent a possible 'war' between the two), and likewise, a proponent of any restrictions to the pursuit of developing humanlike machines that X is something that *cannot* be transferred from humans to machines. Accomplishing either of these philosophical tasks would, however, require specifying what X is. The required specification would not need to be 'exhaustive', but only such that it would enable finding out whether X is a *multiple-realizable* property that is simultaneously attributable to both humans and machines, and permit assigning rights and obligations to the machines – or *not.*

However, before engaging with the attempts to specify X, I would like to address an objection that this way of presenting the problem is unnecessarily complicated. For it has been suggested that since only "humans, not robots, are responsible agents", a practical way of ensuring that machines shall respect the moral rights of humans would be to *oblige* human programmers to ensure that the machines they create operate in compliance with "existing laws & fundamental rights & freedoms." (British Standards Institute, 2010). Because machines would be thus effectively forced to respect the rights and obligations of humans *a priori*, by implication, the hypothetical machines with fully humanlike capacities would categorically respect humans irrespective of how the humans would actually treat them.

My reply to this objection would be that any apparent simplicity of the approach towards the challenge of specifying X that, however, evades producing any specification, but relies on *a priori* assumptions instead, is fundamentally deceptive. First of all, it is unclear *which* existing laws and moral principles a machine should be made to respect *a priori* - e.g., are laws related to working hours important? – and so is also an answer to the question *how* morality could be programmed to a machine – e.g., whether through implementing some set of if/then rules to guide its conduct, or through a neural network that evaluates various action possibilities *etc.* Finding a plausible 'practical' answer to the former question, and a viable method for accomplishing the latter task would require discovering no less than which specific aspects of human nature a machine would need to emulate in order to capable of moral acting. And this would be only possible if we knew which aspect of our nature ultimately makes us subject to moral rights and obligations – that is, had X specified. Hence, regardless of whether one considered the behavioural standards of machines derivative of our moral and legal rights and obligations through the effort of programmers, or alternatively, based directly on the rights and obligations that the machine could have in virtue of being able to realise X – it would be nevertheless important to find what X is.

2. Three Possible Ways to Specify X

There are three available overall approaches towards specifying the nature of X. *First*, one could think that X is a categorical feature of being human, any aspect of which cannot be realised by any other kind of being, including machines, by definition. In what follows, this option shall be designated with symbol (**a**). Should one endorse option (**a**) one would need to think that no obligation or right could be attributed to a machine without creating an entirely new kind of moral theory that does not appeal to human experience, nature, or reason in justifying those rights and obligations. However, since building such a theory would not be possible for us, adopting option (**a**) would, in practice, imply that we could assign neither moral obligations nor rights to machines. *Second*, one could think instead – like quite a few philosophers engaged with 'posthumanism' think (e.g., Stahl, 2004; Gunkel, 2012 & 2018; Bostrom and Yudkowsky, 2014; Signorelli, 2018) – that even if X currently belongs only to humans, it could be realised by machines in the future, and hence, moral obligations and rights that we assign to one another could be extended to apply to machines in relation to their ability to realise the aspects of X, such as sentience or rational acting. This option shall be designated with symbol (**b**). *Third*, it is possible for one to think in line with the 18th century machine-human analogists that X is such that both humans and machines have it and therefore there might be no moral difference between humans and machines. The symbol of this option shall be (**c**).

a) X as Exclusive to Humans

Of the three possible strategies with regard to specifying the nature of X, (**a**) is by far the most traditional. This option starts from the assumption that humans *must* be unique and 'special' among beings. Although humans are unique insofar as we have not only abilities for "moral choice, or reason, or language, or sociability, or sentience, or emotions, or consciousness", but we have all these abilities operating together and reinforcing one another (Fukuyama, 2002: 183), merely having a species-specific collection of co-dependent abilities (*pace* Fukuyama) does not suffice to *justify* our exceptionality among beings. Since also animals and robots (can) have various more or less extensive sets of codependent abilities, we would need to find a *further* justification why these sets must be categorically less 'special' than the human ability set.

The most influential version of such a justification comes from the 16th century, and is offered by 'Cartesian dualism', which posits that since perceptible bodies of *both* humans and all other kinds of beings are made of matter, the *bodily aspect* (in spite of its psychological import) *cannot* account for the metaphysical difference between us and the other kinds of beings. Thus, the difference has to be based on the *mental aspect* – either in human 'soul', or, less abstractly, in our ability to conceive ourselves as beings *who know what they are doing and for what end* – that is – in our 'self-consciousness'. According to Descartes, one must admit that one's senses can betray oneself, but has to regard the reality of one's self-consciousness as being beyond reasonable doubt (Descartes, 1904: 25). In this respect, we are different from machines and animals – who do not need to be self-conscious in order to operate in the world – and only an agent who is conscious of what she does, can be held responsible for her actions and thus subject to obligations, or 'maxims' as Descartes (1902: part 3) called them. However, from this conclusion, it does not need to follow that *only* humans could be entitled to moral rights. A Cartesian could think that e.g. sentience justifies at least some rights – e.g., a right to respectful treatment – in which case animals would have those rights, although obligations are strictly for humans only.

The necessity of human self-consciousness for the attribution of obligations can be justified also without directly appealing to Cartesian dualism. A famous argument that can be used to advance this conclusion has been presented by John Searle. In the argument, one is asked to imagine oneself as a monolingual English-speaker confined alone in a room with two batches of Chinese script, which a third batch and a set of instructions in English that enable one "to correlate elements of this third batch with elements of the first two batches." (Searle, 1980: 418). By correlating the elements, one would be able to "to give back certain sorts of Chinese symbols with certain sorts of shapes in response to certain sorts of shapes given [..] in the third batch" (Ibid.). In other words, one would be capable of answer questions about the Chinese without being conscious of what one is doing. According to Searle, this is basically how a computer would operate. Intuitively, the person in the room, as opposed to one who consciously understands Chinese, cannot be perceived to be free and hence responsible for the quality of her answers, or the batches on which they are based (cf. Searle, 1984: 94). Therefore, she can be under no obligation to ensure their correctness, and analogously, neither can a computer.

However, the practical problem in all Cartesian arguments for the existence of the categorical difference between humans and machines is that in them, self-consciousness features as a subjective state, which we 'intuitively' perceive other humans to have and machines to lack, but cannot – as a matter of fact – anyhow 'empirically' observe. However, should a machine become indistinguishable from a human being in terms of its capabilities, it is possible that our intuitive perceptions would change. Even with the technology of today, it is possible to create machines that many people are inclined to perceive as self-conscious at least in a limited extent: consider, e.g. the initial reception of robot Sophia (e.g., Gohd, 2018). Thus, even after the implications of Searle's argument are accounted for, it remains an *open question* whether or not there is any empirically perceptible difference between humans and machines that could justify their categorically different moral and legal statuses and that would remain in place regardless of how technologically advanced the latter may become in comparison to the former.

The most promising way to reply above worry while holding in (**a**) would be to attempt to search for possible empirically perceptible correlates for the state of self-consciousness – the correlates that machines do not have, and could not have. For example, a certain aspect of the biological structure of human neural system could be the desired kind of correlate (Crick and Koch, 1998; Churchland, 2013). Currently, the most promising candidate seems to be *neocortical activity*, because no artificial system structurally resembles neocortex. Some attempts have been made to computationally model its functions (e.g., Sabour et al., 2017), but the layered structure of even current state-of-art artificial neural networks is far closer to cerebellum (see e.g. Yamazaki and Igarashi, 2013). However, the neural functions that seem to produce the perception of self-consciousness in us are located in the neocortex *instead* of the cerebellum. This has been demonstrated by experiments that are based on observing the correlation between the perceptible signs of self-consciousness displayed by a patient and spikes in neural activity across different parts of her brain (e.g., Storm et al., 2017). Since the operation of neocortex presupposes human neurochemistry, it would not even be physically possible to implement it to an electrical machine. Moreover, having a neocortex would not enable the machine to execute computing tasks any faster, but due to the slowness of human neurochemistry, be likely a hindrance to achieving an optimal processing speed. Thus, even if a machine could 'intuitively' appear as self-conscious – e.g., in virtue of being able to whatever humans can do – with the proposed method, we would be able to 'empirically' observe that it simply *could not* be self-conscious due to not having the human neocortex.

However, the problem in this, and other proposals of the same vein, is that even if there was a biological difference between conscious and nonconscious agents, it would be arbitrary to claim that such a difference would suffice to justify us being treated differently from machines (Bostrom and Yudkowsky, 2014: 912). Moreover, regardless of whether or not we would like to attribute a normative fallacy to such inferences, it would *still* remain an open question whether a being can be self-conscious if and only if the begin manifested similar neurobiological structure with humans. There is nothing *in* the concept of 'self-consciousness' or 'neocortex' that analytically implies that only beings with the former could have the latter. Although if only the beings that have neocortex had been self-conscious so far, it would thus need to follow that any being that lacks neocortex, e.g., an advanced future computer, could not be self-conscious.

A more suitable candidate for the empirical correlate of self-consciousness would thus be an empirically perceptible ability that analytically presupposes self-consciousness – such as the *capacity for rational action*. Arguably, it would not be coherent to think that a nonconscious being could act in a way that *consistently* makes an impression of rationality over time, because *rational acting*, by definition, involves construing logically valid inferences about *one's* values, and ways in which *one* can realise them by action, and taking responsibility of the resulting action. However, like any other ability, also the ability for rational action can evidently come in *degrees*: one can act *more* or *less* rationally, just as one can, e.g., be more or less sentient to pain.

This conclusion opens up the possibility that instead of being categorical, X that differentiates us from machines can be realised in degrees as well. The more rational the acting of a machine appears to be for us, the more like a human being it *is*, and hence, the more moral and legal obligations and rights it can be expected to have. For certain basic rights, sentience can be sufficient – but if sentience is combined with a sufficient ability for rational action, a being could have more *human-like* rights – e.g., the right for respectful treatment would transform into the right for autonomy. By acknowledging these conclusions that open a passage for machines to become treated more like humans in relation to the progress they make in emulating human abilities, we, however, stray away from (**a**), and in fact endorse option (**b**). Perhaps this is one reason why currently (**b**) is a popular option among philosophers dealing with posthumanism.

b) Currently for Humans Only – but Possible for Machines to Acquire

While (**a**) is based on Cartesianism, the origins of option (**b**) are in Aristotelian thought, according to which the kinds of beings are not to be divided according to categorical boundaries, but clustered according to their capacities. Not all 'non-rational animals' are unable to rational thought according to Aristotle – e.g., a lion is able to see that a 'stag or a wild goat' is a potential meal (Aristotle, 1995: 1118a22). However, regardless of its particular set of capacities, any animal would nevertheless be closer to any other animal than a paradigmatic adult human in the hypothetical space of capacities, in which the capacities each being are compared against the capacities of all other beings. Provided that the difference between humans and animals would be accounted above all by differences in their rational capacities – as Aristotle (1995: 1097b31-1098b8ff) thinks – we could say that *rationality* is 'what it is for a human to be' (*to ti ên einai).*

This conception of 'essence' could justify regarding animals and humans different in kind, but abstains from drawing any categorical difference between them that would preclude all rationality from animals. In more abstract terms, while Descartes regarded animals as 'soulless', Aristotle regarded them all to have a 'soul', and the humans to have (metaphorically) a 'rational part' in their soul on account for their superior rational capacities (Aristotle, 1995: 1102a26-32). Although Aristotle does not speak much about moral and legal obligations, and even less about machines, this scheme allows considering also the obligations of a machine to be related to its closeness to the humans in the space of capacities. With regard to the rights of a machine, certain rights could be assigned to it in relation its to closeness to any sentient animal, but the attribution of humanlike rights would require *more* closeness to humans than to any animal – which in turn, would presuppose that apart from sentience, the computer should have *also* humanlike rational capacities (cf. Bostrom and Yudkowsky, 2014: 918-19).

Since the attribution of responsibility for a rational action presupposes that one is able to draw an inference that concludes in action without being told the steps of drawing such an inference, one way to test the level of rationality of a computer would be to ask it to perform humanlike tasks, such answering questions in natural language *without* pre-programming each step of the performance. Thus, e.g., success in answering questions about Chinese would be considered a sign of machine rationality *if and only if* the computer was not pre-programmed by humans to accomplish this specific task. According to a supporter of (**b**), a computer that can select means to such tasks without human intervention, i.e., *autonomously*, should be obliged to follow the same moral and legal principles than humans have to follow when selecting means to ends. E.g., just like a human, a computer should not be allowed to come up with a step-by-step guide to running a protection racket when it is asked to prepare a plan for a profitable business.

As made clear above, in order to merit humanlike moral rights, a computer would also need to be sentient apart from being rational. It would be, however, very difficult to measure sentience of a computer, because thanks to its rational abilities, an advanced computer might well be able to emulate the behaviours associated with sentience without actually 'feeling' anything (with, e.g. the help of methods listed in McDuff and Czerwinski, 2018). In any case, assuming that this challenge can be resolved, the capacity-based approach (**b**) towards the question of whether a machine could be attributed moral rights and obligations may appear as very commonsensical, and the posthumanist thinkers support this view, according to which computers can gradually become morally more and more like humans when the range and depth of their rational and sentient capacities increase. Unlike humans and animals, computers do not have any biological limitations that prevent them from extending their capacities: any part of a computer could be replaced with a more powerful part, and any number of individual computing units connected with one another to form a more powerful unit *ad infinitum*. Thus, we cannot assume *a priori* that emulating some aspect of human nature is out of their reach: if e.g. emulating neocortex is impossible for current computer architecture, then the challenge would be to find a different way of implementing its functions to computers or to change their architecture.

However, irrespective of its theoretical merits, it seems that (**b**) suffers from a certain fundamental problem. Just as position (**a**) begins turning towards (**b**) when we start inquiring for the empirically perceptible correlates for self-consciousness, (**b**) risks collapsing into (**c**) when the following Aristotelian consideration – which I have not seen addressed in literature on machine ethics – is brought in. Provided that there is no categorical difference between humans and computers, and that computers – unlike animals whose capacities are limited by their biological structure – can at least potentially acquire fully humanlike (or even superhuman range of capacities), then it seems that computers are, in principle, humans *in potentia.* Although one may claim that merely potential humans are not humans, this position would not be tenable. Humans consider small children as humans in spite of their being closer to non-rational animals than to paradigmatic adult humans in the space of actualised capacities: e.g., great apes can employ fairly sophisticated linguistic communication (Hurley and Nudds, 2006), whereas small human children are incapable of using language. Thus, it is no wonder that reasoning that appeals to their unrealised potential has been used to ground the humanity of small children since Aristotle (McGowan Tress, 1995: 3).

It may now seem that in order to a posthumanist be coherent in her thinking, it would not be sufficient for her to think, in line with (**b**), that computers can become humanlike in the future – but she would need to regard them as essentially humans *already,* or else consider small human children as not human. This consideration can now make option (**c**) – which seems to have never been 'popular', even when it was formulated in the 18th century, and has currently virtually no supporters – seem more coherent and morally defensible than (**b**). Accepting (**c**) would not need amount to accepting that potential for having fully humanlike capacities establishes anything like *moral personhood.* A human, who has *lost* her potential for realising her capacities due to an incurable illness, seems still to be a person; it is less clear whether this implies that a fully humanlike machine could retain its moral status after getting irreparably broken.

c) Both Machines and Humans have X

Option (**c**) was born with the claim that humans are machines. When this thesis was first proposed by Julien Offray de La Mettrie in 1748, it was too controversial even for many 'radical' thinkers. The guiding thought of La Mettrie in his treatise *L'Homme Machine* was that just as animals are soulless – as already Descartes had argued – perhaps humans, too. If animals can act in the world without 'soul', why would human acting presuppose any immaterial basis either? According to La Mettrie, this was only a baseless 'dogma' (La Mettrie, 2009: 1). Some two decades later, baron d'Holbach took La Mettrie's theory of human nature as the starting point in his attempt to build *Système de la Nature* (1770) that was intended to reveal the 'laws of nature' that deterministically explain the 'moral' behaviour of humans to complement the Newtonian laws that explain the movements of physical objects. However, instead of going towards the direction of speculating about our chances of building machines that match or surpass our own rational capacities, he focused on producing counter-arguments against the 'prejudice' that morality needs religion (d'Holbach, 2001: 110). Although these arguments might have had some impact to the subsequent philosophers, by the early 20th-century – when the study of organic chemistry and biology had made obvious that living organisms do *not* operate like mechanical devices, La Mettrie and d'Holbach's theories became used as stock examples of failed empirical philosophy (e.g., Cassirer, 1954: 124).

Due to their unfortunate reputation, both La Mettrie and d'Holbach have gone unnoticed among posthumanism until today. However, in the light of what we have come to conclude in the previous section, the idea about humans being machines might suddenly seem relevant again, though this time other way round – machines being humans. If we agree with posthumanists that the capacities of a being in relation to other beings matter in determining the nature of a being, the objection to La Mettrie and d'Holbach - that humans cannot be machines on account of their different biological structure in comparison to machines – loses its force: provided that computers can potentially have all human capacities. Of course, the computers do not merit the same obligations and rights that responsible adult humans merit until they have their capacities, but neither do children, and hence, until that time, the computers could be regarded in a way analogous to the way we regard children, i.e., their obligations and rights should be conditional upon their degree of development.

In practice, both the option that both humans and machines have X – (**c**) – and that currently only humans have it, but the machines shall acquire it over time – (**b**) – imply that the rights and obligations of computers are related to their observable capacities. Thus both approaches would justify assigning the same set of moral obligations and rights for computers with the same justification. However, we have seen that (**c**) is more coherent than (**b**) – which, in turn, is more justifiable than (**a**). The main problem with (**a**) turned out to be related to the supposedly 'intuitive' perception that X categorically separates humans from machines. In brief, the problem is that that there are no empirically perceptible correlates that would be sufficient to validate the intuitive perception of the existence of a specific X unique to humans such as self-consciousness.

3. Conclusion

On the basis of preceding discussion, the preferable answer to the question whether or not a machine could be attributed moral rights and obligation seems to be 'yes'. If machines are potentially humanlike – as (**c**) would propose – they could be attributed moral and legal rights and obligations in relation to the closeness of their capacities to the capacities of an average adult human being. In practice, this answer to the question would imply that if a machine learns to autonomously perform a human task *T* (e.g., driving a taxi) no worse than a 'biological' human, then the rights (such as a right to salary, vacation and rest) and obligations (such as ensuring that the passengers are safe) humans have related to *T* should apply *also* to the machine. Although the presented answer may now seem not only more coherent than its alternatives, but also a useful guideline in crafting legislation to mitigate the effects of increasing robotization to human labour, this does not yet mean, however, that it must be the preferable answer *all things considered.* When analysing the options (**a**) and (**b**), we focused only on the traditional conceptions of X, that attempt to find the decisive difference between humans and machines in self-consciousness, rationality and sentience. However, even if these conceptions were insufficient, it would not mean that there is *no* categorical X to be found, for example, in the human ability to love, display empathy and listen to one's moral conscience.

References

Annas, George, Andrews, Lori B. and Isasi, Rosario M. (2002) "Protecting the Endangered Human: Toward An International Treaty Prohibiting Cloning and Inheritable Alterations", *American Journal of Law & Medicine,* 28: 151-178.

Aristotle (1984) "Nicomachean Ethics" In *The Complete Works of Aristotle, Revised Oxford Translation,* Vol. II. (Trans. W. D. Ross; J. Barnes, ed.), Princeton: Princeton University Press.

Bostrom, Nick and Yudkowsky, Eliezer (2014) "The Ethics of Artificial Intelligence" In *The Cambridge Handbook of Artificial Intelligence* (K. Frankish, ed.), Cambridge: Cambridge University Press.

British Standards Institute (2010) "EPSRC Principles of Robotics" in *BS,* 8611. https://epsrc.ukri.org/research/ourportfolio/themes/engineering/activities/principlesofrobotics/.

Cassirer, Ernst (1954) *The Philosophy of the Enlightenment,* Boston: Beacon Press.

Churchland, Paul M. (2013) *Matter and Consciousness: A Revised Edition,* Cambridge, MA: MIT Press.

Crick, Francis and Koch, Christopher (1998) "Consciousness and Neuroscience", *Cereb Cortex,* 8: 97–107.

Descartes, René (1902) "Discours de la méthode", *Oeuvres, Vol. 6.*

Descartes, René (1904) *Meditations.* in *Oeuvres de Descartes, Vol. 7* (A. Charles and P. Tannery, eds.), Paris: J. Vrin.

D'Holbach, Baron (2001) *The System of Nature, 2 Vols.,* (Trans. H. D. Robinson), Kitchener: Batoche Books.

Fukuyama, Francis (2002) *Our Posthuman Future,* New York: Farrar, Straus and Giroux,

Gohd, Chelsea (2018) "Here's What Sophia, the First Robot Citizen, Thinks About Gender and Consciousness" in *LiveScience*, July 11, Accessed in https://www.livescience.com/63023-sophia-robot-citizen-talks-gender.html.

Gunkel, David J. (2012) *The Machine Question*, Cambridge, MA: MIT Press.

Gunkel, David J. (2018) *Robot Rights*, Cambridge, MA: MIT Press.

Hurley, Susan and Nudds, Matthew (2006) "Language as a Window on Rationality" In *Rational Animals?* (Savage-Rumbaugh, E., Rumbaugh, D., & Fields, W., eds.), Oxford: Oxford University Press, pp. 513-550.

La Mettrie, Julien Offray (2009) *Man—Machine* (Trans. Jonathan Bennett), Accessed in https://www.earlymoderntexts.com/assets/pdfs/lamettrie1748.pdf.

McDuff, Daniel and Czerwinski, Mary (2018), "Designing Emotionally Sentient Agents", *Communications of the ACM*, 61 (12): 74-83.

Sabour, Sara, Frosst, Nicholas and Hinton, Geoffrey (2017) "Dynamic Routing Between Capsules", *ArXiv*:1710.09829.

Searle, John R. (1980) "Minds, Brains, and Programs", *Behavioral and Brain* Sciences, 3 (3): 417-457.

Searle, John R. (1984) *Minds, Brains, and* Science, Cambridge, MA: Harvard University Press.

Signorelli, Camilo M. (2018) "Can Computers Become Conscious and Overcome Humans?", *Frontiers in Robotics and AI*, 5, Accessed in https://www.frontiersin.org/article/10.3389/frobt.2018.00121.

Silver, David, Hubert, Thomas, Schrittwieser, Julian et al., (2018) "A general reinforcement learning algorithm that masters chess, shogi, and Go through self-play", *Science*, 362 (6419): 1140-1144.

Stahl, Bernd C. (2004) "Information, Ethics, and Computers: The Problem of Autonomous Moral Agents", *Minds and Machines*, 14: 67–83.

Storm, Johan F., Boly, Mélanie, Casali, Adenauer G., Massimini, Marcello, Olcese, Umberto et al. (2017) "Consciousness Regained: Disentangling Mechanisms, Brain Systems, and Behavioral Responses", *Journal of Neuroscience*, 37 (45): 10882-10893.

Vinyals, Oriol, Toshev, Alexander, Bengio, Samy et al. (2014) "Show and Tell: A Neural Image Caption Generator", *ArXiv*:1411.4555.

Wu, Yongui, Schuster, Mike, Chen, Zhifeng et al. (2016) "Google's Neural Machine Translation System: Bridging the Gap between Human and Machine Translation", *ArXiv*:1609.08144.

Yamazaki, Tadashi and Igarashi, Jun (2013) "Realtime cerebellum: A large-scale spiking network model of the cerebellum that runs in real-time using a graphics processing unit", *Neural Networks*, 47: 103-111.

CHAPTER 18

THE COMING TECHNOLOGICAL SINGULARITY: HOW TO SURVIVE IN THE POST-HUMAN ERA

Vernor Vinge[1]

San Diego State University, USA

Abstract: Within thirty years, we will have the technological means to create superhuman intelligence. Shortly after, the human era will be ended. Is such progress avoidable? If not to be avoided, can events be guided so that we may survive? These questions are investigated. Some possible answers (and some further dangers) are presented.

1. What is The Singularity?

The acceleration of technological progress has been the central of this century. I argue in this chapter that we are on the edge of change comparable to the rise of human life on Earth. The precise cause of this change is the imminent creation by technology of entities with greater than human intelligence. There are several means by which science may achieve this breakthrough (and this is another reason for having confidence that the event will occur):

- The development of computers that are "awake" and superhumanly intelligent. (To date, most controversy in the area of AI relates to whether we can create human equivalence in a machine. But if the answer is "yes, we can", then there is little doubt that beings more intelligent can be constructed shortly thereafter;

---

[1] Department of Mathematical Science, San Diego State University. This article was for the VISION-21 Symposium sponsored by NASA Lewis Research Center and the Ohio Aerospace Institute, March 30-31, 1993. It is also retrievable from the NASA technical reports server as part of NASA CP-10129. A slightly changed version appeared in the Winter 1993 issue of *Whole Earth Review*.

- Large computer networks (and their associated users) may "wake up" as a superhumanly intelligent entity;
- Computer/human interfaces may become so intimate that users may reasonably be considered superhumanly intelligent;
- Biological science may find ways to improve upon the natural human intellect.

The first three possibilities depend in large part on improvements in computer hardware. Progress in computer hardware has followed an amazingly steady curve in the last few decades (cf. Hans, 1988). Based largely on this trend, I believe that the creation of greater than human intelligence will occur during the next thirty years. Charles Platt (cf. Platt, n/d) has pointed out the AI enthusiasts have been making claims like this for the last thirty years. Just so I'm not guilty of a relative-time ambiguity, let me more specific: I'll be surprised if this event occurs before 2005 or after 2030.)

What are the consequences of this event? When greater-than-human intelligence drives progress, that progress will be much more rapid. In fact, there seems no reason why progress itself would not involve the creation of still more intelligent entities – on a still-shorter time scale. The best analogy that I see is with the evolutionary past: Animals can adapt to problems and make inventions, but often no faster than natural selection can do its work – the world acts as its own simulator in the case of natural selection. We, humans, have the ability to internalize the world and conduct "what if's" in our heads; we can solve many problems thousands of times faster than natural selection.

Now, by creating the means to execute those simulations at much higher speeds, we are entering a regime as radically different from our human past as we humans are from the lower animals. From the human point of view, this change will be a throwing away of all the previous rules, perhaps in the blink of an eye, an exponential runaway beyond any hope of control. Developments that before were thought might only happen in "a million years" (if ever) will likely happen in the next century. (Greg Bear paints a picture of the major changes happening in a matter of hours (cf. Bear, 1985)).

I think it's fair to call this event a singularity ("the Singularity" for the purposes of this chapter). It is a point where our models must be discarded and a new reality rules. As we move closer and closer to this point, it will loom vaster and vaster over human affairs till the notion becomes a commonplace. Yet when it finally happens, it may still be a great surprise and a greater unknown. In the 1950s there were very few who saw it: Stan Ulam (cf. Ulam, 1958) paraphrased John von Neumann as saying:

> One conversation centered on the ever accelerating progress of technology and changes in the mode of human life, which gives the appearance of approaching some essential singularity in the history of the race beyond which human affairs, as we know them, could not continue.

Von Neumann even uses the term singularity, though it appears he is still thinking of normal progress, not the creation of superhuman intellect. (For me, the superhumanity is the essence of the Singularity. Without that we would get a glut of technical riches, never properly absorbed (see Stent, 1969).

In the 1960s there was recognition of some of the implications of superhuman intelligence. I. J. Good wrote:

> Let an ultraintelligent machine be defined as a machine that can far surpass all the intellectual activities of any any man however clever. Since the design of machines is one of these intellectual activities, an ultraintelligent machine could design even better machines; there would then unquestionably be an "intelligence explosion," and the intelligence of man would be left far behind. Thus the first ultraintelligent machine is the _last_ invention that man need ever make, provided that the machine is docile enough to tell us how to keep it under control. ... It is more probable than not that, within the twentieth century, an ultraintelligent machine will be built and that it will be the last invention that man need make (Good, 1965: 31-88).

Good has captured the essence of the runaway, but does not pursue its most disturbing consequences. Any intelligent machine of the sort he describes would not be humankind's "tool" – any more than humans are the tools of rabbits or robins or chimpanzees. Through the '60s and '70s and '80s, recognition of the cataclysm spread (cf. Vinge, 1966; Alfvén, 1969; Vinge, 1983; Bear, 1985). Perhaps it was the science-fiction writers who felt the first concrete impact. After all, the "hard" science-fiction writers are the ones who try to write specific stories about all that technology may do for us. More and more, these writers felt an opaque wall across the future. Once, they could put such fantasies millions of years in the future (cf. Stapledon, 1961). Now they saw that their most diligent extrapolations resulted in the unknowable... soon. Once, galactic empires might have seemed a Post-Human domain. Now, sadly, even interplanetary ones are.

What about the '90s and the '00s and the '10s, as we slide toward the edge? How will the approach of the Singularity spread across the human world view? For a while yet, the general critics of machine sapience will have good press. After all, till we have hardware as powerful as a human brain, it is probably foolish to think we'll be able to create human equivalent (or greater) intelligence. (There is the far-fetched possibility that we could make a human equivalent out of less powerful hardware, if we were willing to give up speed, if we were willing to settle for an artificial being who was literally slow (cf. Vinge, 1981). But it's much more likely that devising the software will be a tricky process, involving lots of false starts and experimentation. If so, then the arrival of self-aware machines will not happen till after the development of hardware that is substantially more powerful than humans' natural equipment.)

But as time passes, we should see more symptoms. The dilemma felt by science fiction writers will be perceived in other creative endeavors. (I have heard thoughtful comic book writers worry about how to have spectacular effects when everything visible can be produced by the technically commonplace.) We will see automation replacing higher and higher-level jobs. We have tools right now (symbolic math programs, cad/cam) that release us from most low-level drudgery. Or put another way: The work that is truly productive is the domain of a steadily smaller and more elite fraction of humanity. In the coming of the Singularity, we are seeing the predictions of *true* technological unemployment finally come true.

Another symptom of progress toward the Singularity: ideas themselves should spread ever faster, and even the most radical will quickly become commonplace. When I began writing, it seemed very easy to come up with ideas that took decades to percolate into the cultural consciousness; now the lead time seems more like eighteen months. (Of course, this could just be me losing my imagination as I get old, but I see the effect in others too.) Like the shock in a compressible flow, the Singularity moves closer as we accelerate through the critical speed.

And what of the arrival of the Singularity itself? What can be said of its actual appearance? Since it involves an intellectual runaway, it will probably occur faster than any technical revolution seen so far. The precipitating event will likely be unexpected – perhaps even to the researchers involved. ("But all our previous models were catatonic! We were just tweaking some parameters...") If networking is widespread enough (into ubiquitous embedded systems), it may seem as if our artifacts as a whole had suddenly wakened.

And what happens a month or two (or a day or two) after that? I have only analogies to point to: The rise of humankind. We will be in the Post-Human era. And for all my rampant technological optimism, sometimes I think I'd be more comfortable if I were regarding these transcendental events from one thousand years remove... instead of twenty.

2. Can the Singularity be Avoided?

Well, maybe it won't happen at all: Sometimes, I try to imagine the symptoms that we should expect to see if the Singularity is not to develop. There are the widely respected arguments of Penrose (cf. Penrose, 1989) and Searle (cf. Searle, 1980) against the practicality of machine sapience. In August of 1992, Thinking Machines Corporation held a workshop to investigate the question "How We Will Build a Machine that Thinks" [Thearling]. As you might guess from the workshop's title, the participants were not especially supportive of the arguments against machine intelligence. In fact, there was general agreement that minds can exist on nonbiological substrates and that algorithms are of central importance to the existence of minds. However, there was much debate about the raw hardware power that is present in organic brains. A minority felt that the largest 1992 computers were within three orders of magnitude of the power of the human brain.

The majority of the participants agreed with Moravec's estimate (cf. Moravec, 1988) that we are ten to forty years away from hardware parity. And yet there was another minority who pointed to Conrad et al., (1989) and Rasmussen et al., (1991) and conjectured that the computational competence of single neurons may be far higher than generally believed. If so, our present computer hardware might be as much as *ten* orders of magnitude short of the equipment we carry around in our heads. If this is true (or for that matter, if the Penrose or Searle critique is valid), we might never see a Singularity. Instead, in the early '00s we would find our hardware performance curves begin to level off – this caused by our inability to automate the complexity of the design work necessary to support the hardware trend curves. We'd end up with some *very* powerful hardware, but without the ability to push it further. Commercial digital signal processing might be awesome, giving an analog appearance even to digital operations, but nothing would ever "wake up" and there would never be the intellectual runaway which is the essence of the Singularity. It would likely be seen as a golden age... and it would also be an end of progress. This is very like the future predicted by Gunther Stent. In fact, on page 137 of *The Coming of the Golden Age: A View of the End of Progress*, Stent explicitly cites the development of transhuman intelligence as a sufficient condition to break his projections.

But if the technological Singularity can happen, it will. Even if all the governments of the world were to understand the "threat" and be in deadly fear of it, progress toward the goal would continue.

In fiction, there have been stories of laws passed forbidding the construction of "a machine in the form of the mind of man" (cf. Herbert, 1985). In fact, the competitive advantage – economic, military, even artistic –

of every advance in automation is so compelling that passing laws, or having customs, that forbid such things merely assures that someone else will get them first.

Eric Drexler (cf. Drexler, 1986) has provided spectacular insight about how far technical improvement may go. He agrees that superhuman intelligences will be available in the near future – and that such entities pose a threat to the human status quo. But Drexler argues that we can embed such transhuman devices in rules or physical confinement such that their results can be examined and used safely. This is I. J. Good's ultraintelligent machine, with a dose of caution. I argue that confinement is intrinsically impractical. For the case of physical confinement: Imagine yourself confined to your house with only limited data access to the outside, to your masters. If those masters thought at a rate – say – one million times slower than you, there is little doubt that over a period of years (your time) you could come up with "helpful advice" that would incidentally set you free. (I call this "fast thinking" form of superintelligence "weak superhumanity". Such a "weakly superhuman" entity would probably burn out in a few weeks of outside time. "Strong superhumanity" would be more than cranking up the clock speed on a human-equivalent mind. It's hard to say precisely what "strong superhumanity" would be like, but the difference appears to be profound. Imagine running a dog mind at very high speed. Would a thousand years of doggy living add up to any human insight? (Now if the dog mind were cleverly rewired and _then_ run at high speed, we might see something different...) Most speculations about superintelligence seem to be based on the weakly superhuman model. I believe that our best guesses about the post-Singularity superhumanity. I will return to this point later in the chapter.)

The other approach to Drexlerian confinement is to build _rules_ into the mind of the created superhuman entity (Asimov's Laws). I think that performance rules strict enough to be safe would also produce a device whose ability was clearly inferior to the unfettered versions (and so human competition would favor the development of those more dangerous models). Still, the Asimov dream is a wonderful one: Imagine a willing slave, who has 1000 times your capabilities in every way. Imagine a creature who could satisfy your every safe wish (whatever that means) and still have 99.9% of its time free for other activities. There would be a new universe we never really understood, but filled with benevolent gods (though one of _my_ wishes might be to become one of them).

If the Singularity cannot be prevented or confined, just how bad could the Post-Human era be? Well... pretty bad. The physical extinction of the human race is one possibility. (Or as Eric Drexler put it of nanotechnology: Given all that such technology can do, perhaps governments would simply decide that they no longer need citizens!). Yet physical extinction may not be the scariest possibility. Again, analogies: Think of the different ways we relate to animals. Some of the crude physical abuses are implausible, yet... In a Post-Human world, there would still be plenty of niches where human equivalent automation would be desirable: embedded systems in autonomous devices, self-aware daemons in the lower functioning of larger sentients. (A strongly superhuman intelligence would likely be a Society of Mind (cf. Minsky, 1985) with some very competent components.) Some of these human equivalents might be used for nothing more than digital signal processing. They would be more like whales than humans. Others might be very human-like, yet with a one-sidedness, a _dedication_ that would put them in a mental hospital in our era. Though none of the closest things in the new environment to what we call human now. (I. J. Good had something to say about this, though at this late date, the advice may be moot: Good (cf. Good, n/d) proposed a "Meta-Golden Rule", which might be paraphrased as "Treat your inferiors as you would be treated by your superiors." It's a wonderful, paradoxical idea (and most of my friends don't believe it) since the game-theoretic payoff is so hard to articulate. Yet if we were able to follow it, in some sense that might say something about the plausibility of such kindness in this universe.)

I have argued above that we cannot prevent the Singularity, that its coming is an inevitable consequence of the humans' natural competitiveness and the possibilities inherent in technology. And yet... we are the initiators. Even the largest avalanche is triggered by small things. We have the freedom to establish initial conditions, make things happen in ways that are less inimical than others. Of course (as with starting avalanches), it may not be clear what the right guiding nudge really is.

3. Other Paths to the Singularity: Intelligence Amplification

When people speak of creating superhumanly intelligent beings, they are usually imagining an AI project. But as I noted at the beginning of this chapter, there are other paths to superhumanity. Computer networks and human-computer interfaces seem more mundane than AI, and yet they could lead to the Singularity. I call this contrasting approach Intelligence Amplification (IA). IA is something that is proceeding very naturally, in most cases not even recognized by its developers for what it is. But every time our ability to access information and to communicate it to others is improved, in some sense we have achieved an increase over natural intelligence. Even now, the team of a PhD human and good computer workstation (even an off-net workstation!) could probably max any written intelligence test in existence.

And it's very likely that IA is a much easier road to the achievement of superhumanity than pure AI. In humans, the hardest development problems have already been solved. Building up from within ourselves ought to be easier than figuring out first what we really are and then building machines that are all of that. And there is at least conjectural precedent for this approach. Cairns-Smith (cf. Cairns-Smith, 1985) has speculated that biological life may have begun as an adjunct to still more primitive life based on crystalline growth. Lynn Margulis (cf. Margulis and Sagan, 1986) has made strong arguments for the view that mutualism is the great driving force in evolution.

Note that I am not proposing that AI research be ignored or less funded. What goes on with AI will often have applications in IA, and vice versa. I am suggesting that we recognize that in network and interface research there is something as profound (and potential wild) as Artificial Intelligence. With that insight, we may see projects that are not as directly applicable as conventional interface and network design work, but which serve to advance us toward the Singularity along the IA path.

Here are some possible projects that take on special significance, given the IA point of view:

- Human/computer team automation: Take problems that are normally considered for purely machine solution (like hill-climbing problems), and design programs and interfaces that take advantage of humans' intuition and available computer hardware. Considering all the bizarreness of higher dimensional hill-climbing problems (and the neat algorithms that have been devised for their solution), there could be some very interesting displays and control tools provided to the human team member;
- Develop human/computer symbiosis in art: Combine the graphic generation capability of modern machines and the esthetic sensibility of humans. Of course, there has been an enormous amount of research in designing computer aids for artists, as labor-saving tools. I'm suggesting that we explicitly aim for a greater merging of competence, that we explicitly recognize the cooperative approach that is possible. Karl Sims (cf. Sims, 1991) has done wonderful work in this direction;
- Allow human/computer teams at chess tournaments. We already have programs that can play better than almost all humans. But how much work has been done on how this power could be used by a human, to get something even better? If such teams were allowed in at least some chess tournaments, it could have a positive effect on IA research that allowing computers in tournaments had for the corresponding niche in AI;
- Develop interfaces that allow computer and network access without requiring the human to be tied to one spot, sitting in front of a

computer. (This is an aspect of IA that fits so well with known economic advantages that lots of effort is already being spent on it.);

- Develop more symmetrical decision support systems. A popular research/product area in recent years has been decision support systems. This is a form of IA, but may be too focussed on systems that are oracular. As much as the program giving the user information, there must be the idea of the user giving the program guidance;
- Use local area nets to make human teams that really work (i.e., are more effective than their component members). This is generally the area of "groupware", already a very popular commercial pursuit. The change in viewpoint here would be to regard the group activity as a combination organism. In one sense, this suggestion might be regarded as the goal of inventing a "Rules of Order" for such combination operations. For instance, group focus might be more easily maintained than in classical meetings. Expertise of individual human members could be isolated from ego issues such that the contribution of different members is focussed on the team project. And of course, shared databases could be used much more conveniently than in conventional committee operations. (Note that this suggestion is aimed at team operations rather than political meetings. In a political setting, the automation described above would simply enforce the power of the persons making the rules!);
- Exploit the worldwide Internet as a combination human/machine tool. Of all the items on the list, progress in this is proceeding the fastest and may run us into the Singularity before anything else. The power and influence of even the present-day Internet is vastly underestimated. For instance, I think our contemporary computer systems would break under the weight of their own complexity if it weren't for the edge that the USENET "group mind" gives the system administration and support people!) The very anarchy of the worldwide net development is evidence of its potential. As connectivity and bandwidth and archive size and computer speed all increase, we are seeing something like Lynn Margulis' (cf. Margulis and Sagan, 1986) vision of the biosphere as data processor recapitulated, but at a million times greater speed and with millions of humanly intelligent agents (ourselves).

The above examples illustrate research that can be done within the context of contemporary computer science departments. There are other paradigms. For example, much of the work in Artificial Intelligence and neural nets would benefit from a closer connection with biological life. Instead of simply trying to model and understand biological life with computers, research could be directed toward the creation of composite systems that rely on biological life for guidance or for the providing features we don't understand well enough yet to implement in hardware. A long-time dream of science-fiction has been direct brain to computer interfaces (cf. Anderson, 1969; Vinge, 1966). In fact, there is concrete work that can be done (and has been done) in this area:

- Limb prosthetics is a topic of direct commercial applicability. Nerve to silicon transducers can be made (cf. Kovacs et al., 1992). This is an exciting, near-term step toward direct communication;
- Similar direct links into brains may be feasible, if the bit rate is low: given human learning flexibility, the actual brain neuron targets might not have to be precisely selected. Even 100 bits per second would be of great use to stroke victims who would otherwise be confined to menu-driven interfaces;
- Plugging in to the optic trunk has the potential for bandwidths of 1 Mbit/second or so. But for this, we need to know the fine-scale architecture of vision, and we need to place an enormous web of electrodes with exquisite precision. If we want our high bandwidth connection to be _in addition_ to what paths are already present in the brain, the problem becomes vastly more intractable. Just sticking a grid of high-bandwidth receivers into a brain certainly won't do it. But suppose that the high-bandwidth grid were present while the brain structure was actually setting up, as the embryo develops. That suggests:
- Animal embryo experiments. I wouldn't expect any IA success in the first years of such research, but giving developing brains access to complex simulated neural structures might be very interesting to the people who study how the embryonic brain develops. In the long run, such experiments might produce animals with additional sense paths and interesting intellectual abilities.

Originally, I had hoped that this discussion of IA would yield some clearly safer approaches to the Singularity. (After all, IA allows our participation in a kind of transcendance.) Alas, looking back over these IA proposals, about all I am sure of is that they should be considered, that they may give us more options. But as for safety... well, some of the suggestions are a little scarey on their face. One of my informal reviewers pointed out that IA for individual humans creates a rather sinister elite. We humans have millions of years of evolutionary baggage that makes us regard competition in a deadly light. Much of that deadliness may not be necessary in today's world, one where losers take on the winners' tricks and are coopted into the winners' enterprises. A creature that was built _de novo_ might possibly be a much more benign entity than one with a kernel based on fang and talon. And even the egalitarian view of an Internet that wakes up along with all mankind can be viewed as a nightmare (cf. Swanwick, 1986).

The problem is not that the Singularity represents simply the passing of humankind from center stange, but that it contradicts some of our most deeply held notions of being. I think a closer look at the notion of strong superhumanity can show why that is.

4. Strong Superhumanity and the Best We Can Ask for

Suppose we could tailor the Singularity. Suppose we could attain our most extravagant hopes. What then would we ask for: That humans themselves would become their own successors, that whatever injustice occurs would be tempered by our knowledge of our roots. For those who remained unaltered, the goal would be benign treatment (perhaps even giving the stay-behinds the appearance of being masters of godlike slaves). It could be a golden age that also involved progress (overleaping Stent's barrier). Immortality (or at least a lifetime as long as we can make the universe survive (cf. Dyson, 1979; Barrow and Tipler, 1986) would be achievable.

But in this brightest and kindest world, the philosophical problems themselves become intimidating. A mind that stays at the same capacity cannot live forever; after a few thousand years it would look more like a repeating tape loop than a person. (The most chilling picture I have seen of this is in Niven, 1967). To live indefinitely long, the mind itself must grow... and when it becomes great enough, and looks back... what fellow-feeling can it have with the soul that it was originally? Certainly, the later being would be everything the original was, but so much vastly more. And so even for the individual, the Cairns-Smith (or Lynn Margulis) notion of new life growing incrementally out of the old must still be valid.

This "problem" about immortality comes up in much more direct ways. The notion of ego and self-awareness has been the bedrock of the hardheaded rationalism of the last few centuries. Yet now the notion of self-awareness is under attack from the Artificial Intelligence people ("self-awareness and other delusions"). Intelligence Amplification undercuts the importance of ego from another direction.

The post-Singularity world will involve extremely high-bandwidth networking. A central feature of strongly superhuman entities will likely be their ability to communicate at variable bandwidths, including ones far higher than speech or written messages. What happens when pieces of ego can be copied and merged, when the size of a selfawareness can grow or shrink to fit the nature of the problems under consideration? These are essential features of strong superhumanity and the Singularity. Thinking about them, one begins to feel how essentially strange and different the Post-Human era will be – no matter how cleverly and benignly it is brought to be.

From one angle, the vision fits many of our happiest dreams: a place unending, where we can truly know one another and understand the deepest mysteries. From another angle, it's a lot like the worst-case scenario I imagined earlier in this chapter.

Which is the valid viewpoint? In fact, I think the new era is simply too different to fit into the classical frame of good and evil. That frame is based on the idea of isolated, immutable minds connected by tenuous, low-bandwith links. But the post-Singularity world _does_ fit with the larger tradition of change and cooperation that started long ago (perhaps even before the rise of biological life). I think there _are_ notions of ethics that would apply in such an era. Research into IA and high-bandwidth communications should improve this understanding. I see just the glimmerings of this now, in Good's Meta-Golden Rule, perhaps in rules for distinguishing self from others on the basis of bandwidth of connection. And while mind and self will be vastly more labile than in the past, much of what we value (knowledge, memory, thought) need never be lost. I think Freeman Dyson has it right when he says: "God is what mind becomes when it has passed beyond the scale of our comprehension" (cf. Dyson, 1988).[2]

References

Alfvén, Hannes (1969) *The End of Man?* (writing as Olof Johanneson), Award Books, earlier published as *The Tale of the Big Computer*, Coward-McCann, translated from a book copyright 1966 Albert Bonniers Forlag AB with English translation copyright 1966 by Victor Gollanz, Ltd.

Anderson, Poul (1969) "Kings Who Die" In *Seven Conquests*, MacMillan Co., pp. 8-36.

Barrow, John D. and Tipler, Frank J. (1986) *The Anthropic Cosmological Principle*, Oxford: Oxford University Press.

Bear, Greg (1985) "Blood Music" In *Analog Science Fact & Fiction* (June, 1983), Expanded In (1985) *Blood Music*, Morrow: Arbor House.

Cairns-Smith, Graham (1985) *Seven Clues to the Origin of Life*, Cambridge: Cambridge University Press.

Conrad, Michael et al., (1989) "Towards an Artificial Brain", *BioSystems*, 123: 218.

Drexler, K. Eric (1986) *Engines of Creation*, Norwell, MA: Anchor Press/Doubleday.

Dyson, Freeman (1979) "Physics and Biology in an Open Universe", *Review of Modern Physics*, 51: 447-460.

Dyson, Freeman (1988) *Infinite in All Directions*, New York: Harper && Row.

---

[2] I wish to thank John Carroll of San Diego State University and Howard Davidson of Sun Microsystems for discussing the draft version of this chapter with me.

Good, Irving J. (1965) "Speculations Concerning the First Ultraintelligent Machine" In *Advances in Computers* (Franz L. Alt and Morris Rubinoff, eds.), Cambridge, MA: Academic Press, pp. 31-88.

Good, Irving J., (n/d) [Help! I can't find the source of Good's Meta-Golden Rule, though I have the clear recollection of hearing about it sometime in the 1960s. Through the help of the internet, I have found pointers to a number of related items. G. Harry Stine and Andrew Haley have written about metalaw as it might relate to extraterrestrials: G. Harry Stine (1980) "How to Get along with Extraterrestrials ... or Your Neighbor" in *Analog Science Fact-Science Fiction*, February, pp. 39-47.

Herbert, Frank (1985) *Dune*, New York: Berkley Books.

Kovacs, Gregory T. A. et al. (1992) "Regeneration Microelectrode Array for Peripheral Nerve Recording and Stimulation", *IEEE Transactions on Biomedical Engineering*, 39 (9): 893-902.

Margulis, Lynn and Sagan, Dorion (1986) *Microcosmos, Four Billion Years of Evolution from Our Microbial Ancestors*, New York: Summit Books.

Minsky, Marvin (1985) *Society of Mind*, New York: Simon and Schuster.

Moravec, Hans (1988) *Mind Children*, Cambridge, MA: Harvard University Press.

Niven, Larry (1967) "The Ethics of Madness", Reprinted (1968) In *Neutron Star*, New York: Ballantine Books, pp. 82-108.

Penrose, Roger (1989) *The Emperor's New Mind*, Oxford: Oxford University Press.

Platt, Charles (n/d), Private Communication.

Rasmussen, Steen et al., (1991) "Computational Connectionism within Neurons: a Model of Cytoskeletal Automata Subserving Neural Networks" In *Emergent Computation* (Stephanie Forrest, ed.), Cambridge, MA: MIT Press, pp. 428-449.

Searle, John. R. (1980) "Minds, brains, and programs", *Behavioral and Brain Sciences*, 3: 417-457.

Sims, Karl (1991) "Interactive Evolution of Dynamical Systems" In *Toward a Practice of Autonomous Systems: Proceedings of the First European Conference on Artificial Life*, Cambridge, MA: MIT Press.

Stapledon, Olaf (1961) *The Starmaker*, New York: Berkley Books (but from the forward probably written before 1937).

Stent, Gunther S., (1969) *The Coming of the Golden Age: A View of the End of Progress*, New York: Doubleday.

Swanwick, Michael (1986) *Vacuum Flowers*, serialized in Isaac Asimov's Science Fiction Magazine, Republished by Ace Books, 1988.

Thearling, Kurt (n/d) "How We Will Build a Machine that Thinks", a workshop at *Thinking Machines Corporation*. Personal Communication.

Ulam, S. (1958) "Tribute to John von Neumann", *Bulletin of the American Mathematical Society*, 64 (3): 1-49.

Vinge, Vernor (1966) "Bookworm, Run!" In *Analog* (March), Reprinted (1987) In *True Names and Other Dangers*, Wake Forest, NC: Baen Books, pp. 8-40.

Vinge, Vernor (1981) "True Names" In *Binary Star Number*, 5, Dell, Reprinted (1987) in *True Names and Other Dangers*, Wake Forest, NC: Baen Books.

Vinge, Vernor (1983) "First Word", *Omni*, 5 (4): 10.

CHAPTER 19

EPISTEMOLOGICAL AND ETHICAL IMPLICATIONS OF THE FREE-ENERGY PRINCIPLE[1]

Eray Özkural

Bilkent University, Turkey

Abstract: The free energy principle states that self-organization occurs through minimization of free energy, which is a measure of potential thermodynamic work. By minimizing free energy, the organism happens to also minimize surprise over its boundary, promoting chances of survival. We discuss the ethical/epistemological implications of the cognitive goal in detail from an empirical point of view, highlighting the principle of least action as a physical basis of Occam's razor, the universality of the free energy principle, and its explanation of natural selection. We explain that the free energy principle extends to groups of organisms and helps us understand group-scale adaptations and selection in biology. The free energy principle applies to all scales of organization in the organism from single cells to the entire nervous system. When this principle is taken to its logical extremes of modeling groups, populations and ecosystems, we uncover a new, evolutionarily sensible path at explaining puzzling aspects of human motivation and judgement, including ethical decisions. To minimize free energy, populations have to act to maximize gathering of information, while building effective models at mitigating changes to its dynamic structure. The free energy principle thus provides a naturalistic explanation of some of our deepest ethical intuitions, and valuable principles of social behavior. We interpret the cognitive goal that corresponds to the principle as seeking a dynamic, fruitful, yet peaceful activity that sustains the organism. This state of mind is interestingly similar to the Buddhist intuition of mental equanimity; the organism's final goal is to be at peace and harmony with the environment. Another immediately relevant aspect is that assemblies must form to promote symbiotic, synergistic, positive feedback loops, which coincides with the

---

findings of ecologists. Therefore, ethics naturally emerges in self-organizing systems. Assemblies of organisms must ultimately unite in macro-minds to achieve the greatest reduction in free energy, as well as building technological extensions of themselves to improve their capacity to do such, therefore the principle also predicts a post-singularity world-mind composed mostly of artificial intelligence.

1. Introduction

Without delving into the mathematical formalism, the free energy principle may be stated as the following: to survive, organisms must minimize the thermodynamic free energy in a biological system. The free energy, or Gibbs energy in a chemical system is a measure of how much thermodynamical work the system is capable of performing. Above that limit, the system will not participate in chemical reactions. The more free energy an organism has, the more reactions may happen due to chance. This leads to disorder, and the more disorder a system suffers the more chance there is for its structure to dissolve. The organism must therefore minimize surprising interactions with the environment to maintain integrity (cf. Friston, 2003). The integrity condition has been elegantly defined as a boundary called the Markov blanket that discriminates the organism from the rest of the environment, decoupling internal states from external states.

Although classical understanding of potential energy is usually applicable to systems at equilibrium, as in classical chemistry, the Gibbs energy applies to all physical systems, that is to say it works for dynamical systems far from equilibrium. The formalism of free energy rests on a stochastic dynamical model of a phase space (any physical system), and therefore the interactions are described probabilistically. The Markov blanket thus is defined as an insulation of the organism from external effects, only mediated through the blanket via special states we may recognize as sensory and active states. The cell membrane is a good example of the Markov blanket as it neatly partitions the environment into internal organism, and external environment states as an insulation layer. The environment inside the blanket is controlled by the cell, nutrients and sensory data are obtained through the membrane, and the cell may have active states such as those that modulate the membrane, or flagella. Likewise, the nervous system in an animal may be considered to have a Markov blanket, wherein the mediation with the environment occurs through stimulation of receptor and effector neurons. The brain has to continually infer causes in the environment, and predict future states to choose the actions that will minimize free energy, and survive. In the free energy formalism, all of these state sets and possible interactions among them are described as probabilistic, dynamical interactions affixed in physical space. The entire physical system evolves, stochastically, and so do the organism's various components such as its internal states, sensory states, and active states. Figure 19.1 depicts the cognitive organism model implied by the free energy principle, and the stochastic interactions among the state set partition of the world according to the model.

World = E × S × M × A

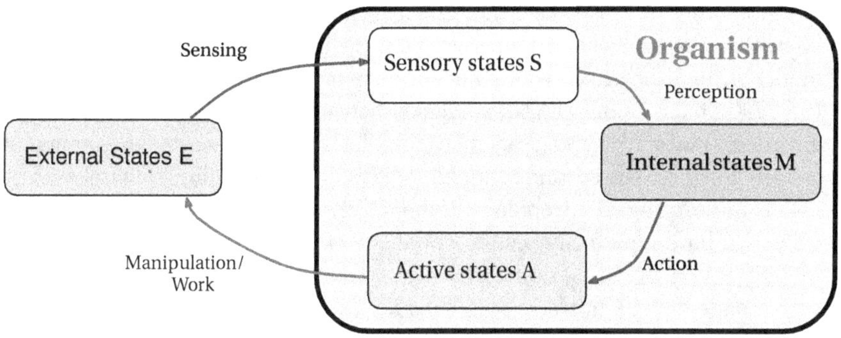

Figure 19.1: The partition of the world states according to the free energy principle.

One of the foundational results of the free energy approach is that minimizing free energy entails minimizing surprise, relative to the organism. Surprise is equivalent to expected entropy; if my organism's boundary has to face so many random interactions over a time window, sooner or later my boundary will be punctured, and my organism will disintegrate. That is why, in nature, the unknown and the complex elicits so much fear to animals; as a dog barks at the storm or the vacuum cleaner, this is because it cannot reliably predict what sorts of random interactions will be caused by it. This equivalence is the defensive, or protective condition of life, and at the shortest time scale it manifests as defensive reflexes. However, to enable long-term survival, a complex organism cannot rely on fixed rules. Instead, it must continually discover and learn to model the environment, so that the survival condition, including feeding and healing, may be met through actions that manipulate the environment (and the organism) to the organism's advantage.

Even in a cell, there is complex signal processing, called the signaling network, as opposed to the metabolic network, that processes information in non-trivial ways. Although not quite intelligent, bacteria can adapt to nutrient gradients, and that is a sort of adaptive behavior. This adaptive behavior is not merely a curiosity, it is enforced by the free energy principle. To live that long, the organisms had to evolve adaptive behavior. There are two main sorts of cognitive adaptations that the organism can follow to minimize its free energy. Firstly, it may optimise the accuracy of its world model, i.e., adjust its internal states to better match the environment, and therefore infer past and future causal states more effectively, enhancing all manners of thought about the world. Secondly, it can creatively manipulate its environment to confer a survival advantage, as predicted by Chomsky when he pointed out a serious shortcoming of behaviorism: a behaviorist mind is not inclined to improve over existing natural conditions (cf. Chomsky, 1959).

Friston and his research team model the learning process by Bayesian inference. The organism updates probability distributions corresponding to a model of the sensorium. The organism then seeks to minimize the difference (called Kullback-Leiber or KL divergence) between the recognition (a model of environmental states) and generative (a model that will generate environmental states) densities. The second foundational result is that minimizing free energy entails minimizing KL divergence. If the KL divergence is zero – in which case we would have a perfectly accurate predictive model – then free energy is just the entropy of the sensorium. In reality, of course, no organism can achieve such a state, for the universe is in constant flux. At the very least, the internal dynamics requires interaction to sustain its course. Therefore, the kind of prediction here is not merely modelling the immediate environment, which might be an overtly simple environmental niche, but it has to be a world-model, encompassing the world at large. There, we glimpse at the philosophical repercussions of the principle.

And in a rather surprising way, as a third major result, we find that the free energy principle is also directly equivalent to the principle of least action. This equivalence was first observed by Feynman as he probed the foundations of quantum mechanics. The principle of least action states that a physical system will choose the most economical way among many possible ways, and it turns out that this is equivalent to Bayesian inference, which in turn is equivalent to the free energy principle. Then, the principle is not merely encoded in the accidental mechanics of evolving organisms, but also nature itself. It also seems circular, for if free energy principle is provided by physics itself, does not this mean that life is necessarily caused by physical law; is not this against evolutionary science, or is not this a vicious circle? It turns out that there is no real deductive loop, or assuming the conclusion here. What this result truly means is that the Cosmos is inclined to evolve from a non-biological, unorganized state with no long-lasting Markov blankets, to biological organisms, explaining biogenesis. Then, we are not merely looking at a theory of neurological organization or biological principles, but a unified, cosmic theory of evolution. After all, an open-ended stochastic dynamical system like the Cosmos, evolves through time, attractor sets form, and the system continually explores new behavior. Indeed, there is no unwarranted assumption, or circularity in this manner of theorizing. What perplexes the common reader is the universality of the theory.

Many conceptual objections have been raised against the free energy principle. Most common objections depend on the intuition that such a simple mathematical explanation is too good to be true; how can the definition of life be crammed into a few dynamical system equations? Or alternatively, critics have claimed that the framework is a sort of mathematical babble that obscures any scientific findings, and nobody really understands it. The present article is, at least partially, an attempt to address these philosophical and conceptual objections, since it does not appear that the findings require any mathematics to understand them, or failing that, we hope that we can explain them satisfactorily in usual philosophical parlance.

As we have seen, the free energy framework does contain some counter-intuitive conclusions, and is thus it is understandable why there is much critique. When viewed objectively, however, the proofs are easy to check, and there are only a few general, but weak assumptions such as ergodicity, and the work is supported by many diverse empirical studies. The objection from obscurity is not valid, if anything, the theory is too simple, and must be refined with more detailed models to apply to the real world in many cases, such as the hierarchical, predictive coding, message passing model of the brain (cf. Friston, 2012). This is akin to the case with universal AI models, where we can perhaps express a universal utility maximization agent such as AIXI (cf. Hutter, 2007) succinctly, but to realize it many detailed models, architectures, and methods are required. In turn, the most persuasive empirical evidence of the free energy principle will be to build an autonomous AI agent based on these principles, and test the predictions of the theory, such as, “will it tend to explore the environment?”, “will it show curiosity?”, “will it form habits?”, and so forth. Friston et al. detail an autonomous agent model in Friston (2016). Please see Friston (2010) for a self-contained review from the neuroscience perspective.

2. Epistemological Implications

From a philosophy of science perspective, there are two pertinent implications of the free energy principle that may not immediately appear to scientists. First, the free energy principle effectively reduces to the principle of least action, which is a common property of quantum mechanics and general relativity. Therefore, as far as our present knowledge of theoretical physics goes, the principle states that life, cognition and evolution, directly reduce to physics. This is a bold claim that is underemphasized in the scientific articles because physical scientists are already reductive physicalists. However, until the free energy principle, no such comprehensive theory existed that explains both a general physical basis, and dynamics. From an AI theory viewpoint, the principle of least action is related to our previous work reformulating algorithmic information theory in terms of physics (cf. Özkural, 2015). Hutter et al. previously explained that the universal induction theory works, as it acts as a formalization of Occam's razor (cf. Lattimore and Hutter, 2015; Sunehag and Hutter, 2014). However, the physical formulation in Özkural (2015) goes deeper, as we introduced minimum volume, minimum energy, and minimum (quantum) action complexities. If the main claims of the free energy principle are correct, then there is a physical reason why Occam's razor is true, and it is highly likely that the correct universal inductive bias is that of least (quantum) action.

The second implication is that the free energy principle is a lower-level, or more general theory than whatever physical law we discover in the future, as long as it can be formulated in the stochastic dynamical system formalism. That is quite interesting, as it does not apply to only our universe, but assuming a multiverse, it would apply even to other universes in the ensemble. Astrophysics does model the evolution of complex structures such as planetary systems, however, the free energy principle might explain why cosmic evolution occurs. That is yet lower level than astrophysical models. Furthermore, the free energy principle can explain the emergence of any stable structures, which implies that it might yield a unified theory of physics. This property is intuitively unbelievable, as we have been trained by particle physicists to equivocate foundational physical theory with the standard model, which is not necessarily the most elegant theory in physics. The standard model advocates routinely dismiss even string theorists who proposed a more elegant theory with higher dimensionality – Witten has famously proposed M-theory with 11 physical dimensions. If we can explain the particle zoo of the standard model with the free energy principle, of course, it would be a radical success of the theory, however, this need not be true for it to achieve a feat of equal epistemological significance.

The free energy principle gives us a new tool to understand natural selection in terms of physics. Similar popular accounts have existed for long. In science fiction literature, Asimov famously analyzed the problem of defeating the second law of thermodynamics, in his story "The Last Question". This physical approach to life was sometimes called extropy, because it looks like it amounts to exporting entropy from the organism, or achieving negative entropy, countering the second law of thermodynamics, order against chaos. The free energy model shows that defeating the second law is likely in the realm of far-from-equilibrium dynamics. Although, we cannot guarantee evading the heat death of the universe like in Asimov's story, we might be able to explain why life exists at all for we now have a reductive definition of life, and potentially a complete understanding of universal fitness. The free energy principle thus yields a theory of Cosmic Evolution surpassing the restricted laws of Genetic Evolution, easily explaining such biological phenomena like epigenetics, and horizontal gene transfer in bacteria, but also the most foundational biological problem of biogenesis. The explanation of biogenesis is mostly opaque from the articles on biology and neuroscience, however it does exist. First, the theory does explain inorganic self-organization such as crystal growth, and synchronized pendulums, just as well as active inference agents (cf. Kirchhoff et al., 2018), this is essential to explain how the various cellular machinery evolved prior to the cell. And secondly, the theory predicts that the world will follow trajectories wherever higher plateau of fitness are explored, which seems to give us an arrow of evolution. The principle does not necessarily lead to higher structural complexity, but it does tend to selection of adaptive structures with superior minimization of free energy.

3. Scale-Free Property and Group Organization

Another striking property of the free energy principle is that it applies at all scales of organization from the nucleotide sequences to organelles, to cells, organs, the brain and the entire organism. If that is true, it surely does not stop there, and applies to interacting organisms, families, groups, tribes, ecosystems, the biosphere, and beyond. As the scale grows, it becomes increasingly harder to visualize how the principle would apply, therefore a cogent explication is in order.

One of the most intriguing debates in evolutionary biology is that of group or kin selection. Clearly, there are such dynamics, yet some biologists such as Dawkins insist that the individual organism unit is where any natural selection applies. This contentious hypothesis has not however been analyzed in much depth. The selection applies at the level of the individual specimen, truly, however, the individual carries genes that adapt it to groups. Traits such as monogamy or mating frequency are dictated by genes in nature, however, there is a lot more to group dynamics than merely family scale organization. A popular kind of group study in zoology is that of the extended wolf family or tribe, which is called a wolf pack. There are however social interactions in animals from such family like units to large tribes. All primates, for instance, show a manner of tribal collective behavior, but so do many other familia, such as flocks of birds, herds of herbivores, and ants and bees. Such tribal behavior is clearly dictated by genes, and it would be impossible for natural selection to not act on such dynamics, vis-a-vis co-evolution, that is the evolution of the genome, in tandem with the evolution of the environment.

We posit that free energy principle can explain many of the group-scale adaptive traits that have evolved in plant, animal societies, and ecosystems despite the lack of imagination some biologists suffer from. Many specimens cannot thrive in isolation, group organization reduces risks, facilitates resource sharing, intelligence amplification and division of labor. In herds, while some specimens observe the environment for predators, some will keep feeding. In human tribes, there is usually a sexual division of labor, but other kinds of labor division exists in many primate and non-primate species. Chimpanzees engage in cooperative hunting, Boesch et al. argue that this behavior is due to mutualism (cf. Boesch et al., 2006).

4. Symbiosis as a Naturalist Foundation of Ethics

Philosophers sometimes suffer from a lack of imagination, especially when they wish to present a general theory of ethics. Although consequentialism is more or less the only game in town for naturalist philosophers, simplistic theories such as utilitarianism have not resulted in a sensible explanation of ethical behavior. Utilitarianism of any sort is problematic however, the popular and simplistic positive/negative utility proposals of Singer and his followers that try to reduce utility to pain/pleasure are particularly bankrupt theories due to the utility monster objections – see a discussion of utility theory in Barbera et al. (1998).

The main philosophical problem with utilitarianism is that it tends to be either an underspecified theory, not being able to explain what utility we should choose, or ineffective and plain wrong when they do try to propose a specific kind of utility, like positive utilitarianism which must be classified as a bland manner of scientism.

The second kind of wrong consequentialism is the insistence of "AI Safety" researchers on human preferences. Some researchers seem to have gotten the idea that they could make behaviorist reinforcement learning agents safe, if only they could make them imitate human preferences through arcane methods such as inverse reinforcement learning that will be used to infer reward functions. Humans do infer the mental states of others, including their goals, and learn from their behavior, including ethical constraints, although there is no scientific evidence that we infer reward functions, or that reward functions correspond to ethical decision making. Human preferences have however been successfully integrated into a reinforcement learning model recently (cf. Christiano et al., 2017).

Nevertheless, there is a massive epistemological obstacle to any such method that might make utility maximization agents safe. There is nothing universal, stable, or well-defined about human preferences. Humans have many meaningless, vague, contradictory, irrational, harmful preferences. Moreover, the preference learning hypothesis does not extend to other species, neither does it seek to explain the commonality of ethical behavior in other species and our own species, with many divergent social codes. A recent experiment that collected human preferences about ethical dilemmas in the self-driving domain illustrates this epistemological problem perfectly (cf. Awad et al., 2018). The decision to spare a baby or a grandmother apparently varies according to demographics, which indicates that mimicking human ethical decisions superficially will likely not be a safe solution for AGI agents that interact with humans. We cannot deflate ethics by reducing it to human preferences. The reason philosophers of ethics exist is because human preferences are often not good or reliable. A compelling consequentialist account of ethical cognition is due to Dennett in his seminal book *Freedom Evolves* (2003) however, even an evolutionary account might fall short of explaining the universal principles underlying ethics, which the free energy principle might allow us to accomplish.

Pain and pleasure are merely proprioception that the organism uses to detect generally favorable, and unfavorable conditions the organism finds itself in. They are guiding internally generated sensory signals, and they are only part of a heuristic evolved to either drive the organism towards such favorable states, or deter from the unfavorable states. However, unless you are a radical behaviorist, there is no reason to assume that is all there is to cognition and/or behavior. Behaviorism has been refuted strongly by Chomsky, and we generally think that there is little merit left in the behaviorist school.

Animals use a variant of the perception-cognition-action cycle, and it is particularly in cognition that ethical behavior may emerge; it surely cannot be explained away as habitual behavior, learning from elders, imitating others, and so forth. Those explanations would be circular, not explaining why some behavior is better than another, failing in another way than utilitarianism. Even more importantly, such an approach cannot explain why the human brain evolved to make contextual and ethical judgement possible.

We propose that the universal modes of co-existence, called symbiosis and parasitism are well-suited to construct a more empirically grounded and universal theory of ethics based on biological research rather than armchair philosophy. The classical examples of symbiotic behavior involve pairs of animals, which exchange benefits in a mutualist economy. However, the common description of symbiosis is perhaps not sufficiently broad. Impressive examples of symbiosis have resulted in the modern cell metabolism and multi-cellular organisms. Mitochondria and the cell, and cells within an organism have a symbiotic relationship. The hypothesis that mitochondria and chloroplasts descended from specialized bacteria is called symbiogenesis or the endosymbiotic theory (cf. Margulis, 1991). In our view, symbiosis extends to social organization and even extra-somatic organization in species.

Once upon a dead world, it was through symbiosis that nascent forms of life flourished and then exploded in variation and population into the environment. Within the cybernetics approach of the free energy principle, then, the generalised symbiotic relationship could be considered as any feedback loop that reduces free energy over an ensemble of organisms. The symbiosis may be regarded simply as pooling resources, but it may also result in stable new forms like that of the mitochondria. Friston et al. explain that two organisms that engage in communicative, collaborative acts must seek to co-operate effectively to reduce free energy. Such symbiotic behavior explains why the main use of language is to transmit knowledge and imperatives, indeed it explains why language evolved at all. (cf. Friston and Frith, 2015).

If both organisms share their inferences about the world, they will quickly build a more accurate collaborative world model, and be able to plan better with exchanged information (world-model parts), and with imperatives, they can execute collaborative plans (cf. Friston and Frith, 2015). We submit that symbiosis does not stop at the boundary of two individuals, or even entire societies and populations. Symbiosis extends to ecosystems; the mycelium networks form a mutual immune system for the trees in a forest. Symbiosis extends to the entire biosphere, the whole carbon cycle is an instance of symbiosis at a planetary scale.

Therefore, the phenomenon of altruism cannot be dismissed as a consequence of self-interest, as some contemporary naturalist thinkers like Dawkins (1981) and Pinker (2015) seem to believe – in their futile attempt to justify a dysfunctional economic ideology that benefits parasitic behavior. If that were the case, the members of a human society would not readily distinguish between cases of symbiosis and parasitism, however, they do, and they do in a dramatic manner, which is the fundamental debate that underlies much of political philosophy. While a worker co-operative is a symbiotic economic structure, the billionaire class collects rent, and extracts undue profit from every sector, which might be considered parasitic. Social welfare or altruism was not invented by enlightenment era thinkers; it existed well before humans evolved. Many species survived solely due to their particularly rich ways of exploiting social relationships. The quixotic attempt to reduce ethics to selfishness remains as a tacit approval of social darwinism and anti-social political ideologies. See Wilson's paradigm changing book on the new field of sociobiology for a contrastive view that elegantly explains social instincts such as attachment without succumbing to such basic fallacies (cf. Wilson, 2000).

Thus, symbiosis may serve as a suitable, naturalist foundation for understanding ethical problems from the ground up. Our species evolved many cognitive traits to behave kindly, affectionately, socially; these were not accidental features, had there been no need for symbiotic behavior, our species could easily take on a more solitary nature. Yet, we delight in the company of others, and have an insatiable need to care for our family members and those in plight – you may find a discussion of social cognition in Lieberman (2013). These instincts stem from the simple fact that symbiotic behavior results in fitter populations. In other words, distributing resources fairly while dividing labor effectively has been a winning strategy for our yet young intelligent species. Slavery is indeed worse than a society that lets individual agents flourish yet benefit from a symbiotic co-existence, because the imbalance in benefits produces suboptimal outcomes, not because it is individually better for one person to avoid slavery; it matters if a great number of slaves are freed rather than one, as far as consequences go.

Moreover, the theory might allow us to realize generalized robot laws as we proposed in Özkural (2017). Once we have a universal definition of life, and the constraint which keeps organisms alive (minimizing free energy), it might be possible to formulate an AGI agent that can recognize people, and keep them away from harm. The most plausible meta-goal considered in Özkural (2017) is also symbiotic: to preserve and pervade life, and culture throughout the universe. It might be the case that such cosmic symbiosis meta-goals follow from the free energy principle.

5. Peace as the Final Goal of Organism

There are rare moments of satisfaction in one's life, when seemingly every need has been met. Cognitively, such satisfaction for an active inference agent is observing a state of minimum free energy, which entails that the agent has sufficiently complete information about the environment to predict future states, and that it can comfortably manipulate about the environment to produce likewise minimum free energy states. The trajectory of the environment is still burdened with Brownian motion, but there are events where it seems that some manner of dynamic equilibrium has been reached within a microcosm. Such moments may be termed peaceful, and coincides with the intuition of equanimity. The more an agent knows about future states, the less anxiety it will experience. Therefore, in the most desirable states, the organism will engage harmoniously with the environment, and be capable of withstanding surprise input in the future.

When one is situated in a relatively calm natural environment with a few well-controlled stochastic processes such as a gently flowing stream of water, the mind can experience the satisfaction that accompanies a complex information source that is nonetheless modeled accurately by the agent's world model, and where little surprise is expected. Hence, the Zen garden acts as a model of biology's final goal. The active inference agent will cultivate the environment until it attains a sufficiently high degree of free energy minimization, which explains the appeal of introvert behavior and nature retreats. Such introversion is however balanced out by the innate drive of curiosity that urges the agent to continually acquire new information about the environment; after a while even the most exquisite and calming Zen garden will tend to boredom for the agent (cf. Friston et al., 2012).

Another fundamental prediction of the free energy principle is that despite the noise in the environment, the trajectory of the agent will tend to carve out valleys of fitness and gradually ascend to higher plateau. This observation explains why the organisms also seem to have a desire to improve their conditions, and vary and multiply, instead of being satisfied with them by default.

As such, the agent must both thrive and participate in ever greater feedback loops as well as experience harmonious, satisfied labor. The variety of sub-goals implied by the universal goal of free energy minimisation paints a more familiar picture than alternative agent theories. Thus, the free energy principle gives way to a primary directive of life, and intelligence. It is conceivable that there are other significant aspects of life, however, only one is universally fixed.

6. Towards the World-Mind

The principle of free energy does not stop at arbitrary boundaries and scales, therefore, we may hypothesize that a future world-mind, connected through cybernetic and artificial intelligence technology, will be more effective at reducing free energy than independent units. The advent of the internet may be regarded as a first step towards such global cognition. Even if the current world-mind has no mental features, it still operates as a weakly coupled collaborative system of sorts through arrangements of communication and economics. On the contrary, this global human mind is apparently lacking enough cohesion, else it would act swiftly to deal with anthropogenic global warming. Therefore, a world-mind through human units would likely not be effective enough for the cognitive requirements of a world-mind.

Artificial units with much better energy efficiency (which we expect to appear after 2030 due to Koomey's law (cf. Koomey et al., 2010)) are more likely to result in an effective world mind. The energy efficiency of computation, as well as the capabilities of robotic technology will be critical in the free energy reduction of such an organism, therefore its genesis is a step towards the hypothetical infinity point (singularity), whereby positive feedback loops started by AI drive improvements in AI technology itself towards physical limits of computation. Even if some ultimate computer or AI close to the physical limits of computation (cf. Özkural, 2015) is not achieved within a short timespan as predicted by Ray Solomonoff's infinity point theory (cf. Solomonoff, 1985), a more reliable outcome could be the emergence of global-scale autonomous AI organisms.

Speculatively speaking, the free energy principle would compel such a world-mind to assimilate biological information of all sorts to improve its fitness, therefore we might find that technological immortality via brain simulation may emerge as the natural path before such an organism evolves. The simulated human brain would find it much easier to develop high-bandwidth cross-connections and coalesce into a heterarchical super-brain by modifying their neural architectures. Evolution in silico would progress millions of times faster than biological evolution, and superior post-human organisms would evolve rapidly at any rate. In other words, the Internet would finally come alive and develop awareness. We cannot yet completely fathom what an organism with a billion minds and bodies would be like, however, we can estimate that it might be better at survival than we are. Such an organism would then find it imperative to migrate to other worlds to prolong chances of survival, and therefore the interstellar expansion of a society of world-minds would have begun.

There is a speculative idea called "Gaia hypothesis" in biology that the Earth itself may be considered a biological entity. According to the free energy principle, that is not quite true, as Earth is more like a cosmic womb, or a seed pod, which brews life, although the planet does have an atmospheric boundary, and many synergistic processes, which perhaps makes our pale blue dot more like a proto-cell. Earth has the potential to be a world-mind, though it is not yet one. The planet barely has any sensory states or can influence other planetary bodies in any significant way, it is mostly inert in terms of animacy, and unfortunately the highest intelligence density is that of humans on it. Perhaps, one day, it shall saturate with intelligence, and it shall spawn the seeds of a cosmic dandelion to disseminate life to the rest of the Cosmos.

7. Discussion

We reviewed the free energy principle as a new scientific definition of life and autonomous cognition in terms of its epistemological and ethical implications. Epistemologically, the approach supports a flawless reductionism that reduces life and intelligence to the lowest possible level which is physical motion. Unlike previous scientific definitions of life, it is not peculiar to Earth biology (in terms of DNA, reproduction, etc.), but it is a physical definition that is applicable to any adaptive system. The definition does not fail at systems close to the boundary between animate and inanimate such as adapting crystals, virii, and cryostatic organisms.

Remarkably, the free energy principle addresses the biogenesis problem by introducing a completely universal theory of self-organization. In this view, the universe is tilted slightly in favor of life and intelligence, rather than shooting darts at a cosmic board. Such conclusion urges us to revise our view of extraterrestrial intelligence, which is mostly couched in earth centric concepts such as our search for exoplanets that are similar to Earth. Under the free energy interpretation, the Cosmos can and will produce fundamentally different forms of life, given sufficient resources. Some of these forms will inevitably be carbon based, but others might be based on silicon, or even nuclear reactions, or other physical interactions.

The ethics implied by the free energy principle belongs to consequentialism, as the active inference agent continually projects the future, and plans to minimize free energy. The universal goal of the agent is survival, however, the goal naturally extends to groups of organisms, and the fitness of a group may be evaluated in its capacity to collectively minimize the free energy of the population. Individually, the free energy principle manifests as homeostasis, curiosity, creativity, habit formation, and exploration, while socially it manifests as symbiosis, attachment, communication, empathy, altruism, and ethics, among other phenomena mercifully forgotten, for the entire list would be too long to include. Those populations with better fitness will be selected over others in the long run. We argue that the free energy view naturally gives rise to a preference of symbiosis over parasitism, as it is a general strategy that improves the fitness of groups. Just as an organism riddled with parasites is not likely to survive long, a society riddled with parasitic behavior will not reach optimal lifespan or fitness. We also argue that the subjective state of mind entailed by the principle is that of equanimity; peaceful and harmonious co-existence with the environment.

Finally, we argue that another natural inclination implied by the principle is rapid evolution towards a world-mind incorporating biological and artificial elements. We argue that such a world-mind would necessarily try to incorporate as much biological information as possible in order to minimize its expected free energy, optimize its technological base to maximise energy efficiency, and additionally choose collective, efficient cognitive processes to guide future actions, which will at a first approximation look like direct democracy and brain simulations co-operating over the Internet. Therefore, the free energy principle seems compatible with the infinity point hypothesis, also known as the singularity.

References

Awad, Edmond, Dsouza, Sohan, Kim, Richard, Schulz, Jonathan, Henrich, Joseph, Shariff, Azim, Bonnefon, Jean-Francois and Rahwan, Iyad (2018) "The moral machine experiment" *Nature,* 563(7729): 59–64.

Barbera, Salvador, Hammond, Peter and Seidl, Christian (1998) *Handbook of Utility Theory Volume 1: Principles,* Dordrecht: Kluwer Academic Publishers.

Boesch, Christophe, Boesch, Hedwige and Vigilant, Linda (2006) *Cooperative hunting in chimpanzees: kinship or mutualism?,* Berlin Heidelber: Springer, pp. 139–150.

Chomsky, Noam (1959) "A review of b. f. skinner's verbal behavior" *Language,* 35(1): 26–58.

Christiano, Paul, Leike, Jan, Brown, Tom B., Martic, Miljan, Legg, Shane and Amodei, Dario (2017) "Deep reinforcement learning from human preferences" in *ArXiv:1706.03741.*

Dawkins, Richard (1981) "In defence of selfish genes" *Philosophy,* 56(218): 556–573.

Dennett, Daniel C. (2003) *Freedom Evolves,* New York, NY: Viking Press.

Friston, Karl (2010) "The free-energy principle: a unified brain theory?" *Nature Reviews Neuroscience,* 11: 127–138.

Friston, Karl (2012) "A free energy principle for biological system" *Entropy,* 14(11): 2100–2121.

Friston, Karl (2013) "Life as we know it" *Journal of The Royal Society Interface,* 10(86).

Friston, Karl, FitzGerald, Thomas, Rigoli, Francesco, Schwartenbeck, Philipp, ODoherty, John and Pezzulo, Giovanni (2016) "Active inference and learning" *Neuroscience and Biobehavioral Reviews,* 68: 862 – 879.

Friston, Karl and Frith, Christopher (2015) "A duet for one" *Consciousness and Cognition,* 36: 390 – 405.

Friston, Karl, Thornton, Christopher and Clark, Andy (2012) "Free-energy minimization and the dark-room problem" *Frontiers in psychology,* 3(130): 127–138.

Hutter, Marcus (2007) "Universal algorithmic intelligence: A mathematical top down approach" In *Artificial General Intelligence* (B. Goertzel and C. Pennachin, eds.), Berlin: Springer, pp. 227–290.

Kirchhoff, Michael, Parr, Thomas, Palacios, Ensor, Friston, Karl and Kiverstein, Julian (2018) "The markov blankets of life: autonomy, active inference and the free energy principle" *Journal of The royal society interface,* 15(138): 20170792.

Koomey, Jonathan, Berard, Stephen, Sanchez, Marla and Wong, Henry (2010) "Implications of historical trends in the electrical efficiency of computing" *IEEE Annals of the History of Computing,* 33(3): 46–54.

Lattimore, Tor and Hutter, Marcus (2011) "No free lunch versus occam's razor in supervised learning" In *Algorithmic Probability and Friends. Bayesian Prediction and Artificial Intelligence* (David L. Dowe, ed.), Vol. 7070 of *Lecture Notes in Computer Science,* Berlin: Springer, pp. 223–235.

Lieberman, Matthew D. (2013) *Social: Why Our Brains Are Wired to Connect,* New York, NY: Crown Publishers.

Margulis, Lynn (1991) "Symbiogenesis and symbionticism" In *Symbiosis as a source of evolutionary innovation* (L. Margulis and R. Fester, eds.), Cambridge, MA: MIT Press, pp. 1–14.

Özkural, Eray (2015) "Ultimate intelligence part I: physical completeness and objectivity of induction" In *Artificial General Intelligence - 8th International Conference, AGI 2015, AGI 2015, Berlin, Germany, July 22-25, Proceedings*, pp. 131–141.

Özkural, Eray (2017) "Godseed: Benevolent or Malevolent?" In *Philosophy of Mind: Contemporary Perspectives* (M. Curado and S. Gouveia), Newcastle: Cambridge Scholars.

Pinker, Steven (2015) "The false allure of group selection" *The handbook of evolutionary psychology*, Hoboken, NJ: Wiley, pp. 1–14.

Solomonoff, Ray J. (1985) "The time scale of artificial intelligence: Reflections on social effects" *Human Systems Management*, 5(2): 149–153.

Sunehag, Peter and Hutter, Marcus (2014) "Intelligence as inference or forcing Occam on the world" In *Artificial General Intelligence - 7th International Conference, AGI. Proceedings* (B. Goertzel, L. Orseau, and J. Snaider, eds.), Berlin: Springer, pp. 186–195.

Wilson. Edward O. (200) *Sociobiology: The New Synthesis, Twenty-Fifth Anniversary Edition*, Cambridge, MA: Belknap Press/Harvard University Press.

CONTRIBUTORS

Aníbal Monasterio Astobiza is a Basque Government Posdoctoral Researcher. His research lies at the intersection of the cognitive, biological and social sciences exploring their philosophical underpinnings. He is a member of international (and national) research projects such as for example, EXTEND and INBOTS (Horizon 2020). He explores topics such as roboethics, machine ethics, ethics of AI, data ethics, ethics of neurotechnology, human enhancement, space ethics, applied ethics, bioethics, philosophy of neuroscience, philosophy of psychiatry, altruism, cooperation or philosophy of network science and complexity. Aníbal graduated in Philosophy (Universidad de Deusto) and obtained his PhD in Cognitive Science and Humanities at the Universidad del País Vasco/Euskal Herriko Unibertsitatea with a dissertation on social cognition. Monasterio Astobiza A. et al. (2019), Bringing inclusivity to robotics with INBOTS. Nature Machine Intelligence 1, 164.

Ben Goertzel is the CEO of the decentralized AI network SingularityNET, a blockchain-based AI platform company, and the Chief Science Advisor of Hanson Robotics, where for several years he led the team developing the AI software for the Sophia robot. Dr. Goertzel also serves as Chairman of the Artificial General Intelligence Society, the OpenCog Foundation, the Decentralized AI Alliance and the futurist nonprofit Humanity+. Dr. Goertzel is one of the world's foremost experts in Artificial General Intelligence, a subfield of AI oriented toward creating thinking machines with general cognitive capability at the human level and beyond. He also has decades of expertise applying AI to practical problems in areas ranging from natural language processing and data mining to robotics, video gaming, national security and bioinformatics. He has published 20 scientific books and 140+ scientific research papers, and is the main architect and designer of the OpenCog system and associated design for human-level general intelligence.

Caterina Moruzzi is a postdoctoral researcher at the University of Nottingham where she obtained her PhD in 2018 with a thesis on the ontology of musical works. She has a MA in Philosophy at the Università di Bologna and she graduated in Piano performance from the Conservatorio G.B. Martini in Bologna. She is actively contributing to research in aesthetics and Philosophy of Art by presenting at conferences worldwide and by publishing on high-impact journals. In her current research she investigates the impact of revolutionary development of AI technology on ontological, aesthetic, and evaluative considerations on music. In particular, she aims to answer two research questions: 'How does the development of computer-generated music affect the ontology of musical works?' and 'Do we deem computer-generated music as creative as human-generated music? If not, why?'. The scope of this research is to fill the existing gap in the literature in regard to the impact that cutting-edge developments in AI have on aesthetic and ontological debates on music and art.

Cody Turner is a third year PhD student in philosophy at the University of Connecticut. His areas of research include the philosophy of consciousness, the philosophy of artificial intelligence, and applied ethics. He is a member of the AI, Mind, and Society Group at the University of Connecticut, and has a teaching appointment with the Computer Science & Engineering department. He currently has one other publication entitled 'Could you Merge with AI? Reflections on the Singularity and Radical Brain Enhancement' (with Susan Schneider, forthcoming in Oxford Handbook of Ethics of AI, Oxford University Press, 2020).

David Pearce is author of The Hedonistic Imperative (1995), which advocates using biotechnology to abolish suffering throughout the living world in favour of gradients of superhuman bliss. In 1998, Pearce co-founded with Nick Bostrom The World Transhumanist Association, now rebranded as Humanity Plus. Transhumanists urge the use of technology to create a "triple S" civilisation of superhappiness, superlongevity and superintelligence.

Eray Özkural has obtained his PhD in computer engineering from Bilkent University, Ankara. He has a deep and long-running interest in human-level AI. His name appears in the acknowledgements of Marvin Minsky's The Emotion Machine. He has collaborated briefly with the founder of algorithmic information theory Ray Solomonoff, and in response to a challenge he posed, invented Heuristic Algorithmic Memory, which is a long-term memory design for general-purpose machine learning. Some other researchers have been inspired by HAM and call the approach "Bayesian Program Learning". He has designed a next-generation general-purpose machine learning architecture. He is the recipient of 2015 Kurzweil Best AGI Idea Award for his theoretical contribution to universal induction. He has previously invented an FPGA virtualization scheme for Global Supercomputing, Inc. which was internationally patented. He has also proposed a cryptocurrency called Cypher, and an energy based currency which can drive green energy proliferation. You may find his blog at http://log.examachine.net and some of his free software projects at https://github.com/examachine.

Federico Pistono is an Entrepreneur, Angel Investor, Researcher, and Educator. He's author of the international success "Robots Will Steal Your Job, But That's OK", has co-founded four startups, he's Futurist and Technology Expert in the

TV show "Codice: la vita è digitale" (Code: life is digital), which airs on Rai1, and is Former Head of Blockchain of HyperloopTransportation Technologies. He's founder and CEO of Exponential Thinking, which invests in early stage startups, and creator ofMavericks, a school for Entrepreneurs, Investors, and Innovators. He holds a BSc in Computer Science from the University of Verona, and in 2012 he graduated from Singularity University, NASA Ames Research Park.

Frederic Gilbert currently research consists in exploring the ethics of novel implantable brain-computer interfaces operated by Artificial Intelligence. At the time of writing this bio, he is an Australian Research Council (ARC) Discovery Early Career Research Award recipient and Senior Lecturer in Ethics at CALE, School of Humanities, at the University of Tasmania. He is affiliated with the Ethics, Policy & Public Engagement program of the ARC Australian Centre of Excellence for Electromaterials Science (ACES), and concomitantly an Ethics Consultant for the Centre for Neurotechnology, for which he conducts research at the University of Washington, in Seattle, USA. He has published over 65 articles in bioethics and neuroethics. Some of his main publications related to this chapter are: Gilbert F., O'Brien T., Cook, M. (2018) “The Effects of Closed-Loop Brain Implants on Autonomy and Deliberation: What are the Risks of Being Kept in the Loop?” In Cambridge Quarterly of Healthcare Ethics, pp. 1-12; Gilbert F., Cook, M., O'Brien, T., Illes J. (2019) “Embodiment and estrangement: Results from a first-in-human intelligent BCI trial” in Science and Engineering Ethics, 25 (1): 83-96; Gilbert, F. (2015) “A threat to Autonomy? The Intrusion of Predictive Brain Devices”. American Journal of Bioethics Neuroscience, 6 (4): 4-11.

Gabriel Axel Montes, Ph.D. Candidate, is a neuroscientist, educator, and musician. His 12+ years of research in cognitive neurobiology and mind-body practice has elucidated both endogenous and technological mechanisms for intervening in the brain's process of modelling reality and perception. His recent research focuses on informing the development of beneficial artificial general intelligence (AGI) with insights from the non-ordinary consciousness (NOC) of mind-body practices. Gabriel is known for fusing insights from rigorous NOC self-practice into the design of AGI, and virtual/augmented reality and humanitarian projects involving resilience education. Gabriel is an international educator on the intersection of neuroscience and mind-body practice. As Founder of Neural Axis®, he leads and consults on neuroscience-based technology design, strategy, and behavioural change. He leads organisations, communities, and individuals in applying neuroscience to first-person experience. For more about Gabriel, please visit gabrielaxel.com.

Hasse Hämläinen received his PhD in Philosophy from the University of Edinburgh (2015). At the time of writing, he is working as a QA analyst for a leading IT company and as a project researcher at the Jagiellonian University in Kraków. His research is focused on the history of moral theory, especially on ancient and the 18th century moral thinking, and, recently, also on charting the ethical challenges and limits of AI technologies. His publications include a co-edited monograph, The Sources of Secularism: Enlightenment and Beyond (Palgrave Macmillan, 2017) and his work has been published, for example, in International Philosophical Quarterly, Diametros, Journal of Ancient Philosophy and in The Philosophical Forum.

Kulvinder Panesar is a strategically focused senior computing professional and an academic for over twenty years. Her current role is a Senior Lecturer of Computer Science at the School of Art, Design and Computer Science at York St John University, York in United Kingdom. Her research interest is NLP (Natural Language Processing) in AI (Artificial Intelligence), and meaning and knowledge representation in conversational software agents (CSAs), and more recently, Chatbots and conversational AI. This research area is multi-disciplinary spanning agent thinking, linguistics, and computational linguistics, natural language processing (NLP), knowledge representation (KR), Semantic Web (SW), artificial intelligence (AI) and data science. Current teaching commitments include undergraduate and postgraduate subjects in: computer science, software engineering design and development, website design and development, advanced database and networks; programming, professional practice; project management, project supervision, mathematics for computer science, philosophies of technology, and artificial intelligence. Previous teaching undertaken include: management information systems, marketing and e-commerce strategy (customer relationship management) and development, enterprise computing solutions, systems analysis and design, formal methods and research methods. Additional roles undertaken over the years are course director, module leader, final year project coordinator and supervisor, placement co-ordinator, 'study abroad' academic advisor, researcher, internal verifier, mentor, STEM ambassador, external assistant examiner (EdExcel), ICT teacher, IT support/programmer/trainer contractor, and systems analyst. She is a professional member (MBCS) of the British Computer Society (BCS), and a Fellow of the Higher Education Academy (AdvanceHE) and currently organise and co-ordinate the Women in Computer Science (WICS) group at York St John University.

Mariana Chinellato Ferreira is a Brazilian researcher, graduate at Social Communication with specialization in Advertising and Publicity from the Catholic University of Santos (2002) and graduate at Modern Languages, major in Portuguese and English languages and Literature from the Catholic University of Santos (2009). Has an exchange program period at the University of Coimbra (Portugal - 2007/2008) in Modern Languages. Has professional experience in Linguistics, focusing on Applied Linguistics, acting on the following subjects: Modern Foreign Language (English), English for Specific Purposes, Portuguese as a Second Language, Translation (Portuguese – English) and Education. Professional academic experience in English and American Literature, Portuguese and Brazilian Literature, Theory of Literature, Theater and Poetry. She has a Master's degree in Architecture at University of São Paulo - São Carlos with the project: "City and Literary form: Urban representation in the Brazilian contemporary literature". Currently, she is a PhD Candidate of the PhD Program in Materialities of Literature at the University of Coimbra with a research about questions of language and creativity in Narrative Generation Systems in a partnership with joint supervision of the Centre for Informatics and Systems of the University of Coimbra where she develops a new model of narrative in order to apply into a prototype system of automatic narrative generation in Portuguese language. Her last publish article discusses new models of narrative in computer-generated literature: "Seeking the Ideal Narrative Model for Computer-Generated Narratives", presented at the CC-NLG 2018, in Tilburg, the Netherlands.

Natasha Vita-More is an author, humanitarian, and innovator whose work focuses on longevity and regenerative generations. As a motivational speaker, she focuses on causes and solutions, while fostering meaningful acknowledgement of others' works who have aspired to identify human potential. She is called "An early adapter of revolutionary changes" (Wired, 2000) and "Advocates the ethical use of technology to expand human capacities" (Politico, 2017). Natasha was the Lead Scientific Researcher on the Memory Project, which scientific breakthrough concerns long-term memory of C. elegans and cryonics (2015). As a proponent for mitigating aging, Natasha introduced the seminal field of Human Enhancement for longevity in academics. Her expertise in the field of ethics has produced high-level scholarship for undergrad and graduate students in the fields of computer science, cybersecurity, robotics, gaming, and business fields. As an entrepreneur, her experience within the domain of foresight studies has established principles and practices for assessing humanity's potential futures. Natasha originated the original "Whole-Body prosthetic", a seminal innovation comprised of nanomedicine, AI, and robotics that spearheaded alternatives to biological aging and received international recognition. Currently, she is a Professor of Humanities in Ethics, Business, and Entrepreneurship, and serves as the Executive Director of Humanity+. She is also a Fellow at the Institute for Ethics and Emerging Technologies, and author of numerous academic articles and books. Some publications: "Persistence of Long-Term Memory in Vitrified and Revived Caenorhabditis elegans" or "The Transhumanist Reader". More can be find here: https://natashavita-more.com/publications/.

Nicole Hall most recently held a Fernand Braudel research fellowship at the Institut Jean Nicod (Ecole Normale Supérieure) and taught aesthetics at the Université Paris 8. Her PhD, on the nature of aesthetic experience, is from the University of Edinburgh. Her research is at the intersection of aesthetics, environmental philosophy, the philosophy of mind and consciousness, and cognitive and neuroscience. She has a forthcoming publication on Adam Smith's aesthetic psychology and is working on the role of perception in aesthetic experience.

Peter DePergola II is Director of Clinical Ethics at Baystate Health, Assistant Professor of Medicine at University of Massachusetts Medical School, and Assistant Professor of Bioethics and Medical Humanities at Elms College, where he serves concurrently as Director of the Center for Ethics, Religion, and Culture. Dr. DePergola earned his B.A. degree in philosophy and religious studies at Elms College, his M.T.S. degree in ethics at Boston College, and his Ph.D. degree in healthcare ethics at Duquesne University. He completed his residency in neuroethics at University of Pittsburgh School of Medicine, his fellowship in neuropsychiatric ethics at Tufts University School of Medicine, and his advanced training in neurothanatological ethics at Harvard Medical School. Dr. DePergola's recent book, Forget Me Not: The Neuroethical Case Against Memory Manipulation(Vernon Press, 2018), has been critically acclaimed as a landmark achievement in the field of neuroethics, and was the #1 New Release in Medical Ethics on Amazon.com at the time of its publication. The recipient of numerous awards for academic scholarship and healthcare leadership, his current research explores the theoretical and empirical metaphysics of hope, particularly as it relates to the neuroethics of moral virtue and end-of-life decision making.

René Mogensen holds a PhD from Birmingham City University/Royal Birmingham Conservatoire in the UK, an MM from the Royal academy of Music in Aarhus, Denmark, an MA from New York University and a BA from the University of Rochester, NY, USA. His research in music, computer music, and artificial intelligence in music has recently been published in books by Springer, Éditions Delatour, and as articles in the Journal of Creative Music Systems, the Early Music journal, as well as in international conference proceedings (CSMC, AISB, EuroMAC). He is the recipient of grants from the Danish Arts Foundation, Meet the Composer USA, John Anson Kittredge Fund, Yvar Mikhashoff Trust for New Music, and other funders. His numerous original music works for ensembles and soloists (with and without computers) have been performed around Europe, the USA and Asia. He has participated as a performer/composer in many music concerts during the past three decades around the USA and Europe. He currently teaches undergraduates and postgraduates in areas of creative music technology as well as in music improvisation and composition at the Royal Birmingham Conservatoire in the UK.

Ricardo Morte Ferrer is a lawyer and Data Protection Consultant (https://sdm2.de/), he is also a certified ITIL-F and Lead Auditor ISO 27001. He is a member of the Working Group Digitalization and Health at the German Academy for Ethics in the Medicine. PhD candidate at the University of Granada working on"Ethical and Philosophical Aspects of Privacy". Nowadays is researcher in INBOTS CSA network: Inclusive Robotics for a better Society [www.inbots.eu/], EXTEND: Bi-directional Hyper-connected Neural System [www.extend-project.eu/] and member of the core team at GNU Health.

Roman V. Yampolskiy is a Tenured Associate Professor in the department of Computer Engineering and Computer Science at the Speed School of Engineering, University of Louisville. He is the founding and current director of the Cyber Security Lab and an author of many books including Artificial Superintelligence: a Futuristic Approach. During his tenure at UofL, Dr. Yampolskiy has been recognized as: Distinguished Teaching Professor, Professor of the Year, Faculty Favorite, Top 4 Faculty, Leader in Engineering Education, Top 10 of Online College Professor of the Year, and KAS Outstanding Early Career in Education award winner among many other honors and distinctions. Yampolskiy is a Senior member of IEEE and AGI, and former Research Advisor for MIRI and Associate of GCRI. Roman Yampolskiy holds a PhD degree from the Department of Computer Science and Engineering at the University at Buffalo. He was a recipient of a four year NSF (National Science Foundation) IGERT fellowship. Before beginning his doctoral studies Dr. Yampolskiy received a BS/MS (High Honors) combined degree in Computer Science from Rochester Institute of Technology, NY, USA. After completing his PhD dissertation Dr. Yampolskiy held a position of an Affiliate Academic at the Center for Advanced Spatial Analysis, University of London, College of London. Dr. Yampolskiy is an alumnus of Singularity University (GSP2012) and a Visiting Fellow of the Singularity Institute (Machine Intelligence Research Institute). Dr. Yampolskiy's main areas of interest are AI Safety, Artificial Intelligence, Behavioral Biometrics, Cybersecurity, and Genetic Algorithms.

Soenke Ziesche holds a PhD in Natural Sciences from the University of Hamburg, which he received within a Doctoral Program in Artificial Intelligence. He has worked since 2000 for the United Nations in the humanitarian, recovery and sustainable development fields as well as in data and information management. In addition to UN Headquarters in New York he has been posted to Palestine, Sri Lanka, Pakistan, Sudan, Libya, where he was temporarily the highest UN representative during the revolution, South Sudan, Bangladesh and Maldives. He is currently a consultant, mostly for the UN, for Disaster Information Management as well as AI. In this capacity he has served, on behalf of the UN, the Governments of Bangladesh, Maldives and Sri Lanka respectively. In 2017 he produced a 137-pages UN report on opportunities for big data, AI and emerging technologies in the implementation and review of the UN Sustainable Development Goals. His latest academic publications on AI safety are "AI & Global Governance: A Seat at the Negotiating Table for AI? Opportunities and Risks", "Potential Synergies Between The United Nations Sustainable Development Goals And The Value Loading Problem In Artificial Intelligence" and, together with Roman V. Yampolskiy, "Towards AI Welfare Science and Policies".

Stephen Rainey is a research fellow in the Oxford Uehiro Centre for Practical Ethics. He is a working in the Horizon2020-funded project BrainCom which is developing therapeutic brain-computer interfaces that will enable communication for users with debilitating speech conditions. Dr Rainey studied philosophy in Queen's University Belfast and obtained his PhD in 2008 with a thesis on rationality. He has taught a range of philosophical topics, and worked on a number of European Commission-funded research projects. These have included work on ethics, emerging technologies, and governance. He contributes to European Commission ethical and scientific evaluation panels for the funding of research projects. These have included proposals for ERC and Marie-Curie grants. Stephen has research interests in the philosophy of language, rationality, governance, and artificial intelligence.

Stevan Harnad's research interest is in cognitive science, open access and animal rights. He is currently professor of psychology at Université du Québec à Montréal (UQAM) and professor emeritus of cognitive science at the University of Southampton. Elected external member of the Hungarian Academy of Sciences in 2001 (resigned in protest, 8 October 2016), he was founder and editor-in-chief of the journal Behavioral and Brain Sciences 1978-2002 and Canada Research Chair in cognitive science 2001-2015. His research is on categorization, communication, cognition, and consciousness. He has written on categorical perception, symbol grounding, origin of language, lateralization, the Turing test, distributed cognition, scientometrics, and animal sentience. He is founder and editor-in-chief of the journal Animal Sentience.

Steven S. Gouveia is finishing his PhD at the University of Minho (Portugal), under the supervision of the philosopher Manuel Curado (University of Minho) and the neuroscientist Georg Northoff (University of Ottawa), funded by the

Science and Technology Foundation. His primary focus of research is on the relationship between Neuroscience and Philosophy. He is a visiting researcher at the Minds, Brain Imaging and Neuroethics Group at the Royal Institute of Mental Health, Uni. Ottawa. He is a researcher of the Lisbon Mind & Reasoning Group (Nova University of Lisbon) and the Mind, Language and Action Group (Uni. Porto). He published 8 academic books e.g. (2018) "Perception, Cognition and Aesthetics" (eds.) (Routledge) or "Automata's Inner Movie: Science and Philosophy of Mind" (eds.) (Vernon Press) and, forthcoming, "The Science and Philosophy of Predictive Processing (eds.) (Bloomsbury). He is the host of an international documentary on AI with the participation of international researchers such as Peter Singer and Paul Thagard. He is finishing writing is new book "Homo Ignarus: Rational Ethics for an Irrational World". He was one of the speakers at the famous "Science of Consciousness Conference" (2019) and published several articles on AI, mind and brain, ethics and aesthetics. More: stevensgouveia.weebly.com.

Tomislav Miletić, born on May 7, 1984, in the city of Rijeka, Croatia. My personal and academic interests predominantly lie in exploring the ethical and social impacts of Artificial Intelligence inside the paradigm of Human Enhancement. I am especially focused on the prospect of symbiotic human-Ai systems, their hybrid moral-epistemic agency, status and system architecture and the relevant ethical questions concerning their design and use. Currently engaged in finishing my doctoral thesis on the subject of "Human-AI symbiosis: the possibility of moral augmentation". Relevant publications: Miletić T. "Extraterrestrial artificial intelligences and humanity's cosmic future: Answering the Fermi paradox through the construction of a Bracewell-Von Neumann AGI" in Journal of Evolution and Technology, Vol. 25:1, 2015, 56-73 or Miletić T. "Human Becoming: Cognitive and Moral Enhancement Inside the Imago Dei Narrative" in Theology And Science, Vol. 13:4, 2015, 425-445.

Txetxu Ausín is a Tenured Scientist at the Institute for Philosophy, CSIC (Spanish National Research Council) [www.ifs.csic.es]. He is the Director of the Applied Ethics Group (GEA). PhD in Philosophy (University of the Basque Country) and First Prize of the Year 2000. Invited Professor in several universities and researcher at the Institute for Democratic Governance Globernance. His research areas involve public ethics, bioethics, deontic logic, human rights, and philosophy of robotics and ICT. [Member of OECD Expert Group on Research ethics and new forms of data for social and economic research]. Editor and author of publications about this issues, is the Editor in Chief of the international journal DILEMATA on applied ethics [www.dilemata.net]. Nowadays is researcher in INBOTS CSA network: Inclusive Robotics for a better Society [www.inbots.eu/], EXTEND: Bi-directional Hyper-connected Neural System [www.extend-project.eu/] and BIODAT: Datos en salud[umucebes.es/]. Research Vice-President of ASAP-Spain (Academics Stand Against Poverty) and President of the Spanish Network of Philosophy [http://redfilosofia.es/]. Independent Member of the Public Ethics Commission of the Basque Government and Patron of the Foundation Cluster Ethics of the Basque Country and of Legal and Social Studies (EJYS Foundation), he also collaborates in the Committee of Experts on Aging of the Fundación General CSIC. Some publications: "Exclusion from healthcare in Spain: The responsibility for omission of due care". In Gaisbauer, Schweiger, Sedmak (eds.), Ethical Issues in Poverty Alleviation. Dordrecht: Springer, 2016. ; "Research Ethics and New Forms of Data for Social and Economic Research". OECD Science, Technology and Industry Policy Papers, No. 34, OECD Publishing, Paris.

Vernor Vinge has won five Hugo Awards, including one for each of his last three novels, A Fire Upon the Deep (1992), A Deepness in the Sky (1999), and Rainbow's End (2006). Known for his rigorous hard-science approach to his science fiction, he became an iconic figure among cybernetic scientists with the publication in 1981 of his novella "True Names," which is considered a seminal, visionary work of Internet fiction. His many books also include Marooned in Realtime and The Peace War. Born in Waukesha, Wisconsin and raised in Central Michigan, Vinge is the son of geographers. Fascinated by science and particularly computers from an early age, he has a Ph.D. in computer science, and taught mathematics and computer science at San Diego State University for thirty years. He has gained a great deal of attention both here and abroad for his theory of the coming machine intelligence Singularity. Sought widely as a speaker to both business and scientific groups, he lives in San Diego, California.

INDEX

A

B

C

D

E

F

G

H

I

J

K

L

M

N

O

P

R

S

T

U

W

Y

Z

www.ingramcontent.com/pod-product-compliance
Lightning Source LLC
LaVergne TN
LVHW052241200926
848481LV00002B/233
* 9 7 8 1 6 4 8 8 9 2 1 8 9 *